ACS SYMPOSIUM SERIES 361

Detection in Analytical Chemistry

Importance, Theory, and Practice

Lloyd A. Currie, EDITOR

Developed from a symposium sponsored
by the Divisions of Analytical Chemistry,
Environmental Chemistry, and
Nuclear Chemistry and Technology
at the 191st Meeting
of the American Chemical Society,
New York, New York,
April 13–18, 1986

American Chemical Society, Washington, DC 1988

Library of Congress Cataloging-in-Publication Data

Detection in analytical chemistry.
ACS symposium series; 361

Includes bibliographies and indexes.

1. Chemistry, Analytic—Measurement—Congresses.

I. Currie, Lloyd A., 1930- . II. American Chemical Society. Division of Analytical Chemistry. III. American Chemical Society. Division of Environmental Chemistry. IV. American Chemical Society. Division of Nuclear Chemistry and Technology. V. American Chemical Society. Meeting (191st: 1986: New York, N.Y.) VI. Series.

QD75.4.M4D46 1988 543 87-30735
ISBN 0-8412-1445-X

PRINTED IN THE UNITED STATES OF AMERICA

Foreword

The ACS SYMPOSIUM SERIES was founded in 1974 to provide a medium for publishing symposia quickly in book form. The format of the Series parallels that of the continuing ADVANCES IN CHEMISTRY SERIES except that, in order to save time, the papers are not typeset but are reproduced as they are submitted by the authors in camera-ready form. Papers are reviewed under the supervision of the Editors with the assistance of the Series Advisory Board and are selected to maintain the integrity of the symposia; however, verbatim reproductions of previously published papers are not accepted. Both reviews and reports of research are acceptable, because symposia may embrace both types of presentation.

Contents

Preface

CHEMICAL MEASUREMENTS ARE CHARACTERIZED by three fundamental processes: detection, identification, and quantification. The first of these relates to the ultimate measurement capability as expressed in the detection limit. The invitation to organize a symposium on this topic carried the suggestion that we address the "true meaning" of detection limits. That charge, in fact, influenced the structure of the symposium and the content of this volume. The objective of this book, therefore, is primarily to explore, from both fundamental and practical perspectives, the meaning of detection in chemical measurement science. It is *not* intended to serve as a compendium of the current "detection limits" for a broad range of analytical methods.

The "meaning" of detection and detection limits has been examined in two senses: (1) the basic scientific meaning as to exactly what is signified by the minimum detectable quantity of a chemical substance, and how that quantity is derived; and (2) the meaningfulness or usefulness of such detection limits in the context of external problems, such as those affecting society, industrial processes, or scientific research. It is timely and appropriate for these two sides of detection to be considered together, for the ability to detect prescribed amounts of chemical substances in the natural environment, foods, or manufactured products can have important effects on our economic or physical well-being. However, unless the technical significance of detection decisions and detection limits is fully defined, misleading or even dangerous conclusions can follow. Equally important is mutual understanding by the lay public and the technical community of their respective interpretations of detection.

This text consists of an overview chapter and four principal sections. The first section addresses the issue of detection from the perspective of the well-informed but nonscientific public, that is, the most extensive user community of detection limits. The authors of Chapters 2 and 3, a former member of Congress and a former congressional subcommittee staff director, respectively, are eminently qualified to present the public view because they have both helped to shape that view and because they respond to the public's technological needs. The second section begins with a tutorial chapter, followed by six contributions treating fundamental characteristics of the chemical measurement process, which must be taken into account to derive meaningful detection limits. Included in this section are expositions on comparative detection limits, a matter of some importance in selecting chemical methods or in interpreting interlaboratory data. Five chapters make up the third section of the book. Selected examples illustrate the care and energy involved in devising methods of extreme sensitivity. Exhaustive method characterization and attention to a host of potential sources of error mark these contributions. These chapters, together with several chapters in the preceding section, also treat important areas of application.

The text concludes with two panel discussions. The first reflects the overall focus of the book, in that it examines with some vigor "real-world" problems and needs for meaningful detection, both in the laboratory and in the regulatory environment. The second panel, comprising members of an international team working on coding environmental analytical data, shares some of the approaches and issues involved in preparing low-level chemical data for computerized data bases. Opportunities for information loss or distortion are quite significant in this regard because of problems of rounding and truncation and interpretation of the meaning of detection.

The objectives of this book will have been met if improved communication on the subject of detection results. Such communication would be beneficial not only between the public and the technical community, but also within the technical community. As indicated in the overview chapter, the history of detection limits in analytical chemistry has been marked by an unfortunate degree of diversity in terminology and meaning and a lack of attention to the probabilities of *both* false negatives and false positives. At the same time, we should help the public understand that all detection limits must allow for these two types of error, and that "zero" detection limits cannot, in principle, be attained.

DISCLAIMER

This book was edited by Lloyd A. Currie in his private capacity. No official support or endorsement by the National Bureau of Standards is intended or should be inferred.

LLOYD A. CURRIE
Gaithersburg, MD 20899

September 1, 1987

Chapter 1

Detection: Overview of Historical, Societal, and Technical Issues

Lloyd A. Currie

Center for Analytical Chemistry, National Bureau of Standards, Gaithersburg, MD 20899

Practical societal needs and basic scientific advances frequently rely on Measurement Processes possessing specified detection capabilities with acceptable probabilities of false positives and false negatives. The first part of this overview introduces the basic concept of (chemical) detection, together with its applicability to selected societal problems such as the detection of natural hazards and the implementation of certain regulations. Basic scientific measurement issues concerning assumptions and their validity, plus hypothesis testing and decision theory as related to analyte detection are next introduced. Part two comprises a brief historical review, highlighting major contributions to the concept and realization of detection in chemical applications. The current state of the art is then considered. Part three is the most extensive, as it seeks to expose most of the technical issues involved in deriving meaningful detection decisions and detection limits, considering the overall Chemical Measurement Process. Those concerned primarily with societal or historical matters may wish to pass over this part. Among the topics discussed are: systematic and model error; non-normal random error; the special problem of the blank; replication vs Poisson variance; issues concerning complex data evaluation, calibration, and reporting -- including pitfalls associated with "black box" algorithms; OC curves; power of the t-test; and quality. The section concludes with some new material on discrimination limits, lower and upper regulatory limits, multiple detection decisions, and univariate and multivariate identification. A brief summary follows, bringing together historical, societal, and technical highlights. A concluding observation is that a meaningful approach to practical societal needs is at hand, but that order must be brought out of the extant diversity of technical views on detection.

The DETECTION LIMIT (L_D) is one of the most important characteristics of any Measurement Process. Recognizing the existence of such limits is crucial both for strictly scientific endeavors, such as the search for a new fundamental particle (1), and for vital societal applications of scientific measurements, such as the detection of a pathological state or a hazardous level of a heavy metal. In this latter regard, important progress has been made in conveying to the public and their policy makers that it is a **law of measurement science** that the detection capability of **all** Measurement Processes **must** stop short of zero, in close analogy with the Third Law of Thermodynamics.

Recognition that L_D may not be zero, has alleviated earlier legislative problems, such as the dictum that no residue of proven animal carcinogens may be present in certain food products (2). The fact, however, that detection limits can, at a cost and with technological advances, be made ever smaller has forced reexamination of regulatory issues in the light of extant and even potential detection capabilities. The consequence has been the consideration of cost/benefit or "acceptable risk" alternatives to "no detectable residue" regulatory policy (3). Such alternatives are mandatory in light of the fundamental principles of detection. Defining acceptable levels of risk (4), whether in a regulatory setting or with respect to medical decisions or even in terms of governmental actions in connection with potential natural disasters, is primarily a sociopolitical matter. Although this issue is of central importance, it transcends the theme of this chapter, which is to examine the historical evolution and current state of the art of detection from the perspective of chemical measurement science.

In order to highlight the importance of Detection Decisions and Detection Limits, and to underline the fact that the probability of detection does not immediately pass from zero to unity at the Detection Limit, we have presented in Fig. 1 several situations where valid detection decisions and adequate detection limits are of considerable practical importance. (The presence of a finite risk of error (false negative) at the detection limit -- i.e., the absence of "certainty" -- is the second aspect of the problem that is somewhat foreign to the common understanding, the first being the fact that zero detection limits are unattainable.) This figure introduces the Hypothesis Testing foundation for Detection, and it demonstrates that it is essential for those of us involved in measurement science to develop a sound, common, and quantitative approach to the formulation of Detection Limits. In addition, this formulation must be communicated in an effective manner both within the scientific community and with those who depend on our measurements for societal decisions and policy making.

As a final introductory note, it should be observed that from the perspectives of basic discoveries in Science and the early discernment of fundamental changes in the Global Environment (e.g.,

H Y P O T H E S E S

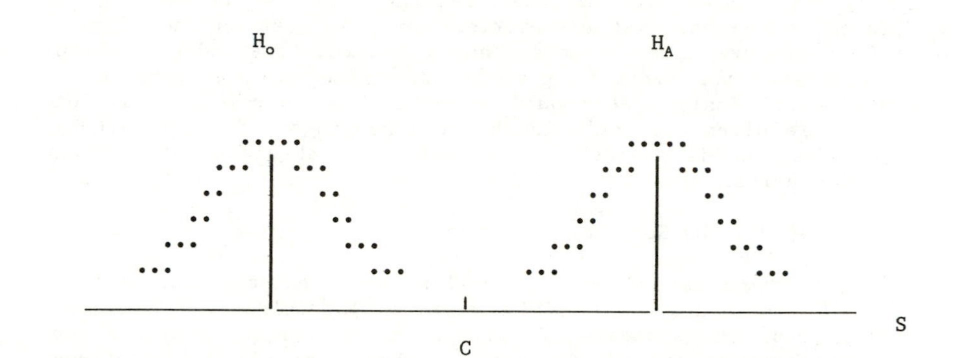

D E C I S I O N S

D_o	$(1-\alpha)$	(β) false negative
D_A	(α) false positive	$(1-\beta)$

Clouds	Aircraft
Background Tremors	Earthquake
Ordinary Luggage	Explosives
Method Blank	Toxic Chemical
Background Radiation	Chernobyl
Medication	Abusive Drug
Volume Fluctuations	Storage Tank Leakage

Fig. 1. Hypothesis Testing and Detection Limits. The upper part of the figure indicates the null [H_o] and alternative [H_A] hypotheses, with the corresponding decisions [D_o, D_A] at the left. Two kinds of erroneous decisions may be made: false positives [probability α] and false negatives [probability β]. (S represents a signal level; C, a decision point or "critical" level.) The lower section contrasts a number of "real world" H_o's and H_A's where adequate detection limits for the H_A's have clear, practical consequences.

stratospheric ozone changes, CO_2-induced global warming), the ability to **design** measurement processes having sufficient detection capability places one at the "cutting edge." Repeatedly in Science, one finds that discoveries are made just as the signals begin to emerge from the noise; and it is the "trained eye" which is generally the first to grasp them. Also, in the context of experimental design, it should be noted that *absolute* detection limits are often the goal, in that the hypothesis (or phenomenon) to be detected is generally conceived of in absolute rather than relative units.

1. BREADTH: The Scope of Detection

Fires, earthquakes and other natural hazards, pathological states, chemical contaminants, new fundamental particles or theories, instigators or sources of pollution or crime, natural or anthropogenic events of the past -- these are all illustrations wherein the basic concept of **Detection**, especially as embodied in the **Statistical Theory of Hypothesis Testing**, occupies a central position. Hypothesis formation -- i.e., specification of the source or system state or phenomenon to be tested [the "null hypothesis"] -- is necessarily the first step. For example, one might wish to test the null hypothesis (H_o) that no earthquake occurred (at a given time and place). To test H_o, one requires a test- or measurement-process (MP), often a Chemical Measurement Process (CMP), the outcome of which yields a **Decision** regarding the validity of the null hypothesis. The "alternative hypothesis" (H_A) which we wish to be able to detect -- e.g., an earthquake of a given magnitude -- must exceed the **Detection Limit** of the Measurement Process employed.

The keys to understanding the meaning of Detection Decisions and Detection Limits in matters of practical importance to science and society are: a) the existence of the two states [or hypotheses] which we wish to distinguish; b) a specified measurement process having an adequate DETECTION LIMIT; and c) a threshold or CRITICAL LEVEL for the measurement variable [Signal] for making the Detection Decision. Unfortunately, no measurement process can be exact, so false positives [α-error, e.g., earthquake erroneously "detected"] and false negatives [β-error, e.g., actual earthquake missed] will occur. Perhaps a more common example is that of the fire alarm. The measurement in this case might be made with a smoke detector, which if set to too low a threshold might give a false alarm [α-error] due to cooking fumes; if the critical level or threshold is set too high, a real fire of some consequence might be missed [β-error]. If an adequate balance between these two types of error cannot be achieved, one needs a **better measurement process** -- i.e., a detector having a lower detection limit. Note that the detection limit is an **inherent property of the measurement process**, whereas the detection decision is made by comparing an outcome or result of measurement with the Critical Level [threshold setting].

Fig. 1 suggests a wide range of situations where adequate detection limits are crucial for the well-being of society. The figure implies that the alternative hypothesis has a unique value on the x-axis. This is sometimes true. For example, the radiocarbon concentration [isotope ratio] for living matter is $^{14}C/^{12}C = 1.18 \times 10^{-12}$, whilst that for fossil fuel carbon is effectively zero (5). A similar situation obtains for population means for chemical concentrations indicative of certain pathological states (e.g., glucose in diabetes (6)), or trace element concentrations characteristic of certain ore bodies. In a great majority of cases, however, the intensity or magnitude variable (x-axis) can take on many discrete (denumerable H_A's), or even continuous values (infinite number of H_A's). Such is the case, for example, with chemical or radioactivity contamination, earthquakes, fires, hurricanes, etc. For a given measurement process a special relation exists among the "distance" between H_A and H_o, and the two kinds of error, α and β. Fixing any two of these quantities determines the third, as will be shown in a later discussion of "Operating Characteristics." (Section 3.2.3.)

1.1 Regulatory Limits and Detection Limits. The practical significance of detection limits is best appreciated in connection with a specific external problem. Thus, based on quantitative assessment of health effects or of a new scientific phenomenon, one may conclude that it is vital to be able to detect a signal or concentration level as low as, say L_R. It follows that a measurement process having L_D no greater than L_R must be selected or, costs permitting, designed to meet the need. This is illustrated in Fig. 2 which depicts the critical level and detection limit schematically for earthquakes. The upper part of the figure presents an hypothetical relation between damage or societal cost and undetected earthquake magnitude, together with a maximum acceptable cost which fixes a "regulatory limit," L_R. (L_R might be defined, for example, by the "balance point" at which the false positive [false alarm] cost -- the cost of evacuation, is equivalent to the false negative cost -- damage incurred or lives lost in the absence of evacuation.) The lower part of the figure indicates the signal detection limit of a measurement process which meets this need. Also shown is the dependence of L_D and the two types of hypothesis testing errors on random measurement error. (The lower portion of the figure, for actual earthquake forecasting, relates to precursor measurement processes. The wealth of physical and chemical precursors utilized are reviewed by K. Mogi in Science, 1986, 233, 324.)

Two observations, perhaps obvious, follow from Fig. 2: first, a zero magnitude earthquake could not in principle be made detectable; second, with improving performance [decreased detection limit] formerly undetectable tremors will be found. Lack of appreciation of these fundamental principles of measurement may lead to regulatory difficulties, such as the requirement that any non-zero quantities of chemical carcinogens should be detectable, or that any detectable amounts should be reported (2). The latter has in effect been equivalent to a moving target, as analytical procedures continue to advance dramatically [Note 1].

A footnote on the matter of regulation, which leads directly to our next topic, relates to the relatively recent cost/benefit basis for regulatory decisions (3) and the emergence of the discipline of Risk Assessment (4). That is, that despite the lack of any explicit incorporation of a dollar value on human life in

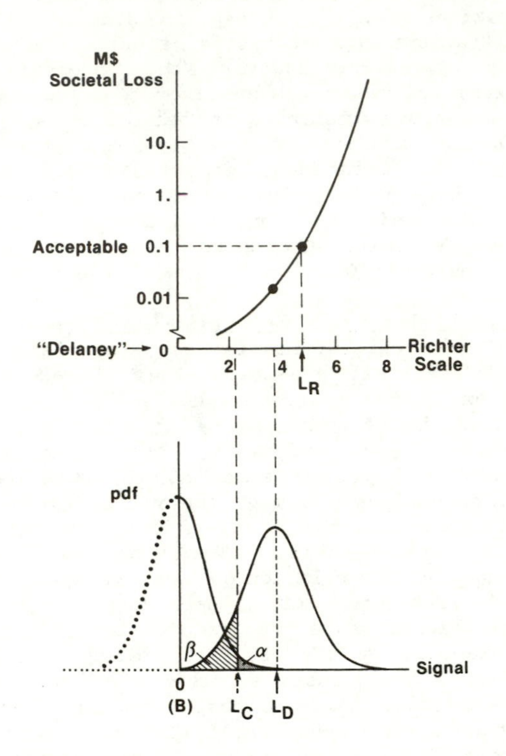

Fig. 2. Regulatory Levels [L_R] and Detection Limits [L_D]. The upper portion of the figure traces a presumed relation between earthquake magnitude [abscissa] and cost to society [ordinate]. The "Delaney amendment" viewpoint (not defined for earthquakes) might be interpreted as requiring zero societal risk and a corresponding L_R magnitude of zero, which of course is scientifically unattainable. Rather, an acceptable cost to society for undetected earthquakes, here imagined to be 0.1 M$, is used to establish the requisite "regulatory" level. The lower part of the figure represents the corresponding earthquake measurement process or precursor alarm (seismograph signal, radon emanations, biological [animal] sensors, etc.). The requisite DETECTION LIMIT [L_D] must now be no greater than L_R, and L_D in turn is related to the probability density functions [pdf] for the null signal [H_o: S=0] and the signal to be detected [H_A: S=L_D], and acceptable false decision probabilities α, β. L_C is fixed by the H_o-pdf and α; L_D is then set by L_C and β, given the H_A-pdf.

the algebra of regulation, a de facto "$2 million unwritten rule" has evolved (7). An analysis of 10 years of regulatory decisions in the US relating to chemical carcinogens showed that this value fairly consistently marked the point above which regulations were classified as too costly to impose, and below which regulations were judged as warranted.

Our use of the symbol L_R, incidentally, is **not** restricted to regulatory matters. Earthquakes, for example, cannot be regulated! Rather, L_R denotes the external limit which drives the design of our measurement process. It could apply as well to the requirements of a high quality production process, or a tracer study of long range atmospheric transport, or the investigation of extremely slow reaction processes, or in fact any of the situations indicated in Fig. 1. Perhaps it might better be labeled "reference limit (or level)" or "requisite limit."

1.2 Decision Theory and Societal Decisions/Actions. The foregoing introduction to detection theory was based strictly on the Neyman-Pearson or "frequentist" approach to significance testing and signal detection (8,9), with the exception of the imposition of an external reference or regulatory limit, L_R, based on sociopolitical and/or scientific considerations. An alternative approach, especially appropriate for (detection) decisions culminating in some kind of action, is provided by the application of Decision Theory (10), or more generally Decision Analysis (11). Although this theory may be of considerable importance for certain societal or business decision-making, its structure is such that it is not generally applied to chemical measurements.

The major advantages of the decision theoretic approach are that it permits one to apply explicit loss functions to the erroneous decisions [α,β-errors], and that it readily incorporates prior (or "subjective") knowledge concerning the probabilities of the respective hypotheses. The ability to utilize loss functions and prior probability is advantageous in that costs and beliefs and values external to the measurement process may be effectively incorporated into the decision making. A complication is that there may not be unanimity concerning the weights to be assigned to these quantities; this is somewhat analogous to the complications in reaching agreement on appropriate values for L_R. [Costs, for example, would doubtless be viewed differently by regulators and regulatees, producers and consumers, physicians and patients, etc. The issue is analogous to the question of "whose experts" are speaking in Court or advising in Congress -- i.e., it is necessarily tempered by advocacy positions.] Except when one is treating a strictly scientific question, however, it is important to realize that the losses and prior probabilities are frequently complex sociopolitical and/or economic matters, best determined by experts in those fields.

Decision theory operates on the basis of an "objective function" which is in some way optimized through the setting of a decision threshold. A lucid presentation to alternative strategies for formulating detection decisions has been given by Liteanu and Rica (8, p. 192). The essence of the matter is that a threshold value k_o for the Likelihood Ratio is derived from a) prior probabilities for the null and alternative hypotheses, b) a cost or loss matrix specifying costs associated with correct and erroneous decisions, and the probability density functions (pdf) for experimental outcomes for each of the hypotheses in question. These data are combined to compute the mean loss (or cost or risk) which is then minimized in order to derive k_o. The decision test is performed by comparing the observed (experimental) value for k with k_o. [k, the likelihood ratio, is the ratio of the pdf for H_A to that for H_o at the signal level in question.] The optimal value, k_o based on the "Bayes Criterion" is given by the product of the net cost of a false positive and the prior probability of H_o divided by the product of the net cost of a false negative and the prior probability of H_A. An interesting illustration leading to the same conclusion is given in Massart, Dijkstra and Kaufman (12, p. 516) in connection with medical diagnoses and selection of the optimal point on the Receiver Operating Characteristic Curve [ROC]. [Operating Characteristic (OC) and ROC curves will be discussed briefly in a subsequent section.] The issue of developing an "optimal" decision strategy based on the prior distributions of both well and ill patients is an interesting one. In the illustration presented in Ref. 12 (pp. 508 ff), for example, there is a presumed preponderance of healthy patients [prior distributions]. By using the distributional crossing point as the threshold, one finds that about half of the abnormal (ill) subpopulation would have been misdiagnosed! (See also Appendix H in Egan (9) for an interesting illustration of the Bayesian approach to medical decision making, and the consequent need for multiple diagnostic tests -- a non-trivial issue in the light of current efforts of major medical insurers to curtail the number of diagnostic tests.)

References (8) and (10) give alternative decision strategies -- Minimax, Ideal Observer, and Maximal Likelihood -- when only partial information is available for prior probabilities and/or costs. [The Minimax approach, for example, cuts one's losses from a wrong guess for the prior probabilities.] The effects of special preferences or aversions [e.g., to extreme cost] are discussed in terms of "Utility Theory" by Howard (11), as well as the use of Decision Analysis for designing sequential experiments and the setting of research priorities.

This brief excursion into Decision Theory is included to indicate the manner in which experimental data can be coupled with external (societal) judgments to form a logical basis for societal decisions and actions. A justification for so complex a strategy for decision making is that "simple" scientific measurements and model evaluations will always be characterized by measurement uncertainty. Yet societal decisions and actions **must** take place even under the shadow of uncertainty. For scientific measurements, as discussed in the following text, however, we shall restrict our attention to the relatively simple Neyman-Pearson hypothesis testing model (8, p. 198).

1.3 Testing of Assumptions. The detection of erroneous assumptions lies at the core of sound measurement science. It is therefore especially appropriate to include reference to Detection Decisions and Detection Limits for key assumptions in our survey of the scope of Detection. Assumptions of principal importance for chemical measurements include those relating to the functional form and parameters for a) the physicochemical (or empirical) model and b) the error model relating the experimental observations to the underlying chemical composition. Among the assumptions, or assumed parameters to be tested, the following are of special importance:

Functional Relation

- o number of chemical components
- o characteristic spectra or chromatographic patterns
- o mathematical relation for the response for each component (includes correct identification, and curve shape)
- o matrix effects and interference [interactions] among components
- o parameters such as the blank, recovery, sensitivity (efficiency)

Error Model

- o cumulative distribution function [type]
- o parameters [variance, higher moments] (variance components for compound distributions)
- o autocorrelation [non-white noise]
- o systematic error or bias [bounds]
- o blunders (discrimination from chance outliers, from discoveries)

Hypothesis testing is applicable to all of the above factors. Detection decisions may be made, for example, using the critical level of Student's-t to test for bias, or the critical level of χ^2 to test an assumed spectral shape or calibration model or error model. For a given measurement design and assumption test procedure, one can estimate the corresponding detection limit for the alternative hypothesis, e.g., the minimum detectable bias. As with analyte detection, the ability to detect erroneous assumptions rests heavily on the design of the experiment; and the study of optimal designs is a field unto itself.

A survey of several of the above model-parameter assumptions, as related to chemical component (or analyte) detection will be presented later. Let us terminate this preview with two observations: a) Tests of assumptions may themselves rest upon assumptions -- an obvious case being the use of Student's t, which rests upon the assumption of normality; b) Detection of an analyte through model failure (lack of fit) -- e.g., evaluating χ^2 when fitting a spectrum with one component missing -- is less sensitive than direct detection using the correct model. This is due to collinearity among spectral patterns (or overlapping chromatographic peaks) (13).

1.4 Analyte Detection. This is a primary focus for this volume, the specification of critical levels or thresholds for analyte detection decisions, and the design of CMP's to achieve requisite analyte detection limits. The following section includes an historical perspective on the topic. A tutorial is provided in the chapter by Kirchmer (14), where a crucial distinction is noted: that is, the detection decision is made in reference to an observed, random experimental outcome (**estimated concentration**), whereas the detection limit refers to the underlying true concentration which the CMP is capable of detecting. The chief reason for interest in the latter is advanced planning and design -- i.e., assessing the capability of the CMP in question to meet the measurement needs.

Because of the broad scope of detection, as outlined in the preceding paragraphs, it is useful to distinguish some of the quantities or events detected with appropriate symbols. For the purposes of this chapter, the following will be used:

	Critical Level	Detection Limit
generic symbol	L_C	L_D
event or system state (earthquake, oil spill)	θ_C	θ_D
analyte concentration (or amount)	x_C	x_D
instrument response (net signal)	S_C	S_D
bias	Δ_C	Δ_D
external random error (non-Poisson; "between")	σ_{xC}	σ_{xD}
model - lack of fit	$\chi^2_c, \ldots$	--

In addition to the above, L_R is used to denote the external limit which drives the design of the Measurement Process (MP). Thus, if successful process control, or early warning (natural or human disasters), or fundamental chemical research depends on achieving a limit L_R, then the MP must be so designed that its $L_D \leq L_R$.

Note that the critical level of the appropriate test statistic ($z_{1-\alpha}$, $t_{1-\alpha}$, etc) can generally be used as a normalized alternative to x_C, S_C, etc. The "detection limit" for a test statistic, however, is meaningless, as x_D, S_D, etc. refer to the true underlying quantity. A corollary is that the term "detection limit" is also without meaning in the absence of an alternative hypothesis. (This is perhaps an obvious philosophical matter, but in principle, the null hypothesis cannot be rejected, except by chance [α-error], if no alternative exists; the β-error is then necessarily undefined. Of course an unexpected rejection can lead to an exciting search for the alternative.)

2. HISTORICAL PERSPECTIVE

The dual questions, "How little can I detect?", and "Has something been detected?" have long caught the attention of analytical scientists. Throughout recent history (i.e., 20th century) a number of responses have been formulated, such as

- The intuitive [formulation]: basing detection decisions and limits on sound, but not readily quantifiable experience

- The ad hoc: selecting a rigid formula, often based on some reasonable limiting condition, via dictum, voting or consensus

- The signal/noise: generally assuming white noise, and addressing primarily testing of an **observed** signal

- The avoidance: only results or measurement processes thoroughly removed from the detection limit deserve our attention

- The hypothesis testing: where explicit attention is given to the risks of **both** false positive and false negative detection decisions.

In reviewing the history of detection limits (in Analytical Chemistry) it is helpful to keep these several, often implicit, differences in mind. If it is agreed that the concept of detection has meaning, then it is essential that the above questions be fully defined and explicitly addressed. In the view of this author a meaningful approach to analyte detection must be consistent with our approach to uncertainty components of measurement processes and experimental results; the soundest approach is probably the last [hypothesis testing] tempered with an appropriate measure of the first [scientific intuition].

Table I has been prepared from this perspective. The authors selected are drawn primarily from those who have contributed basic statements on the issue of detection capabilities of chemical measurement processes ["detection limits"], as opposed to simply addressing detection decisions for observed results ["critical levels"]. In fairness to those not listed, it is important to note that a) a selection only, spanning the last several decades has been given, and that b) there also exist many excellent articles (15,16) and books (12,17,18) which review the topic. It is immediately clear from Table I that the terminology has been wide ranging, even in those cases where the conceptual basis (hypothesis testing) has been identical. Nomenclature, unlike scientific facts and concepts, can be approached, however, through consensus. The International Union of Pure and Applied Chemistry [IUPAC], which appears twice in Table I, is the international body of chemists charged with this responsibility. At this point it will be helpful to examine the position of IUPAC as well as the contributions of some of the other authors cited in Table I.

Fritz Feigl (19), the father of "Spot Tests," heads the list primarily as one who suggested lower limits for chemical measurement, here translated (from the german) as "identification limits," which represented the best experience [or chemical intuition] of the day. Such limits, typically in the microgram range, were scarcely ad hoc, but they of course lacked the statistical sophistication of latter day limits. Feigl's limits, however, deserve our attention even today, in that they recognize the **overall** capability of the measurement process including that which cannot be readily treated by statistics [Note 2].

Table I. Historical Perspective -- Detection Limit Terminology

Feigl ('23)	-	Identification Limit (19)
Altshuler ('63)	-	Minimum Detectable True Activity (21)
Kaiser ('65-'68)	-	Limit of Guarantee for Purity (20)
St. John ('67)	-	Limiting Detectable Concentration (S/N_{rms}) (24)
Currie ('68)	-	Detection Limit (23)
Nicholson ('68)	-	Detectability (25)
IUPAC ('76)	-	Limit of Detection (29)
Ingle ('74)	-	("[too] complex...not common") (27)
Lochamy ('76)	-	Minimum Detectable Activity (92)
Grinzaid ('77)	-	Nonparametric Detection Limit (26)
Liteanu ('80)	-	Frequentometric Detection Limit (8)
NRC ('84)	-	Lower Limit of Detection [28]
IUPAC ('86)	-	Detection Limit (30)
IAEA ('87)	-	Detection Limit (93).

Among the others cited in Table I, Kaiser (20) deserves major credit for introducing the hypothesis testing concept into spectrochemical analysis, as does Altshuler (21) in radioactivity measurement. Wilson (22) championed its use for water analysis, and Currie (23) provided an approach for detection and quantification in analytical and radiochemistry. The reference by St. John (24) has been one of the most cited of those based on signal/noise, though it does not address the error of the second kind (false negative). Nicholson (25) gave one of the earliest treatments for extreme low-level (Poisson) counting data, and Grinzaid (26) offered a robust treatment not requiring the assumption of any specific distribution. Liteanu's frequentometric method (8) was also distribution-free, in the sense that an experimental estimate of the detection limit was derived from the **observed** fraction of false negatives, using a regression technique. The paper by Ingle (27), which was obviously designed to be tutorial (published in the J. Chemical Education) is noteworthy in that it suggested that the concept of the error of the second kind (which is intrinsic to the statistical theory of hypothesis testing) was simply too complex for ordinary chemists to grasp! Regrettably, there seems to be some support for such a statement; but Hypothesis Testing is one of the keystones of every elementary course in Statistics, so its formal introduction into the education of the analytical chemist would seem not too esoteric a step.

An exhaustive review of the definition and application of Detection Limits for nuclear and analytical chemical measurements was published in 1984 (28). The reader may wish to scan the titles of the papers there cited, to gain further insight regarding basic principles and terminology, counting statistics, non-counting and non-normal random errors, random and systematic variations in the blank, Bayesian approaches, reporting, averaging and censoring treatments, optimization, influence of alternative spectrum deconvolution techniques, etc. In the body of Ref. 28 special attention is given also to topics such as simple and multicomponent nuclear spectrum fitting and extreme low-level counting.

With respect to IUPAC, both the position published in 1976 (29), which addressed nomenclature, symbols and units in analytical optical spectroscopy, and the more general analytical nomenclature document, now in review (30), treat detection from the hypothesis testing viewpoint. A non-conceptual difference lies in the choice of the risk level (false positives and negatives). The 1976 report, which grew out of Kaiser's work, used a fixed value of 3.00 for the standard deviation multiplier (S/N) for detection decisions. This would correspond to a false positive risk of 1-0.9986, or 0.14% (1-sided test), if the population were normal, and σ known. (The false negative risk β was not explicitly treated.) The current IUPAC Nomenclature Document recommends risk levels (α, β) of 5%, corresponding to a multiplier of 1.645 for σ known, normal population. Both documents recognize the effects of varying degrees of freedom in estimating the variance of the blank; the latter document specifically recommends the use of Student's-t to compensate, just as is done in the construction of normal confidence intervals.

The historical evolution of this topic has resulted in some very unfortunate and needless confusion in both terminology and concept. **Awareness of the nature of this confusion is crucial**, if we as analytical scientists are to arrive at a common and meaningful approach to detection, an approach that can serve society rather than add an extra level of confusion to a topic which the public regards as already complicated, albeit important.

The facts are that for at least the last decade or two there has been broad international support for the hypothesis testing framework for making analyte detection decisions, and evaluating -- especially for purposes of design and planning -- the inherent detection capabilities of measurement processes. In this context, a number of authors and institutions have employed terms like "detection limit" (or "limit of detection") to denote the latter, inherent detection capability, generally in units of concentration or amount (8,12,17, 21-23, 25,30,36). In the earliest work of some who most strongly came to support the hypothesis testing model, however, the notion of the false negative [β - error] did not appear (31). Kaiser in particular labeled his threshold level "Die Nachweisgrenze," or Detection Limit. In 1965 Kaiser treated the second kind of error (β), and introduced "Die Garantiegrenze für Reinheit" as the corresponding true concentration level for the alternative hypothesis (32). Kaiser's impact on the field of Analytical Chemistry has been extremely significant, and it is not surprising that many chemists have adopted his terminology for the Detection Limit. It has been adopted, however, in many cases to indicate not just the signal/noise level for making detection decisions, but also as a measure of the inherent detection capability of the measurement process in question. Since the error of the second kind [β] exists whether it's recognized or not, this practice has led to a de facto false negative risk of 50% -- a value which is totally out of balance with a false positive risk of 0.14%, or even 5%! Thoughtful and lucid critique of this matter may be found in Ref's. 8 (p. 263) and 14. A curious footnote to this discussion is that one scarcely ever encounters Kaiser's second term, "Limit of Guarantee for Purity," in the scientific literature.

On the subject of nomenclature, a word concerning historically used terms for the detection *decision point* or level is in order. As stated immediately above, a number of analysts, following Kaiser, use "Limit of Detection" or "Detection Limit" as *both* the measure of (true concentration) detection capability and as a statistical critical level or threshold to make detection decisions. Following established practice in Statistics, the term "Critical Level" was recommended in (23). "Criterion of Detection" has been employed by Wilson (22); and Liteanu (8), who speaks of the "decision criterion" as a strategy, terms the numerical comparison level the "Decision (or Detection) Threshold."

The great majority of the authors cited in the foregoing discussion emphasized that the detection limit **must** refer to the entire analytical measurement process. In many cases one finds that not the case -- i.e., workers may refer (sometimes appropriately and intentionally) to just the instrumental measurement step, or to ideal, pure solution detection limits -- both of which may be far too optimistic for real, complex samples. Some compilations indicate "typical" detection limits, an acceptable practice provided the measurement process and sample nature (including matrix and interference effects) are rigidly controlled and subjected to appropriate ruggedness testing.

2.1 Present State of the Art. A perusal of the analytical literature two decades ago revealed considerable disparity in the specification of detection limits. This is shown in Fig. 3 which is reproduced from (23). Then current definitions spanned nearly 3 orders of magnitude when applied to the **same** measurement problem! Concern for such definitional (and/or conceptual) disparity has led a number of national and international organizations to address the need for a common, rational basis for treating this matter. Because of concentration related effects of trace chemical species on health, properties of high purity materials, and even global climate, relative detection limits for different measurement processes are not enough; detection capabilities must be assessed in absolute units. Awareness of the confusion surrounding detection limit practices, by organizations such as IUPAC, IAEA, ACS, a number of US regulatory agencies, and more recently CODATA (Committee on Data for Science and Technology) is a very positive thing. The difficulty and importance of the task is highlighted by several of the authors in this volume, notably: a) Crummett (33) ["In spite of extraordinary efforts (on the part of scientific societies to properly define detection limits) analysts continue to present their results in forms which cause the credibility of the data to be questioned or the meaning to be misinterpreted"]; b) Brossman (34) ["Attempts by our task force on low-level data to make a rigorous conceptual and statistical comparison ... have been unsuccessful. Even similar terms are defined in different, non-comparable ways..."]; and c) Currie and Parr (35), where it was observed that international interlaboratory comparisons involving the **same** bioenvironmental reference materials resulted in mutually exclusive results. For example, quantitative results for arsenic (in horse kidney) were reported by some laboratories at levels which exceeded the "detection limits" of other laboratories (which detected no arsenic) by as much as 4 orders of magnitude!

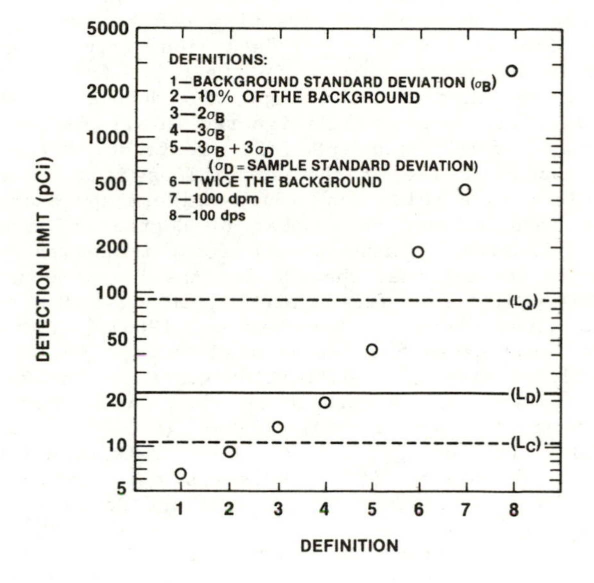

Fig. 3. Ordered detection limits -- 1968. (Reproduced from Ref. 23. Copyright 1968 American Chemical Society.)

Incidentally, the Brossman task force epitomizes a very serious problem: the coding of data into computerized data bases. Such data bases will doubtless have significant distributions and lifetimes, so the effects of possible distortions and information loss will be unfortunately amplified.

Understanding and acceptance of the hypothesis-testing position taken by IUPAC (29,30), the US Nuclear Regulatory Commission [28], the UK Water Research Centre (36), the IAEA, and reflected in many of the recent texts in Analytical Chemistry and Chemometrics (37), promises to resolve the needless, current disarray. Some of the current diversity can be seen in Fig. 4, which presents four of the principal detection limit definitions in vogue (and/or in regulatory guides) in the U.S. Comparisons among the statements, together with the supporting documents, show that: (1) the β error (false negative) is ignored in all but one, causing it to assume a de facto value 50%; (2) treatment of the blank is ambiguous or absent in two of the definitions, and restricted to the reagent blank in a third; and (3) a uniform approach to the α error, taking into account the number of degrees of freedom and Student's t is lacking. There is some irony in the fact that the fourth definition states that the LOD is "the lowest concentration ... statistically different from a blank", in view of a comment in the reference cited (Long and Winefordner, 1983). These authors note that the "well-based but seldom used concept in the calculation of detection limits ... the limit of guarantee for purity, c_G, described by Kaiser ... [represents] the lowest statistically discernable signal." Long and Winefordner go on to show that the original IUPAC definition (29) and the LOD of reference (16) indeed yield 50% false negatives (β). Kaiser's c_G, incidentally, is conceptually identical to Currie's L_D (23) and Boumans' "Limit of Identification" (15).

In addition to the above, one continually finds varying ad hoc or even undefined usage in the peer-reviewed analytical literature. For example, three recent papers, examined because of interest in their chemical content, all deemed detection limits of sufficient importance to include numerical tabulations. However, the first author stated that his detection limit represented a signal to noise ratio of 10; the second defined it as twice the standard deviation of the background signal; and the third gave no indication as to his meaning.

In conclusion, it is urgent that the analytical community adopt a uniform and defensible approach to the concept of detection. Apart from ad hoc or unstated procedures, failure to recognize the error of the second kind [β] -- i.e., failure to distinguish between detection decisions and detection capabilities -- is the most serious conceptual fault, placing false negatives at the level of coin-flipping accuracy. Failure to take into account all major sources of error, especially the nature of the blank, is the most serious measurement fault. A review of some of the more critical assumptions and technical issues related to valid detection limits follows.

Lower Limit of Detection (LLD). "The LLD is defined, for purposes of these specifications, as the smallest concentration of the radioactive material in a sample that will yield a net count, above system background, that will be detected with 95% probability with only 5% probability of falsely concluding that a blank observation represents a "real" signal" (94).

Instrumental Detection Limit (IDL). "The concentration equivalent to a signal, due to the analyte, which is equal to three times the standard deviation of a series of ten replicate measurements of a reagent blank signal at the same wavelength" (95).

Method Detection Limit (MDL). "The method detection limit (MDL) is defined as the minimum concentration of a substance that can be measured and reported with 99% confidence that the analyte concentration is greater than zero and is determined from analysis of a sample in a given matrix containing the analyte" (96).

Limit of Detection (LOD). "The limit of detection (LOD) is defined as the lowest concentration level that can be determined to be statistically different from a blank. The concept is reviewed in [ref. 38] together with the statistical basis for its evaluation. Additional concepts include method detection limit (MDL), which refers to the lowest concentration of analyte that a method can detect reliably in either a sample or blank, and the instrument detection limit (IDL), which refers to the smallest signal above background noise that an instrument can detect reliably. Sometimes, the IDL and LOD are operationally the same. In practice, an indication of whether an analyte is detected by an instrument is sometimes based on the extent of which the analyte signal exceeds peak-to-peak noise" (16).

Fig. 4. Four Definitions for Detection Limits Related to Current U.S. Regulatory Practice.

3. DEPTH: Limitations, Assumptions, and Technical Issues

The foregoing text represents a brief overview of some of the societal, historical, and broad conceptual issues relating to detection and chemical measurements. Here we offer an overview, in catalog or dictionary format, of a series of technical issues directly related to the estimation and validity of analyte detection limits. Balanced coverage has been the intent, but special attention has been given to topics not covered elsewhere in this volume, and to questions arising in discussions or put by "users" of detection limits. In some cases, this led to the introduction of new material such as multiple decisions and probabilistic pattern detection, utilization of physical constraints (on variance), and some effects of varying probability density functions [pdf] as related to experimental design and σ variation. The discussion is divided into three parts: the first, considering issues affecting the validity of detection decisions [null hypothesis testing]; the second, considering the type II error [β] and detection for the alternative hypothesis; the third, considering multiple detection decisions and Discrimination Limits for chemical species and chemical patterns. A guide to the topics presented in this section is given in Fig. 5 [Note 3].

3.1 Null Hypothesis Testing -- Assumptions and Conclusions. When an experiment is performed, we test the experimental result ($\hat{x}$) by comparison to the critical level or threshold (x_C) to decide whether or not analyte has been detected in excess of the blank or background level. Quite apart from questions involving the alternative hypothesis or detection limit, two crucial points must be kept in mind concerning the nature and validity of such a test. The first is that the **Statistical Test for Significance**, at significance level α, is based on **exactly** the same principles as the more fashionable calculation of **Confidence Intervals**, at confidence level 1-α (39). (In both cases, of course, one must pay attention to 1- vs 2-sided tests or intervals.) The second point follows: that assumptions affecting the validity of experimental confidence intervals are just as important for the validity of significance tests. Assumptions which demand our attention include the following: control [i.e., existence] of the measurement process; possible systematic model [functional] or measurement error; and properties of the random component (or components) of error -- i.e., form of the distribution [pdf or cdf], parameters of the distribution [σ^2,...], "color" of the noise (or noise power spectrum), and non-stationarity (changes with time) and heteroscedasticity (changes with concentration or other experimental parameters). Errors may arise also from uncompensated changes in the measurement process itself, such as alteration of the calibration [functional] model or random error model [cdf] due to chemical matrix effects or interference. No less important are the computational and data reporting strategies; these represent intrinsic parts of the **overall measurement process**. A brief catalog of selected hypothesis testing and experimental result related issues follows.

<u>Detection Decisions, α-error [3.1]</u>
(assumptions, validity)

—BASIC ERROR ISSUES
- systematic error [3.1.1]
- normal random error [3.1.2-.3]
- non-normal [3.1.4-.5]
- paired comparisons [3.1.6]

—MEASUREMENT PROCESS ISSUES
- background, baseline, blank [3.1.7]
- error components, truncation [3.1.8-.10]
- evaluation process, calibration [3.1.11-.12]
- reporting low-level data [3.1.13]
- artificial thresholds [3.1.14]

<u>Analyte Detection Limit, β-error [3.2]</u>
(estimation, power)

—DETECTION LIMITS AND POWER
- ignorance of β-error [3.2.1]
- lower, upper L_D's [3.2.2]
- α, β connection (ROC) [3.2.3]
- power of the t-test [3.2.4]

—UNCERTAINTY IN L_D [3.2.5-.6]

—SPECIAL TOPICS
- optimization [3.2.7]
- multicomponent detection [3.2.8]
- random error variation [3.2.9]
- quality (algorithms, controls) [3.2.10-.11]

<u>Discrimination Limit; Multiple Decisions [3.3]</u>

—DISCRIMINATION LIMITS
- lower and upper regulatory limits [3.3.1]
- impurity detection [3.3.2]

—MULTIPLE DECISIONS, IDENTIFICATION
- multiple detection decisions [3.3.3]
- multichannel identification [3.3.4]
- multivariable patterns [3.3.5-.6]

Fig. 5. Topical Guide to Technical Overview.

3.1.1 <u>systematic and model error.</u> Bounds for uncompensated, non-random errors must be allowed for by a corresponding increase in the critical level or confidence interval. This makes α an upper limit if the systematic bounds, which need not be symmetric, are given as upper limits. The validity of the corresponding uncertainty interval clearly depends heavily on the Chemical Intuition or scientific expertise employed, for example in identifying the range of possible alternative models.

3.1.2 <u>normal (white) random noise.</u> If σ is known, $L_C = z_{1-\alpha}\sigma_o$, where σ_o represents the standard deviation of the estimated net signal (23). If "simple" detection [gross signal - blank] is involved, where the blank is estimated from n equivalent observations, and the gross signal from one, then $\sigma_o = \sigma_B \sqrt{(n+1)/n}$ -- σ_B being the standard deviation of the blank. [Note that zero adjustment, as for the null level or baseline of a (recording) galvanometer, chromatograph, spectrophotometer, etc., does **not** eliminate the need for this blank or baseline estimate error propagation. Of course, σ_o may be scarcely greater than σ_B when a large span of linear baseline is quite precisely adjusted, for example, by least squares fitting or by graphic or even "eyeball" subtraction. A related point: the (detection) test must be applied to the **net** signal, or equivalent concentration estimate, because only that has an expected value under the null hypothesis of exactly zero. Imprecise knowledge of the mean value for the blank distribution prevents a rigorous test being applied to the gross signal. See comments below on the background, baseline, and blank.]

3.1.3 <u>σ unknown (normal).</u> Student's t replaces z when σ is estimated by replication. L_C now equals ts. [Note that L_C here, unlike L_D, is no longer a constant, because s (estimate of σ) is a random variable. Note also that σ (or s) as used in this text refers to the standard deviation of the "final" signal; if signal averaging or least squares fitting is employed to arrive at the final signal, then this should be interpreted as the standard error.]

3.1.4 <u>non-white noise.</u> The autocorrelation function (or spectral power density) must be taken into account in calculating critical levels or confidence intervals. This is not a trivial matter, and is remarkably often ignored. Its importance is seen most often for time dependent phenomena [e.g., in chromatography], and where "flicker noise" is found. Note that noise of this sort sets limits to the gains which may be achieved through signal integration or averaging. Note also that detection limits based on the Signal to Background Ratio derive from the assumption of background-carried flicker noise dominance. See especially Smit (40) and Epstein (41) for important discussions of this topic. In a broader sense the underlying issue relates to the limit in the information content of sets of observations which are not fully independent. One encounters it also when interpreting uncertainties for count rate meters (RC signal averages - (28, p. 96)), and uncertainties in functions of partially correlated random variables [error propagation, including covariance: (42)]. The topic is of special relevance when considering instrumental baselines [see below].

3.1.5 non-normal distributions. This problem can be dealt with rigorously if one knows the form of the random error distribution. A notable example of this occurs in "counting" experiments (e.g., radioactivity), where the physics of the process implies Poisson statistics. As the Poisson distribution is discrete, y_C (the critical level for gross counts) takes on integer values only, and α is generally in the form of an inequality -- ie, $\alpha \le 0.05$ (28). Distribution-free techniques, especially those based on order statistics (such as the median and its confidence interval), and transformation techniques (eg, for log-normally distributed errors), are often appropriate (26,57). So-called non-parametric techniques -- the Gauss or Chebyshev inequalities, give (2-sided) α's as no greater than $(2/[3k])^2$ and $1/k^2$ respectively, where the standard deviation multiplier k replaces z of the normal distribution. Note that the Gauss Inequality is applicable for random variables having unimodal, continuous, and symmetric density functions, whereas the weaker Chebyshev Inequality is valid for any distribution having finite mean and variance. A small problem in applying the inequalities is that k must multiply σ, and σ is not generally known. Although s^2 is an unbiased estimate for σ^2 even for non-normal distributions, bounds for s/σ are distribution dependent and therefore also not generally known. [Note 4.] Difficulties are compounded when the measurement process consists of two or more steps comprising different kinds of pdf's. [See for example, Johnson, Ref. (44).]

Recommended solutions for the non-normality problem are: 1) use the percentage points of the actual pdf, if known; b) transform to normality; c) use order statistics; d) design the experiment to take advantage of "pairing" and the Central Limit Theorem. The last approach, which looks very attractive for chemical research, will be discussed below. Information on the other approaches may be obtained from specialized statistical texts (45).

3.1.6 paired comparisons; Central Limit Theorem. The Central Limit Theorem makes quality control charts work. Here, one charts sets of **averages** of observations and checks for excursions beyond Normal control limits. The averaging is done **not** primarily for standard error reduction, but to assure (approximate) normality. It can be shown that averages (or sums) derived from of a sequence of mutually independent random variables having a common distribution tend toward normality, often rather quickly (by the time n = 3 or 4). This "Central Limit Theorem" is valid regardless of the shape of the initial distribution, so long as it has finite variance. The rate of approach to normality, however, depends on the initial shape, being faster for symmetric distributions (45). For low-level chemical measurements, all too often the blank, which forms the basis for the detection decision, is **neither** symmetrically nor normally distributed -- especially when the blank is due to environmental or particulate contamination (46). Very wrong trace analytical confidence intervals and detection decisions may result. By designing the measurement process so that proper paired comparisons can be made (45, Chapt. 4), one can at the same time achieve the best statistical sensitivity and force symmetry (for the estimated net signal), and thus set the stage for approximate normality for averages of such estimates. Two other major reasons for forcing symmetry in this way are: a) to take advantage of the median and its confidence interval for robust estimation, and b) to make possible the use of the Gauss Inequality for distribution free interval estimation. This approach has been suggested, for example, as a possible solution to the severe detection discrepancies obtained in IAEA intercomparisons (35). Specifically, the recommendation is to make detection decisions by comparing the averages of at least n=4 paired comparisons (gross signal - equivalent blank signal) with $ts/\sqrt{n}$, where s^2 is the estimated variance of the n net signals. (Note the direct analogy between the chemical blank measurement, and the "control" observation which plays a central role in clinical and psychological null hypothesis testing.) In anticipation of the next two topics, two related advantages of proper pairing (or equivalent fitting) may be stated: a) bias associated with systematic B-changes is minimized by taking "local" differences (y - B) and using local values for recovery and instrumental detection efficiency; b) effects of imprecision associated with certain "external" random variations (e.g., "between-day") may similarly be avoided.

3.1.7 background, baseline, blank. The variability of the null signal [B] is **the determining factor** in making valid detection decisions (H_o tests) or in deriving valid confidence intervals for low-level signals. In the ideal interference free, "pure solution" measurement environment, the instrumental background is the ultimate limiting factor. In this sense an "instrumental critical level" (decision level, threshold) and the corresponding detection limit mark the best possible performance of a system. Two cautions are in order, however. First, the instrumental noise (background variability) may not be white -- i.e., there may be long- or short-term variations which must be compensated for by appropriate modeling and/or astute paired comparisons. Second, when the chemical, physical, or geometric configuration of the final sample changes, there may be corresponding changes in the effective background (for example due to changes in external scattered radiation). Instrumental detection efficiencies or responses may also be perturbed by such sample-related factors (47), but that is a "calibration" matter, to be taken up separately.

Multicomponent instrumental responses, unless totally selective, generate spectral or chromatographic **baselines** which arise from complex physicochemical phenomena ranging from multiple (sample) particle or radiation scattering to component tailing, depending on the specific analytical technique involved. Such baselines generally subsume any instrumental background, and thus become the limiting factor. Valid net signal estimates and detection decisions then become critically dependent on accurate modeling of the baseline functional shape and noise structure. Empirical baseline shape models are common, the linear model being most used. Deviations from linearity may be modeled using low order polynomials or splines, but since the modeling is empirical one must be alert to possible model errors, such as unanticipated fine structure (48). The noise structure of the empirically modeled baseline deserves special attention in the case of drift, preferred periodicity, or more general autocorrelation such as "1/f noise" (40,41,49).

For all real chemical measurements, the chemical **blank** is the actual limiting factor. To assess its magnitude and variability, there is no better approach than to apply the **entire** measurement process to an adequate number of real blanks. Unfortunately, this ideal may not be realizable, as it requires samples which are identical to those of interest in all respects except for the absence of the target analyte. The alternative approach is to attempt to "propagate" the components of the blank for each step of the CMP, taking into account the points of introduction, and subsequent recoveries and CMP-induced variations. This topic is enormously important and enormously complex. One must consider simultaneously: the effects of multiple blank sources; analyte, blank, and interferant recoveries for each CMP step; and instrumental detection efficiencies plus matrix effects for each (35, 50-52). A small complication arises from the fact that different types of blanks may exhibit different pdf's. Reagent and sample preparation blanks tend to be normally distributed, while environmental blanks are frequently log-normal (44,46,53). In the final analysis, of course, actual variations in the blank are convolved with instrumental noise. Lack of independence or normality for either will be reflected in the final, effective blank distribution. A final observation: systematic or random error in the estimated blank affects not only detection limits and confidence intervals for "low-level" samples; it may also limit the accuracy of high precision, "high-level" samples.

3.1.8 internal vs external error: propagation vs replication. The uncertainties of low level concentration estimates may be derived from error propagation for each stage or step of a compound CMP, or they may be deduced from replication and comparison with low-level SRMs for the overall measurement process, as in laboratory intercomparisons. Consistency between the two approaches is essential for the uncertainty estimates to be considered valid. Among the error characteristics that may be exposed through such ANOVA type testing are: "excess" random error [σ_x], systematic error [Δ], and covariance among internal errors. The first (σ_x) might represent, for example, a between day or between lab variance component, or in the case of Poisson counting statistics, an excess or non-counting component of random error. The second (Δ) could be manifest as the difference between the limiting mean for an intralaboratory measurement and the true (e.g., SRM) value. The third might be seen if internal, non-white noise or other error correlation effects were improperly accounted for in error propagation; the external estimate, derived from independent replicates would automatically compensate for such (internal) behavior. Comparison of internal and external estimates has its limitations, however. This is shown in Fig. 6 [$(\sigma_x,\Delta)_D$ vs n] which displays the detection limits for σ_x and Δ as a function of the number of replicates (54). Thus, in order to detect bias equal in magnitude to the standard deviation, one needs at least 12 degrees of freedom (13 replicates). To detect an extra variance component equal in magnitude to the known internal precision [$(\sigma_x)_D=\sigma_i$], one needs 46 degrees of freedom (55). These error components [σ_x, Δ] **must**, however, be taken into account if valid detection decisions are to be made.

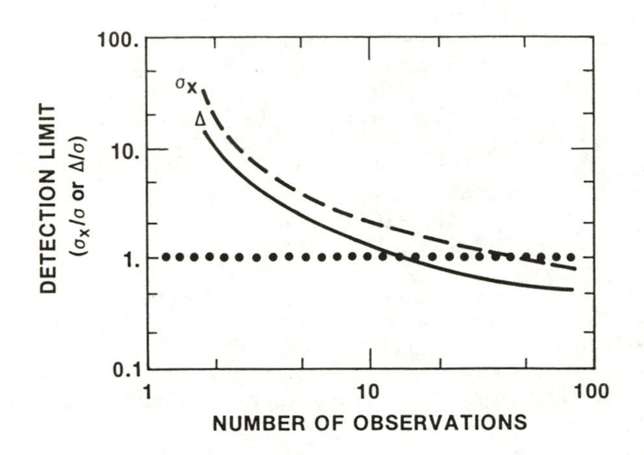

Fig. 6. Detection Limits for Bias [Δ] and excess (external) random error [σ_x] vs the Number of Observations. (Adapted from Ref. 54.)

3.1.9 <u>internal error precisely known: improved detection decisions (and confidence intervals) using inequality constraints.</u> If the excess random error component is well known, then obviously error propagation can be applied -- e.g., $V_t = V_i + V_x$ -- to calculate the total variance V_t, which is the quantity that must be used for calculating the critical level or confidence interval. (For notational simplicity, V is used here to denote variance, in place of σ^2. If multiplicative rather than additive relations are involved V would represent relative variance.) Quite frequently we find it relatively inexpensive to obtain precise estimates of V_i (internal variance) whereas external replicates -- e.g., between laboratories or between days, etc., -- tend to be more costly, hence fewer. Perhaps the extreme case of this sort occurs with Poisson counting statistics, where $V_i \approx N$ where N is the observed number of counts. If expected value of N is sufficiently large (e.g., >60) the expression yields a reasonably good estimate for V_i, as good as would result from 100 or more replicates. If V_x is taken to be zero, the critical level may then be calculated directly from V_i which here is estimated as the number of counts [<u>Note 5</u>].

If V_x is not known, we have three alternatives. One of extreme conservatism would be to use the lower and upper limits for V_x, based on replication [s^2=est(V_t)] and knowledge of V_i. Intermediate, and most common, is simply to calculate L_C as t•s, where t is based on the external number of df (number of replicates minus one). More interesting is the use of our knowledge that $V_x \geq 0$, and a **variance weighted t**. (A fourth alternative, ignoring the possible existence of V_x is all too common; the unsupported assumption that counting statistics or other internal instrumental variance fixes the overall imprecision can generate L_C's and CI's that are too small and correspondingly large false positive probabilities.)

The merit of the variance weighted t technique is that it permits us to use our excellent knowledge of V_i together with the fact that V_x cannot be negative to obtain a significantly smaller L_C or CI than we would using s^2 directly. It provides protection against unanticipated external random error with little penalty if that error component is in fact negligible. The technique suggested here is tentative and approximate, but it appears to be conservative and asymptotically correct [<u>Note 6</u>].

To illustrate, let us consider triplicate measurements of a sample using a counting technique, such as ion counting mass spectrometry or photon counting in optical spectrometry, X-ray fluorescence analysis or gamma ray spectrometry. Internal variance derives from Poisson counting statistics [V_i] where the appropriate value of t_i equals its normal limit z_i or 1.645 for $\alpha = 0.05$. Total variance [V_t] for the 3 replicates is estimated as s^2, where t_t is 2.92 for 2 df. Excess variance [V_x] is $V_t - V_i$, and estimated as $s^2 - V_i$, with the constraint that V_x may not be negative. L_C' (or CI) is calculated as $t's'/\sqrt{n}$ where:

$$s' = \sqrt{V_i + V_x} \geq \sigma_i, \text{ i.e., } s' = \max(\sigma_i, s) \quad (1)$$

$$t' = t_i (V_i/V_t) + t_t (V_x/V_t) \quad (2)$$

For the example at hand, we estimate V_x as $s^2 - V_i$. (Eq. 1) thus yields $s' = \sigma_i$ if $s \le \sigma_i$, or $s' = s$ if $s > \sigma_i$. (Eq. 2) becomes $t' = 1.645\ (k) + 2.92\ (1-k)$ where $k = \sigma_i^2/s^2$. For example, if σ_i is equivalent to 1.75 ng-Ca, and s, to 3.04 ng-Ca, k would equal $(1.75/3.04)^2 = 0.331$. As a result, $s' = 3.04$ ng-Ca, $t' = 2.50$, and $L_C' = 4.39$ ng-Ca. Investigation of the properties of L_C' for $V_x = 0$ to $V_x >> V_i$ shows it to be conservative [$\alpha' \le 0.05$] with a limiting value when $V_x = 0$ of approximately 0.03. Also in this limiting case, L_C' on the average is only slightly greater (< 10%) than it would be if one **assumed** V_x was identically zero, whereas for the conventional approach [$L_C = 2.92(s/\sqrt{n})$] it would be 78% larger. When the stakes are higher -- e.g., CI's or L_C's for $\alpha = 0.01$ -- the contrast becomes even greater. In effect, we have used our knowledge of V_i to exclude very small values for estimated total σ, and gained smaller CI's, L_C's, and detection limits in return.

3.1.10 <u>effects of rounding and truncation.</u> Premature rounding of experimental data distorts its error distribution, resulting in erroneous conclusions regarding the shape of the distribution, its parameters [mean, variance], and results of statistical tests (e.g., detection decisions, quality of fit) and confidence intervals. The most obvious distortion is that an inherently continuous distribution is made discrete; the effect is analogous to "discretization noise" which is often found with multichannel and multidetector array techniques involving windows in time, space, energy, wavelength, etc. (56). The tolerable degree of rounding depends on the distribution. For normally distributed data, there is about a 10% chance of finding results within $\sigma/8$ of the mean. Scale divisions much smaller than $\sigma/4$ are therefore required if one is to avoid false coincidences, and fits that are "too good", etc. In fact, clues to excessive rounding or truncation may be found in χ^2 or F statistics which are unusually small, or in pdf's exhibiting unexpected deviations from normality (<u>57</u>). Abnormality is noted also by Cheeseman and Wilson for constrained balance-point measurements, such as the galvanometer needle which is physically confined to non-negative scale readings (<u>36</u>). The importance of these considerations for databases incorporating low-level results is discussed in (<u>34</u>).

3.1.11 <u>the evaluation process [data reduction: fitting].</u> The data evaluation process [EP] is an integral part of the CMP, and as such it helps define σ_o and the critical level. It is perhaps obvious then that L_C, CI's, and the detection limit will differ for the **very same experimental data**, depending on the EP applied. A simple illustration is found in the fitting of spectral or chromatographic peaks. One may use the peak height as the quantitative signal measure, or a model-independent peak area may be used, or a more sophisticated technique such as linear or non-linear least squares may be employed to estimate the peak size according to a selected functional model such as a Gaussian or skewed Gaussian (<u>58</u>). The point is that without explicit specification of the entire CMP, **including the EP employed**, the detection characteristics of the measurement process are undefined. Because of this, a slight problem occurs when the EP is given as a "black box", or algorithm whose characteristics are unclear. (This issue, including the common availability of executable software without source code, will be treated further in the discussion of detection limits in the next section.)

When the EP comprises linear computations (linear in the observations) such as simple differences, y - B, or linear least squares or linear multivariate computations, initial normality (of the observations y) is preserved for the estimated quantities. Non-linear computations, such as arise commonly in iterative model selection and peak search routines, produce estimated parameters having non-normal distributions (59). Caution is in order, in those cases, in applying "normal" values of test statistics to calculate L_C and CI's. (Other factors to consider are the extent of non-linearity, the level of confidence or significance [1-α], and the robustness of the statistic in question.)

Finally, it should be noted that an erroneous model will give erroneous results. This seeming truism is important because models which pass statistical tests [e.g., χ^2 test of fit] are consistent with the data but not necessarily correct. Because of multicollinearity, model error may go undetected, while producing significant bias in the results (48).

3.1.12 calibration error. A number of different approaches may be taken to incorporate the uncertainty in the calibration factor A into the critical level. To illustrate, let us consider the simplest functional relation for the Evaluation Process:

$$\hat{x} = (y - \hat{B})/\hat{A} = \hat{S}/\hat{A} \quad (3)$$

Unfortunately, this is already a non-linear relation, so we cannot expect $\hat{x}$ to be normally distributed. If the relative error in $\hat{A}$ is small (e.g., < 10%) its influence on L_C is likewise small, and deviations from normality are minimal. If the relative uncertainty in A is not necessarily small, or if it includes possible systematic error, a straightforward approach is to use the lower bound for A to calculate an upper bound for L_C (here x_C) which can be used to make conservative detection decisions [$\alpha \leq 0.05$]. (Incorporation of bounds for systematic error is discussed more fully in the section on detection limits.)

Error propagation from the fitting of a calibration curve can be used to treat detection and interval estimation (almost) rigorously provided the model is correct -- a caution being that the intercept-B may not represent the blank-B (60,61). An interesting alternative is to estimate σ_o and the corresponding detection characteristics directly for $\hat{x}$ [i.e., in units of concentration] by full replication of the CMP at the levels of concern, observing y, B, and A for each replicate. (This is the "paired comparison" concept extended to calibration, where a blank and standard is run for every sample.) The statistical properties of the observed $\hat{x}$ distribution can then be used to directly calculate x_C [as ts_o, if A-variation is not too great] and estimate the detection limit. An added benefit of this scheme is that direct observation of the blank decouples it from the calibration curve fitting process, so that an assumed straight line model [constant sensitivity A] can be tested by fitting a line [S=Ax+e] through the origin (62). For valid conclusions, of course, due attention must be given to interference and matrix effects on both parameters, B and A. An illustration of an observed $\hat{x}$ distribution for the null hypothesis [x = 0] is shown for ^{131}I in (44).

3.1.13 reporting of low-level data. Problems associated with data rounding and truncation extend to the reporting of final results. Also, just as in the case of the data evaluation step of the CMP, reporting must be treated as an integral part of the overall CMP. Bias and information-loss are the prime considerations. At the lower extreme, where x = 0 [null hypothesis] suppression of negative estimates forces a positive bias, on the average. Other biases arise when all non-detected results are reported as zero or as equal to (or less than) the detection limit. The difficulties are evident as soon as one attempts to: develop a database comprising large amounts of low-level data (34); to compute temporal or spatial averages for higher order detection decisions (28); or to compute average concentrations across different materials as in the USDIET-1 exercise (63). This last example illustrates the point: composite samples of the U.S diet were prepared for measurement of a broad range of essential and toxic chemical constituents, including the trace element Se. Comparison of the result for Se in the composite sample [128 μg/g] with the weighted average from the large number of individual contributing foods [100 μg/g], showed a significant negative bias for the latter. This was a result of setting all "trace" observations (defined as those below a quantification limit, L_Q) to zero. Adjusting these upward to $L_Q/2$ led to an improvement [110 μg/g], but negative bias was still apparent. These kinds of problems can be completely circumvented if concentration estimates, **even if negative**, are always reported together with their uncertainties (64, 65). Detection decisions can be made by comparison with L_C, and upper limits may be given as $\hat{x} + ts/\sqrt{n}$.

3.1.14 thresholds. The threshold for discriminating "real signals" from blanks may be set in various ways. The only way that is consistent with the relation between confidence intervals and significance tests is the one described, $L_C = ts$. Other techniques include the use of a constant multiplier ks or $k\sigma$, with k = 3 a popular choice; and use of a fixed threshold signal or concentration, such as 1 mV or 2 ng. A drawback of these alternative techniques is that they seldom recognize the existence or magnitude of the α-error, which, however, does exist, and which will take on varying values depending on the number of degrees of freedom or the magnitude of the fixed threshold in comparison to σ_o. For confidence intervals α is conventionally taken as 0.05, so there seems little justification for depressing it by a factor of forty to 0.0013 (corresponding to 3σ) for detection decisions [σ-known], or by more than a factor of ten (corresponding to 3s) for 20 degrees of freedom. Instruments having hardware or software discriminators may have resulting dead zones which correspond to vanishingly small α's. In one case recently, the threshold was set so high that α was beyond the range of any of the statistical tables, $L_C/\sigma_o \approx 34$; in fact the threshold was so high that the critical level exceeded even the conventional limit of quantification -- i.e., the RSD at L_C was but 5% (61, p. 76, case-e)! Such high and varying thresholds for detection decisions lead both to needless confusion and to measurement processes operating far short of their inherent capabilities.

3.2 <u>Matters Concerning the Error of the Second Kind, and the Analyte Detection Limit.</u> The concern in the preceding section was the validity of detection decisions, based on comparisons of experimental outcomes [**estimated signals or concentrations**] with appropriate critical levels or decision thresholds. Here, we turn to issues concerning the inherent detection capability of the CMP in question -- that is, the **true signals or concentrations** which can be detected with, for example, a 95% probability [β=0.05], **given** the critical level (or equivalently α) to be used for testing observed results. L_D is thus tied intimately to β, and to α or L_C. Although a significance test may be performed with no consideration of H_A or the detection limit, L_D is ambiguous without the specification of β and α (or L_C). That is, for a given L_D there is an infinite set of possible α, β pairs. Passing a significance test -- e.g., $\hat{x} \leq x_C$ -- is commonly said to mean "acceptance" of the null hypothesis -- i.e., x = 0. This is unfortunate terminology, for only consistency with the null hypothesis has been demonstrated. "Proof" of the null hypothesis (within certain fuzzy bounds) demands attention to all possible alternative hypotheses H_A; that is the test in use must be sufficiently powerful to "detect" [$\beta \leq 0.05$, given $\alpha = 0.05$] H_A. A major reason for interest in detection limits is thus to allow us to select or design a measurement process having the capacity to detect signals or analytes at prescribed levels of importance. An overview of selected technical issues follows.

3.2.1 <u>ignorance of the error of the second kind (β).</u> False negatives occur whether their existence is recognized or not. The common practice of making detection decisions at the so-called detection limit, or LOD, etc., has the effect of setting $L_D = L_C$, with the result that β = 50% -- equivalent to the proverbial flip of the coin. With a σ-coefficient of 3, α may be as small as 0.0013, resulting in an imbalance [β/α] of a factor of nearly 400! Ignorance of this matter makes possible inadvertent or even intentional misrepresentation of detection capability. For example, the subtle trade-off between α and β could be employed to avoid penalties for false positives associated with an inadequately controlled blank.

3.2.2 <u>lower and upper detection limits.</u> For certain types of chemical measurements there are dual null hypotheses and consequently dual L_C's and L_D's for concentrations differing from these null levels. Examples are found where a lower limit is set by background noise, and an upper limit, by some type of maximum signal limitation such as instrumental detector saturation. A dual illustration is shown in Fig. 7 for two exponential phenomena, radioactive decay and radiation absorption. In each case the lower L_D is given by the smallest detectable difference from a comparator (zero age standard or blank solution), and the upper L_D is given by the smallest detectable difference from an infinitely old sample (no net emitted radioactivity) or an infinitely absorbing sample (no net transmitted radiation.

3.2.3 <u>the α - β connection: OC and ROC curves, and detection power.</u> A convenient way to visualize the relationship between false positive [α] and false negative [β] errors and the normalized difference [d] between the means of two populations for a given statistical test has come to us from signal detection theory [9].

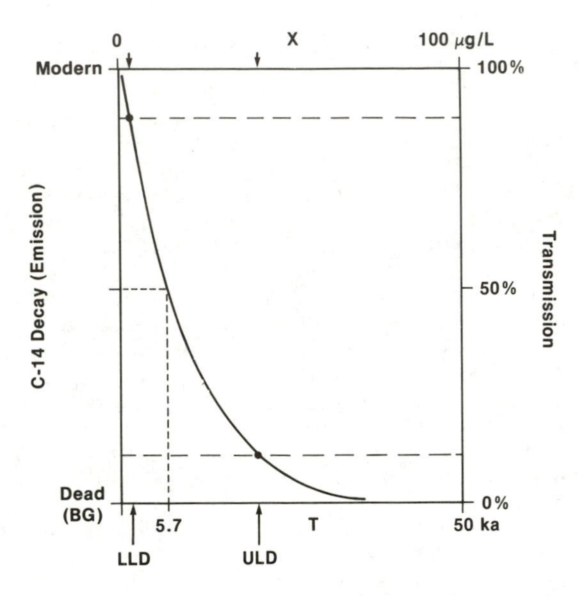

Fig. 7. Lower and Upper Detection Limits. When a measurement process has both minimum and maximum signal bounds, as in radioactive decay and optical absorption spectrometry, LLD and ULD must both be considered. Dashed line **signal** lower and upper detection limits map onto the age and concentration lower and upper limits (arrows) via the exponential function.

In this theory the Receiver Operating Characteristic [ROC] curve for a given test traces the relationship between the true positive probability [1-β] and the false positive probability [α] for a given mean normalized signal difference d. Another representation of the relationship, denoted the Power Curve in statistics, is the curve which traces the relation between Detection Power -- which is synonymous which the true positive probability [1-β] -- and the difference d, for a given value for α. (The complementary relation: β vs d, given α, is described in statistics as the Operating Characteristic [OC] curve.) Fig. 8 shows the normal ROC curve for d = 3.29 in units of σ_o [i.e., the detection limit], and the power curve for α = 0.05. The former [ROC] representation is the more convenient for the comparison of tests and the selection of alternative α, β pairs, for a given difference in population means. For this reason, it is used to compare the diagnostic power of alternative tests in clinical chemistry, where there are two discrete populations [d - fixed] (66). It may be useful also for examining the value of a test or the selection of an "optimal" α, β pair in a regulatory setting where, for example, a specified difference [d_R] is of concern. Also, if the sources or identities of chemical species are characterized by unique element or isotope ratios, an ROC curve could be used to represent the discriminating power of selected measurement techniques. The second [Power curve] representation is more appropriate when one is interested in the detection power as a function of (net) signal level or concentration. Thus, it is clear from the curve that the power is but 50% when d = 1.645. A second scale on the abscissa makes it convenient in this representation to see the relation in units of concentration.

OC and power curves are regularly used in the evaluation of statistical tests (67,68). Similarly, one finds ROC curves employed in medicine and psychology [12, discussion & references in Chapt. 25]. They appear to be little used in Analytical Chemistry, though Liteanu and Rica have proposed the use of different two dimensional projections of the three dimensional relationship [α, β, d] as representations of the "detection characteristic" (8).

3.2.4 power of the t-test. The three dimensional relationship described above is expanded to four for Student's t, with the addition of the number of degrees of freedom. If we restrict our attention to the detection limit, by fixing α and β both to 0.05, the remaining two dimensions can be viewed as a curve, d vs df -- i.e., the detection limit (in units of σ_o) as a function of the number of degrees of freedom, where $t_{1-\alpha}$ is used for making detection decisions. In this case, the value of d is determined by requiring a 95% probability (1-β) that the estimated net signal divided by its estimated standard deviation [$(y-B)/s_o$] will exceed the critical level for Student's t. This ratio is called the non-central t, with non-centrality parameter d, because it is displaced from zero by this amount. The net signal detection limit is given by $d\sigma_o$. Another important application of the non-central t distribution has been to test the validity of presumed detection limits, for example, in connection with medical diagnostic devices (69).

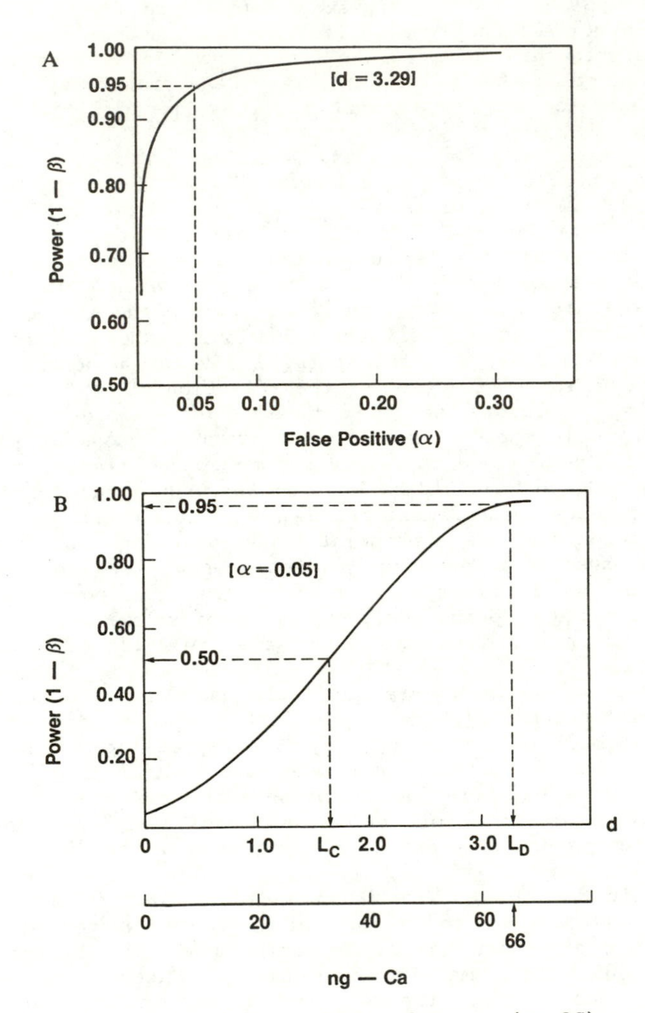

Fig. 8. Detection Power. ROC and Power (or OC) curves yield a graphical display of the relations among detection limits (detectable differences, d), and errors of the first (α) and second (β) kinds. Fig. 8A is the ROC curve corresponding to two normal populations differing by L_D -- i.e., the separation equals 3.29 σ_o, and the curve passes through the point α=β=0.05. Fig. 8B is the corresponding power curve, where now α is fixed, and the power of the test is given as a function of the normalized distance d. The lower abscissa shows the equivalent concentration scale for a hypothetical measurement process for Ca, where σ_o equals 20 ng, and the detection limit is 66 ng.

When df is very large, d is simply 2z or 3.29. For fewer degrees of freedom, 2t yields a conservative estimate, but a still better estimate derives from the following expression (68, p. 252):

$$d \approx 3.29\ (1 + 0.71/df) \qquad (4)$$

This formula is accurate to about 1% or better for df ≥ 8. For 4 - 7 degrees of freedom the correct values are 4.07, 3.87, 3.75, and 3.68. To illustrate, let us suppose that 5 paired y, B observations were made and the mean difference and estimated standard error were 1.8 ± 1.2 mV. The critical level for 4 degrees of freedom would be ts = (2.13)(1.2) = 2.6 mV, so the conclusion would be "not detected." The detection limit would be $d\sigma_o = 4.07\sigma_o$. Using s as an estimate for σ, we would estimate L_D as (4.07)(1.2)=4.9 mV.

3.2.5 uncertainties in detection limits. The previous example raises an extremely important point, namely, that unless σ_o is known without error, the detection limit cannot be exactly known. This is in contrast with the critical level, which can always be explicitly calculated from Student's t and the **estimated** standard error. We can, however, derive a confidence interval for L_D from the bounds for σ, given s and df. For normally distributed errors these bounds can be derived from the χ^2 distribution. (s^2/σ^2 is distributed as χ^2/df.) One finds, for example, that at least 13 replicates are necessary to obtain s within 50% of the true σ (90% confidence level).

For practical application of detection limits -- e.g., in meeting a research or regulatory requirement -- a "safer" procedure is to quote the upper limit for L_D. This in effect casts the uncertainty onto β, in that a specific value (rather than a range) can be given for the detection limit, but with the proviso that $\beta \leq 0.05$ (with 95% confidence). A straightforward, conservative treatment for detection decisions and detection limits when s is estimated from replication is thus: to use L_C = ts (α=0.05) for detection decisions; and to use $L_D = 2ts(\sigma/s)_M$ (α=0.05, $\beta \leq 0.05$) for detection limits. From the brief table of the relevant quantities which follows, we see for example with n = 10, L_C = 1.83 s, and L_D=2(1.83)(1.65)s = 6.04 s (to be compared with 3.29 σ for df = ∞.) [Table II; from Ref. 28, p. 80].

Table II. L_D Estimation by Replication: Student's-t and (σ/s) - Bounds vs Number of Observations

No. of replicates:	5	10	13	20	120	∞
Student's-t:	2.13	1.83	1.78	1.73	1.66	1.645
σ_{UL}/s:	2.37	1.65	1.51	1.37	1.12	1.000

A second source of uncertainty is associated with the quantities comprising the overall calibration factor A, such as recovery, instrumental detection efficiency, matrix absorption or scattering, etc. If A is determined as a random variable each time x (concentration) is estimated, then there is no problem; its random error is automatically taken into account through error propagation or replication when σ_x is estimated. If the same estimate for the calibration factor is repeatedly used, its random error has become a bias, and the bounds (confidence interval) for this bias combined with other possible sources of A-bias produce an uncertainty interval in the concentration detection limit. The recommended approach, again taking into account the practical applications for detection limits, is to transfer the uncertainty to β by taking the upper limit, S_D/A_m, as the concentration detection limit with $\beta \leq 0.05$.

3.2.6 uncertainty bounds (systematic error) for the blank. If the possibility of significant bias in the estimated value for the blank is not taken into account, the resultant detection decisions and limits may be much too optimistic. An upper limit for this bias component can be incorporated into S_C and S_D estimation just as it is in total uncertainty interval estimation, by extending the random uncertainty (confidence) limit by the upper bound for bias Δ_M. Thus, S_C becomes $S_C' = 1.645\ \sigma_o + \Delta_M$, and $S_D' = 2\ S_C'$. The detection limit increases therefore by $2\ \Delta_M$. The rationale for this procedure is indicated in Fig. 9. For a number of measurement disciplines, experience dictates reasonable values for **relative** limits for blank and calibration factor bias $[\Delta_B, \Delta_A]$. Default values of 5% and 10%, respectively, have been suggested (28) and tentatively confirmed (57) for radioactivity monitoring, for example. In this case, $S_C' = 1.645\ \sigma_o + 0.05\ B = S_C + 0.05\ B$, and $x_D' = 1.1\ (2\ S_C')/A = 1.1\ x_D + 0.11$ BEA, where $x_D = 3.29\ \sigma_o/A$ and BEA is the background equivalent activity. Thus for paired measurements with B = 500 counts ($\sigma_o = \sqrt{2B} = 31.6$ counts), and A = 5.0 count/pCi, $S_C' = 52 + 25 = 77$ counts, and $x_D' = 22.9 + 11.0 = 33.9$ pCi. Clearly, detection in this case is neither fully statistical nor fully non-statistical. Balancing the limits imposed by the statistics of signal detection with those derived from our knowledge (or ignorance) of the measurement process is essential for meaningful decision making. Historical use of a multiple of the blank is perhaps more readily understood also, through the formal incorporation of the term 0.11 BEA.

3.2.7 optimization and iteration: figure of merit. Optimal detection limits are sometimes treated through the maximization of some sort of figure of merit (FOM) such as $S/\sqrt{B}$, etc. Simplistic FOM's tend to ignore complex dependence of L_D's on measurement conditions, systematic error components, and the explicit nature of the sample. As shown in (35) for example the variation of radioactivity detection limits with counting time may range from t^{-1} to t^{+1}. Since detection limits may be sample-dependent, because of interference and matrix effects, iterative estimation of the detection limit is sometimes required. Changes in the measurement process may also be necessary if such sample dependence forces the actual detection limit above the corresponding regulatory limit.

3.2.8 multicomponent detection limits. When one leaves the realm of "simple" y - B net signal estimation, modeling and linear or non-linear least squares computations are generally required for component estimation. For the linear multicomponent model it is possible to estimate the detection limit as a closed expression, provided that all interfering analytes are included and the errors (variance), constant. Weighted least squares calculations involving Poisson or other concentration-dependent statistical

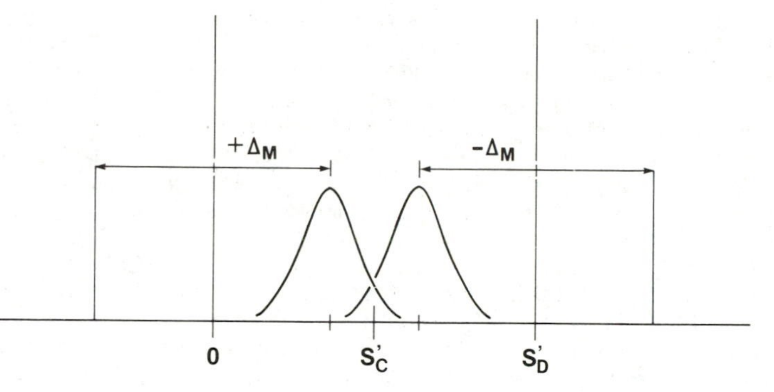

Fig. 9. Effect of Bias on Detection Limits. Allowance for bounds for bias [Δ_M] increases the critical level by Δ_M, and the detection limit by twice that amount (simple detection), taking the sign for the uncompensated bias as unknown. α and β are now inequalities -- ie, α, $\beta \leq 0.05$.

weights require an iterative L_D estimate (13,61). A relatively simple inequality expression for L_D can be given, however, by using the results of a linear model fit, where the component of interest is present at or above its detection limit (28). That is, $x_D \leq 3.29\ \sigma_x$ where σ_x is the standard error for the least squares estimate of the component in question. The basis for this inequality is that the standard error for any specific component will be approximately constant or increase with its increasing concentration in a mixture, all else remaining equal. Special multicomponent problems associated with selected types of pattern matching and multiple independent decisions, as in the detection of several isolated spectral or chromatographic peaks, will be treated in the next section.

3.2.9 effects of gradually changing distribution functions and/or σ's. For $\alpha = \beta$, normally distributed errors with constant, known variance, and simple paired estimates (y - B), $L_D/L_C=2$. This simple ratio does not obtain, however, whenever any of these conditions are not fulfilled. The matter of **high thresholds** (α << 1) has already been noted. In the example discussed in the last paragraph of the preceding section, α was vanishingly small when L_C was set to $34\sigma_o$. The ratio L_D/L_C in this case was 1.09 (61). Similarly, the ratio is unity when the conceptual difference between L_C and L_D is overlooked, such that $\beta = 0.50$.

Changing cdf's is another matter. Because the overall detection process in effect relates to the discrimination of net signals at the detection limit from null signals, one is faced with the possibility of two different distributions at the two levels, S=0 and $S=S_D$. This problem does not arise in making detection decisions, however, for S_C depends only on σ_o, the standard deviation of $\hat{S}$ when its true (mean) value is zero. Two cases will be considered, a) the Poisson counting distribution, which changes shape and (relative) discreteness with increasing signal level; and b) the normal distribution, where σ increases with concentration, a common occurrence in analytical chemistry. A third case of some importance for environmental measurements is the distributional perturbation which occurs as one adds normal measurement errors to log-normal blank variations.

The **Poisson distribution** is decidely asymmetric and discrete (in a relative sense) at the lowest levels. In fact, when the expected (mean) value of the Poisson parameter -- in this case, the blank -- is smaller than 0.05 counts, the critical level (y_C, gross counts) equals zero. (This quantity is necessarily an integer for the Poisson distribution, and α must be treated as an inequality [$\alpha \leq 0.05$].) The detection limit (y_D, gross counts) then equals 3.00 (not necessarily an integer), so y_D/y_C in this case is infinite. With increasing signal level (counts) the Poisson distribution approaches normality, so "the usual equations apply," and L_D/L_C approaches 2.0. For B = 1.0, $L_D/L_C \approx 3.4$; for B≥5, $L_D \approx 2.7+2 \cdot L_C$, with $L_C = 1.645\sqrt{B}$ is a good approximation. See (28) for a more extensive treatment of extreme, low-level counting statistics.

Poisson σ's increase with $\sqrt{(S+B)}$. **A linear increase of σ with concentration** [$\sigma(y) = \sigma_B + mS$], however, is common for many analytical methods. Since $S_C = z\sigma_o$ and $S_D = S_C + z\sigma_D$, the increase in σ in passing from S=0 to $S=S_D$ means that $S_D > 2\ S_C$. A closed expression, however, may be derived using the above linear model. (To simplify the notation in the remainder of this paragraph, S_C and S_D will be represented as C and D respectively.) Let us consider first the case where a precise estimate $\tilde{B}$ is available for B.

Standard deviations are given by:

σ, assumed variation: $\sigma_y = \sigma(S+B) = \sigma_B + mS$
σ, null signal [S = 0]: $\sigma_o = \sigma(B-\tilde{B}) \approx \sigma_B$
σ, detection limit [S = D]: $\sigma_D = \sigma(D+B-\tilde{B}) \approx \sigma(D+B) = \sigma_B + mD$

For normal random errors and $\alpha = \beta = 0.05$, the critical level and detection limit are defined as:

$$C = z\sigma_o \approx z\sigma_B = 1.645\ \sigma_B$$
$$D = C + z\sigma_D \approx C + z(\sigma_B + mD)$$

The last equation may be solved for D:

$$D = 2C/(1-zm)$$

For $\alpha=\beta=0.05$, i.e., z=1.645, an important conclusion emerges: **the detection limit does not exist** for $m > 1/z = 0.61$. This may be academic, however, since so large a slope is unlikely for any reasonable analytical method. A slope of 10%, however, would result in D/C = 2.39. To illustrate, let us take the blank standard deviation for the measurement of toluene in air, by a fully specified method of sampling and gas chromatographic analysis, to be 0.21 µg/L. The critical level for detection decisions, assuming normality, would then be 1.645 (0.21) = 0.34 µg/L. The corresponding detection limit would be $2.39(0.34_5) = 0.83$ µg/L.

For the general case, where B is estimated from n replicates, the algebra is only slightly more complicated. The variance of the estimated net signal is now given by:

$$V_S = (\sigma_B + mS)^2 + \sigma_B^2/n = V_o + mS\ (2\sigma_B + mS)$$

thus, defining $\eta=(n+1)/n$:

$$V_o = \sigma_o^2 = \sigma_B^2\ \eta \quad \text{and} \quad V_D = \sigma_D^2 = V_o + mD\ (2\sigma_B + mD)$$

From these relations and the definitions for C [$C = z\sigma_o$] and D[$D=C+z\sigma_D$], it is relatively straightforward to show that:

$$D = 2C\ [1 + zm/\sqrt{\eta}]/[1 - (zm)^2] \text{ and } \sigma_D/\sigma_o = D/C - 1$$

Thus for $\eta=2$ (paired comparison), and m and z as before (0.1, 1.645), D/C = 2.29 and $\sigma_D/\sigma_o = 1.29$. The asymptotic result (D=2C) follows of course when the slope m is negligible.

3.2.10 black boxes and hidden algorithms. With the advent of "user friendly" (and proprietary) software and automated data reduction and even automated instrument systems which yield final results only, a cautionary note must be sounded. That is, when the computational scheme is not fully and explicitly described, and when the software is not exhaustively studied and tested, erroneous results may emerge. Worse still, there may be no way of recognizing such results as erroneous, particularly if the instrumental system is designed in such a manner that the raw experimental data cannot be retrieved for alternative methods of computation. It is inappropriate in this chapter to document the problems arising, but it may be helpful to glimpse at their nature. Inept or unlucky programming and inaccurate stored parameters will always cause difficulties, but this of course is not restricted to the domain of low-level measurement. Problems of special concern for reliable and efficient detection which have come to the attention of this author include: a) Thresholds which are automatically set so high that detection power is seriously eroded; b) Algorithms (and component models) which are data dependent. This is especially a problem when peaks are marginally discernable, with peak estimation algorithms switching rules depending on the magnitude or apparent presence of a peak; c) estimation and search routines based on inadequate models or inadequately accounting for the effects of non-linear estimation; d) decision or detection algorithms for which assumptions or parameters used are unclear (and possibly incorrect); e) inaccessibility of raw data, especially when peaks are not found, and the consequent inability to investigate extra sources of variability or errors in assumptions, models or data.

Peak search or "model search" and more generally optimization routines that are sometimes heuristic and operate strictly in an empirical fashion on the data at hand, deserve another comment. That is, at the lowest levels and especially at the $S = 0$ extreme [null hypothesis] such generally non-linear routines may provide no S estimates, especially when negative, thus producing biased and skewed S distributions. Once a peak or model is automatically chosen from the noisy data, the algorithm switches, frequently to a linear estimation algorithm. The problem is that the switch point varies, being noise controlled; also the estimation algorithm seldom gets to operate when the null hypothesis [S=0] is true. σ_o is not obtained, and the normal distribution hypothesis testing apparatus cannot be applied at the lowest signal levels. Perhaps this is why the international gamma ray peak detection exercise organized by the International Atomic Energy Agency found the "visual" method of peak detection more successful than all others, including the most sophisticated computer based schemes (58).

The "model search" issue is more profound. In multicomponent chemical analysis, optimal models for estimation (number and nature of components) are often chosen automatically and empirically, for example by applying iterative, non-linear optimization routines, and quite frequently non-negativity constraints. Such automatic chemical model building, accomplished by suppressing (often legitimate) negative estimates, deserves careful scrutiny. It may be even more misleading than zero suppression with simple measurements, especially when noise and multicollinearity are large.

Illustrations of some of these limitations, which are unique for low-level data and therefore meaningful detection limits, may be found in references 35, 57, and 70. Fig. 2 in the first reference illustrates an extreme, yet not uncommon problem; quite visible spectral peaks have failed to be detected by the software.

3.2.11 quality. One solution for inadequate or incorrect approaches to detection -- including control of both false positives and false negatives -- is the incorporation of known and blind standard reference samples and reference simulated data. Such means for control are well established for trace analysis, but they have rarely been brought to bear on the detection problem. Interlaboratory low-level test data, though quite rare, have proven most informative (58,70). Direct validation and control of α and β errors should be made routinely with blind interlaboratory samples and/or data representative of blanks and samples at or near detection (or regulatory) limits, respectively. Evaluation of sets of results via ROC curves could, in turn, be quite fruitful, for the quality of the low-level measurements would be reflected in the loci of the ROC curves, *independent of the particular decision rules employed*.

3.3 Discrimination Limits, Multiple Detection Decisions, and Patterns. When the null hypothesis is defined as zero analyte concentration beyond the blank, or zero signal above the baseline or background, it is appropriate to refer to an analyte (or signal) **detection** process. In many practical cases, however, it is interesting to consider the ability to **discriminate** concentrations from a fixed non-zero reference level, or discriminate patterns or "chemical fingerprints" from a reference pattern. Multiple hypothesis testing decisions form a natural link between these two types of discrimination, and it becomes clear that both fixed level "recognition" and chemical pattern recognition fall under the same statistical frame-work as zero level analyte detection. Both aspects of Qualitative Analysis (detection, identification) share the same probabilistic foundation, including hypothesis specification [H_o, H_A], decision criteria, and type I [α] and type II [β] errors [Ref. 8; pp. 233, 239]. In all cases, it is extremely important to recognize that the respective discrimination or detection limits characterize the measurement process, **not** a particular result. (As always, results are tested by comparison to the corresponding critical level.) Our objective is to evaluate the intrinsic capabilities of CMP's, often shaping these capabilities to meet specific practical or research needs.

3.3.1 lower and upper regulatory limits: balancing risks and costs. We have noted that detection limits dictated by regulatory concerns have been surrounded by considerable confusion, discrepant statistical and ad hoc formulations, ignorance, and even mild deception. The apparent deception is related to the lack of general understanding or agreement concerning the appropriate nature and magnitude of the error of the second kind (β, false negative). By ignoring its presence, whether intentional or not, those who must meet regulatory demands generate a β/α imbalance where, at 50%, false negatives may exceed false positives by nearly a factor of 400. One justification is that identically zero concentrations cannot exist anyway, and very small concentrations cannot be effectively distinguished from the blank. A related, very important observation is that small non-zero concentrations will be "detected" on occasion, necessarily more frequently than the false positive constraint [α] placed on the blank. To reduce the penalty which might be associated with the occasional detection of such small concentrations, it is of course helpful to reduce α for the blank still further -- but this should be done openly, not by subterfuge.

To meet such legitimate concerns, while at the same time keeping an open, realistic, and balanced view of false positives and negatives, we recommend the substitution of lower and upper regulatory limits -- whose difference is the discrimination limit (Δ_D) -- in place of the null limit [zero] and the single, analyte detection limit. To illustrate the suggested approach, Fig. 2 has been modified in Fig. 10 to indicate a non-zero lower limit (L_O), and an upper limit (L_D). As before, the upper limit in the societal or regulatory setting would be established at such a level that the concentration or event of concern would be reliably detected [β=0.05] when its net "cost" to society crossed the limit of acceptability. The lower limit from which the upper limit must be reliably discriminated, is new: its level is established such that the penalty for "detecting" a very small concentration is likewise acceptable. Such penalties can be quite real, especially in terms of intangibles, such as public alarm (71), or indirect long-term negative perceptions affecting the business of a regulatee. In Fig. 10 this concept is presented, again in the context of earthquake detection, with the aid of hypothetical positive and negative cost differentials which would define the "trigger points" for L_O and L_D.

It is important that the (regulatory) level-setting process for these limits be decoupled from their estimation from the characteristics of the measurement process. The former is a sociopolitical matter involving complex risk assessment issues (4), whilst the latter lies in the domain of the scientist. The scientific responsibility is met once the discriminable limits lie within those desired by society. Note that the discrimination limit Δ_D is here defined as the difference L_D-L_O such that α, β each equal 0.05. It is interesting next to consider precision requirements, e.g., at the upper limit, as compared to those for the conventional detection limit. Taking L_O to be 50% of L_D, the relative standard deviation at this L_D would be about 15%, in contrast with 30% for the conventional detection limit. (The change is entirely due to the introduction of the non-zero L_O; the magnitude of L_D is unchanged.) The precision (RSD at L_D) would be "quantitative" (10%), once L_O equals 2/3 of L_D. **Quite possibly the (subconscious) need for such discrimination capability is the underlying motivation of those who call for abandoning detection limits and hypothesis testing in favor of "quantitative" measurements.**

The discrimination limit as depicted in Fig. 10 has two other important applications, one in business and one in science. In business matters involving trade or regulation, one may face the task of "proving" the product or waste stream level exceeds or does not exceed some prescribed value, such as the (upper) regulatory limit. Because of measurement error, the ability to accomplish this is limited, and in fact it is set by the size of the discrimination limit. Balancing of costs will again generally fix the magnitude of Δ_D. Penalties will likely increase with greater **apparent** departures from specifications; and the ability to defend departures as small, or attack departures as large depends upon the producer's or consumer's discrimination limit. The discrimination limit, hence precision of analysis, can only be improved with increased analytical costs. In the socioeconomic arena decision

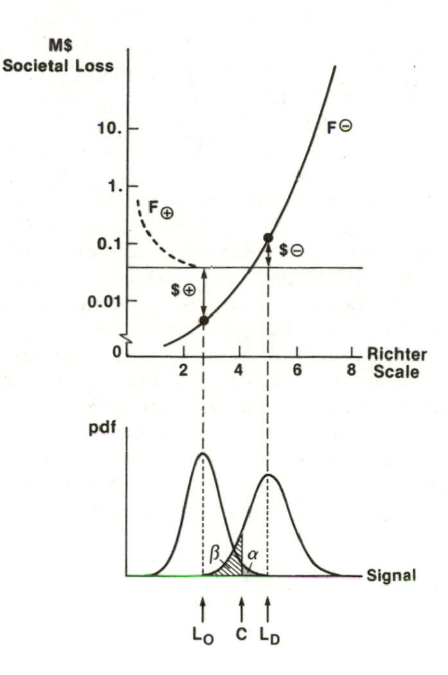

Fig. 10. Discrimination Limits. Curve F^- represents the loss to society as a function of earthquake magnitude; F^+ represents the cost of avoidance (evacuation, etc.), the dashed portion simulating *indirect* costs associated with false alarms -- eg, mental anguish, damaged credibility, lawsuits, etc. Points of imbalance between F^+ and F^- which exceed what is acceptable to society are taken as lower and upper regulatory limits, which must be matched by corresponding lower (L_o) and upper (L_D) measurement limits whose difference is the Discrimination Limit (Δ_D). A non-zero lower limit forces an improved precision requirement in comparison to the "simple" L_D of Fig. 2.

theory may be helpful for deriving the appropriate balance between penalties and analytical costs (therefore the requisite Δ_D), particularly taking into consideration which party has the burden of proof. (Note that the "cost" *differential* associated with the *burden of proof* is equivalent to the size of the [measurement] *dead zone* around the regulatory or specification limit -- i.e., the sum of the "producer's" and "consumer's" discrimination limits.)

The scientific application involves "**identification**" in its simplest sense. That is, if L_0 and L_D are treated as unique or identifying concentrations or isotope ratios, or characteristic energies or wavelengths, etc., then the measurement process must be designed so that Δ_D is sufficiently small to distinguish between these two classes. In analogy with the detection power $(1-\beta)$ characterizing the detection limit (given α), one finds the power associated with Δ_D described as "discriminatory power" (12, p. 517) or "resolving power." This univariate, statistical approach to identification shares much in common with detection. For example, OC and ROC curves are just as appropriate for balancing false positives and negatives, and for comparing capabilities of alternative measurement (and computational) techniques. In addition, the difference between **design** of the measurement process to achieve a given detection or identification capability and **outcome** (specific result) is still manifest in an uncertain region -- i.e., results falling within the RUD [region of uncertain detection] or RUI [region of uncertain identification] **may** be detected or identified, respectively, by chance but this cannot be "assured" (α, β = 0.05) a priori. (See "multichannel identification, below, for further discussion.)

3.3.2 impurity detection. A special issue involving discrimination limits in analytical chemistry, having broad importance, is the detection of impurities or contamination. Conceptually, this can be treated as a direct outgrowth of the "identification" or discrimination of singular classes characterized by unique values of a continuous variable, as described in the preceding paragraph. In Fig. 11 class-0 and class-A are shown at separate unique (identifying) locations of a continuous measurement variable x_i. As depicted, the separation of these two classes far exceeds the discrimination limit Δ_D, so identification of a pure component (in this 2-component universe) will present no problem. If component-0 is contaminated by a small admixture of component-A, however, there exists a limit [Δ_D] below which a contaminated sample will be indistinguishable from the pure component-0. The **minimum detectable contamination** is numerically equal to Δ_D, when Δ_D is expressed relative to the class separation $(x_A - x_0)$ -- i.e., as a mole fraction or mixing ratio. (Note that "mixing" can occur as physical mixing of miscible chemical species, or it can arise from superposition of signals from different sources within the same detector.)

Two fundamental observations follow. First, class separability and impurity detection power degrade with increasing variance of the $\hat{x}_i$ distributions, which in turn, depends on the measurement precision and therefore the detection limits for the two components. This direct, and quantifiable coupling between pure component detection limits, component identification and resolution, and impurity detection is most important, though scarcely surprising. Second, if one has sufficient knowledge of the "chemical universe" -- i.e., x_i locations for the entire population of H_A's -- then for any H_0 of interest, one can deduce the **maximum systematic error due to undetected contamination** by estimating Δ_D for the "closest" impurity source. If this discrimination limit is unacceptable, redesign of the CMP is in order. Reliable estimation of systematic error bounds deriving from undetectable contamination, or undetectable model error is one of the important needs for **accurate** analytical results. Thoughtful consideration of the coupling between experimental design, and component detection and discrimination limits, supported by excellent **scientific knowledge** concerning the H_A universe offers one of the most reliable and objective solutions to this problem. An astute examination of these issues, emphasizing the universe of potential contaminants has been provided by Rogers (6). For simplicity, the discrimination problem was presented here in one dimension (one measurement variable). Multivariable detection and discrimination are obvious extensions, leading generally to increased detection and discrimination power, as one compares or "matches" unique multivariable patterns in place of characteristic values of a single variable. (See below.)

3.3.3 multiple detection decisions. If a number of detection or discrimination decisions are made in the course of a measurement, the overall probabilities of false positives and false negatives are accordingly altered. We consider two cases: first, where the individual tests are unrelated or "serial", and second, where "parallel" tests are made, as in pattern recognition. Independent, serial tests characterize the detection of isolated spectral peaks, as in multichannel gamma ray spectroscopy, as well as residuals following data analysis, and even replication experiments and control charts. In all of these cases, the **overall** probability of false positives and false negatives necessarily exceeds that for the individual peak detection (or outlier detection) test. For example, if a large gamma ray spectrum containing no actual radioactivity were scanned with the equivalent of, say, 50 detection decisions [$\alpha = 0.05$], there almost certainly [>92% chance] would be at least one false positive peak. Similar considerations apply to false negatives, so false alarms and missed radioactivity would be the consequence. (Ignoring this issue has led to some difficulties in the evaluation of low-level gamma ray spectra; see Ref. 28 for further discussion.) The solution is to follow the rules for combining probabilities; namely, adjusting the significance levels so that the **overall** probabilities of correct non-detection [$1-\alpha'$] and correct detection [$1-\beta'$] remain 95%. The probability that all decisions are correct is simply the continued product: $(1-\pi') = 0.95 = \Pi(1-\pi) = (1-\pi)^n$, where π represents α or β, and n, the equivalent number of tests per spectrum. Adjusted values for α and β are then given by (Eq. 5),

$$\alpha \text{ (or } \beta) = 1 - (0.95)^{1/n} \qquad (5)$$

If the total error level is to be held at 5% [α', β'] for a multitest experiment in which H_o is actually true 50 times and H_A, 3 times, then Eq. 5 gives adjusted values of $\alpha = 0.00103$, $\beta = 0.017$ with corresponding critical levels and detection limits of $3.1\ \sigma_o$

and 5.2 σ_o, respectively. Monitoring of mostly empty spectra thus provides justification for unequal α, β, and thus for $L_D/L_C < 2$.

3.3.4 <u>multichannel identification.</u> The linkage between detection and identification was brought up earlier, where "identification" was formulated in terms of the statistical estimation of the characteristic value (e.g., element concentration or ratio, gamma ray energy) for the identifying variable. Linear estimation was at least implied in that discussion, so that initially normal data would lead to normal (though possibly correlated) errors for the estimated results. For example, the frequency distribution of events (counts) along the energy axis [identifying variable] could be used to estimate the mean energy ("centroid") and its variance for a gamma ray peak, and the peak magnitude or "area" could be simultaneously estimated with a simple filter function to compensate for a linear baseline. The decision space is now two dimensional, so contours of the bivariate area (detection) - energy (identification) distribution would be used for significance testing. When multichannel data are intrinsically non-normal, or when they are subjected to non-linear operations as in certain peak search and peak fitting algorithms, normality is not preserved, so caution is in order in making detection decisions and in deriving confidence intervals.

"Non-statistical" identification is important in many facets of analytical science, where signal location or "identity bin" predetermines species identity. Detection and identification are then uncoupled, and **any** signal detected in the characteristic bin simultaneously conveys detection and identity. Classical analytical chemistry (e.g., gravimetry) relied heavily on this model, where unique chemical separations would guarantee identity. Modern instrumental or chromatographic methods similarly succeed when the resolving power (discrimination power) far exceeds the "density" of pure components along the informing variable.

3.3.5 <u>pattern discrimination limits [multivariable identification].</u> We considered the discrimination of chemical components or classes earlier from a univariate perspective, including the paired comparison for a single alternative or contaminating component which would necessarily lie to one side of the null class (known component against which the sample is to be compared). Before considering discrimination with multiple chemical variables (compositional or spectral patterns), let us broaden the univariate problem to two-sided discrimination, since unlike analyte detection, characteristic or identifying variable values generally may be larger or smaller than that of H_o. The H_o discrimination limit test would then be 2-sided -- i.e., $z_C = 1.96$ instead of 1.645 for $\alpha = 0.05$. If a paired comparison of a test sample (unknown) with the control sample (known, H_o) falls within $\pm 1.96\ \sigma_o$, we then conclude that there is no statistically significant difference. This is not, however, **proof** that the patterns are the same; it is only a test of consistency. **It is necessary, but not sufficient.** To establish a real match, or "identification," we must demonstrate that the universe of alternative patterns will not match (statistically). Design of a measurement process for the successful identification of a particular chemical species or compositional state thus requires consideration of both α and β errors, as depicted in Fig. 12.

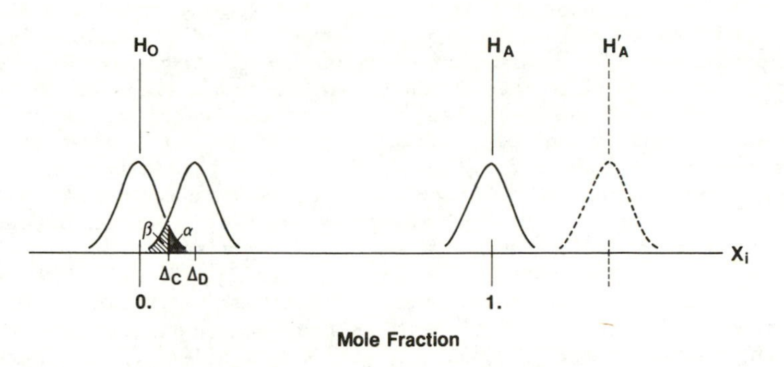

Fig. 11. Impurity Detection. Δ_D represents the minimum detectable concentration of substance-A [H_A] in the "null" substance [H_o]. The abscissa represents mole fraction or mixing ratio. Individual impurity detection limits would obtain for each impurity type, e.g., A' [H_A'].

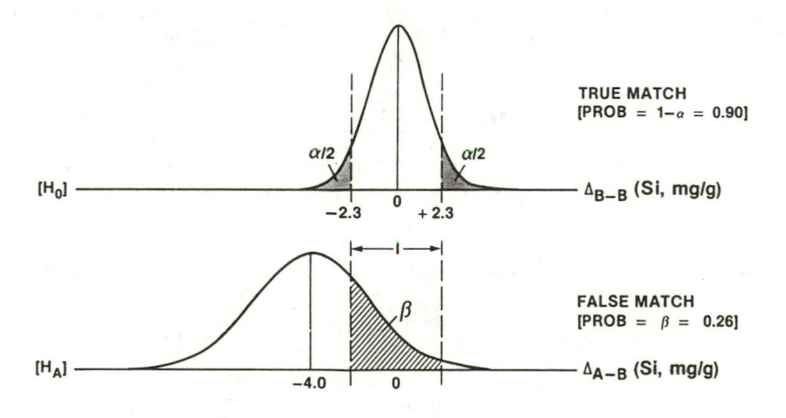

Fig. 12. Single Species Matching; Univariable Identification. For a given location on the abscissa [identifying variable: isotope ratio, X-ray energy,...], unique identification requires that **none of the possible** H_A's overlaps (probability β or less) the two-sided H_o window [I]. That is, all separations must exceed the corresponding discrimination limits. (From the design perspective, since identifying variable separations are generally fixed by Nature, we must design the CMP to achieve corresponding Δ_D's -- cf, Fig. 8A [ROC curve].) [Illustration constructed using $\alpha = 0.10$, $\Delta = -4.0$ mg/g, $\sigma_B = 1.0$ mg/g, and $\sigma_A = 2.6$ mg/g.]

It is a small step to take from univariable identification to multivariable or pattern matching. If we are concerned with just a single alternative pattern [**A**], but several (n) measured variables, then the consistency test requires that all n variables match statistically when the identity of test sample is the same as that of the control sample [**B**]. Combining probabilities as before, $(1-\alpha') = 0.95 = \Pi(1-\alpha_i) = (1-\alpha)^n$. **Proof** of identity, as before, includes consideration of sufficiency -- i.e., we require in addition that [**A**] **not** match (statistically) [**B**] simultaneously for **all** measured variables. Probabilities are combined a little differently in this case; the overall probability of an erroneous match is given by $\beta' = \Pi(\beta_i)$. The product is **also** taken over all n variables, whose individual β_i's will generally differ. Unless $\beta' \leq 0.05$, matching of patterns cannot establish identity. At the same time, it is this multiplicative feature, when individual β's are themselves small, that gives multivariable or pattern discrimination its enormous power.

To illustrate, let us consider matching of trace element patterns in two pure source materials, where the origin of one (control sample, **B**) is known, as is the composition of the possible alternative **A**. Given the characteristics of the measurement process and the compositions of the two known sources, we can tell a priori whether the sources are discriminable as indicated above. If not, the capability of an unknown test sample to match proves nothing. Absence of a match under these conditions, however, would deserve scrutiny; it could indicate either faulty measurements or faulty assumptions. Illustrative data are given in Table III.

Table III. Multivariable Identification

Input data for estimating the discriminability (identifiability) of particle emissions from steel plants A and B [a,b,c]

	H_o: B vs $\hat{B}$			H_A: A vs $\hat{B}$		
	Al	Si	Ca	Cr	Mn	Fe
Concentration (mg/g)						
steel-B	10	12	45	3.2	22	160
steel-A	13	8	70	3.3	16	120
σ	1.1	1.0	5.8	0.32	1.9	14
window [I ±]	4.00	3.63	21.1	1.16	6.91	50.9
distance [Δ]	3.0	-4.0	25.	0.10	-6.0	-40.
β	0.74	0.40	0.32	0.98	0.63	0.71
Δ/σ_o	1.93	-2.83	3.05	0.22	-2.23	-2.02

(a) Based on data from Ref. 72.
(b) Values of I and β are given for n=5.
(c) Fig. 13 depicts the windows [I] and variable separations [Δ].

Concentrations for six elements characterizing two steel aerosol samples (72) are given in the first two rows. Steel-B is taken as the control, and steel-A as the alternative source. H_0 is represented by the vector or pattern difference, $(\mathbf{x}_B - \hat{\mathbf{x}}_B)$; H_A, by $(\mathbf{x}_A - \hat{\mathbf{x}}_B)$. The last five rows of the table indicate, respectively: the standard deviations [σ] for the elements in question, the matching intervals [I], the concentration differences [Δ] under H_A, the probability of false matches [β], and the ratios of concentration differences [Δ] to the paired measurement standard deviations [σ_o]. $1-\beta$ and Δ/σ_o both serve as measures of individual element discriminating power. The quantity I is computed by requiring $1-\alpha'$ to be 0.95; for n=5, this means α = 0.0102 or z_c (2-sided) = 2.57. (For 6-member patterns, z_c increases to 2.63.) Then $I = \pm z_c\sigma_o$, where $\sigma_o = \sigma\sqrt{2}$. Pattern differences [$\Delta$], indicated by the open circles, are shown in comparison with matching intervals in Fig. 13.

For this example, pattern identifiability (H_o "provability") has been approached in two ways. First, β' has been calculated as the product of the individual β_i's, reflecting the series of individual element matching decisions. (For n = 5, omitting Ca, this product equals 0.13.) Second, the vector difference represented by H_A is examined through the use of the non-central χ^2 statistic, where $\Sigma(\Delta/\sigma_o)^2$ is the non-centrality parameter (73). In this second case the test of the vector match (i.e., H_o test) is carried out by comparing the sum of squares of the n observed normalized differences with the critical level for the central χ^2 for n - degrees of freedom. The rms value from the sum of squares -- $(\Delta/\sigma_o)_{rms}$ -- represents the multivariable generalization of the univariate normalized differences. It is a convenient single parameter measure (index) for the vector discrimination power ($1-\beta'$), as β' is uniquely determined by this quantity, given α' and the number of degrees of freedom.

Table IV gives results for the two types of test and several choices of element patterns. Important dual pattern identification conclusions follow: (a) Discrimination power (identifiability) differs according to the type of test, χ^2 being significantly better and becoming more so with increased dimensionality. (b) Optimal feature selection (e.g., for n=5) gives optimal discriminating power for the number of variables selected. (c) There exists an optimal number of dimensions (variables). The most powerful variable (here, Ca) is used for n=1; a second discriminating variable yields increased power with n=2; but eventually addition of poorly discriminating variables "dilutes" the discrimination power -- e.g., n=6 compared to the best set of 5. (d) Increased dimensionality gives enormous leverage to modest improvements in precision, through the product $\Pi\beta_i$. (See bottom line, Table IV.) These four conclusions directly indicate the way toward improved discrimination power, the last being the most influential. (χ'^2 in the table denotes the non-central χ^2.)

3.3.6 <u>generalization.</u> The foregoing considerations of hypothesis testing and pattern identification limits were necessarily simplified, an extended discussion being beyond the scope of

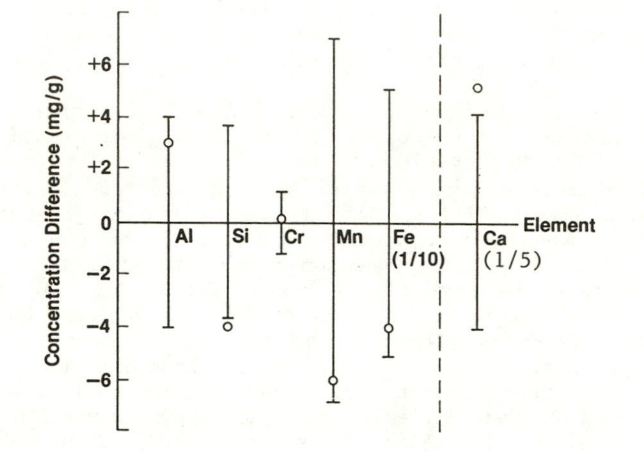

Fig. 13. Multivariable Identification. H_o windows [I] and H_A concentration differences [open circles] for the multi-variable (element) patterns characteristic of particle emissions from two steel plants. [*See* Table III.]

Table IV. Pattern Discrimination Power [1 - β']

(Steel-A particles vs Steel-B particles [control]; α' = 0.05)

	n=1	n=5		n=6
	(Ca)	(-Ca)	(-Cr)	
Sequential matching[a]	0.86	0.87*	0.958	0.949
χ^2 - test[b]	0.88	0.967	0.997	0.996

(a) criterion: $x_i \le I_i$, all i — power: $1 - \Pi \beta_i$
(b) criterion: $\chi^2 \le \chi^2_c$ — power: $1 - P(\chi'^2)$

*8% incr. precision (σ_i's) increases the power to 0.95 [target].

this chapter. The principal generalizations that should be considered, however, are the following:

(1) For the first ("matching") strategy, the requirement of homogeneous variance may be relaxed with the use of individual σ's: i.e., $\sigma_B\sqrt{2}$ for the evaluation of α and I, and $\sqrt{(\sigma_A{}^2 + \sigma_B{}^2)}$, to recalculate the β_i (See Figure 12).

(2) For variances estimated as s^2's, t and F would replace z and χ^2, respectively, for hypothesis testing. To estimate the power of the tests, the corresponding non-central distributions would be employed. The non-centrality parameter for the F distribution is the same as for χ^2. This means that even in the best of circumstances (orthogonal variables) this approach to the identification limit or power requires homogeneity of variance and knowledge of σ. (See reference (74) for a discussion of these issues, as well as an in-depth treatment of multivariate hypothesis testing and classification.)

(3) If the H_A universe contains more than one member, its membership and composition **must be known** for identification to be meaningful. Such knowledge, of course, is in the domain of disciplinary ("scientific") expertise. **Proof** of H_o [identification] comes only when discriminating power is adequate with respect to **all** H_A's. For a given control pattern B, only that region of variable space within the discriminating volume need be explored, however. For sequential matching, this means only A - patterns for which the distribution of the difference spectrum A - B significantly overlaps the I-hypercube; for the alternative approach, the discriminating volume derives from the critical value for χ^2. Multiclass discrimination may be performed, for example, through a series of binary tests (12, 17, 74).

(4) Impurity detection for the multivariable case may be treated as a direct extension of the single variable case. For two patterns, the impurity detection limit (component A contaminating control component B) can be calculated from $(\Delta/\sigma_o)_{rms}$ corresponding to β' = 0.05, where χ^2 (α' = 0.05) is used to test the null hypothesis [H_o: $\mathbf{x}_B$ - $\hat{\mathbf{x}}_B$]. For mixed impurities, a "worst case"

limit may be derived from the "pseudopattern" of closest approach. That is, the two pattern discrimination limit is recalculated substituting A′ for A, where pseudopattern A′ is the linear combination of alternative vectors which lies closest to B. At trace levels, observed patterns become increasingly fuzzy, because of measurement imprecision or baseline noise. Clearly, under such circumstances "detection" and "identification" become entwined. (See reference (75).)

(5) Covariance among variables (e.g., elemental concentrations) **within** individual classes is by far the greatest complication. It may be treated in one of four ways: (a) Select just one variable or function of variables (e.g., first principal component); covariance is then undefined. (b) Select only the most powerful, uncorrelated (or nearly independent) variables, discarding others showing significant correlation (12, Chapt 20). (c) Transform the original variables into a reduced, orthogonal set, as in Principal Component Factor Analysis and SIMCA (76,97). (d) In the absence of a very large sample for testing the multivariate normal assumption and estimating the within class covariance matrices, the fourth alternative is daunting: taking into account the full covariance structure through critical contours $[\alpha,\beta]$ of the hyperellipsoids corresponding to H_o and H_A. Considering just two variables, the treatment would be analogous to the confidence ellipse for the estimated slope and intercept of a fitted calibration line. (Hypothesis testing of calibration curve parameters is far more amenable to this multivariate "parametric" approach, however, since the correlation matrix is known from the design of the experiment (77).) For two variables, the matching intervals I and the respective probabilities $(1-\alpha')$ would not be greatly affected by the lack of rigorous knowledge of the covariance matrix, since $(1-\alpha') \approx (1-\alpha)^2 \approx 1$. The false match probability β' could be significantly in error, however, because $\beta_1 \cdot \beta_2$ must now be replaced by $\beta_1 \cdot (\beta_2|1)$, where $(\beta_2|1)$ is the conditional probability of a false match for variable-2 given a false match for variable-1. If the variables are perfectly correlated, $(\beta_2|1) = 1$, and the second variable lends no incremental discriminating power. Higher dimensions lead to increasing complexity, and estimates of higher order correlations become increasingly imprecise as one runs out of degrees of freedom.

4. CONCLUSIONS AND OUTLOOK

The ability to detect specified (absolute) levels of chemical species in environmental, biological, and physical (material) systems is crucial for the well-being and advancement of our society. Because of the practical importance of reliable detection in the societal setting, on the one hand, and its technical complexity, on the other, we face a "Two Cultures" type situation. We scientists lack the expertise to fully comprehend or effectively influence the sociopolitical issues; experts in that domain, similarly cannot be expected to fully comprehend the technical issues involved. Effective communication and mutual education -- one of the aims of this volume and this overview chapter -- is therefore essential. With this objective in mind, let us re-consider briefly some of the observations and suggestions of this tripartite overview.

4.1 Sociopolitical Perspectives

- Adequate detection capabilities are important to society, both for natural or anthropogenic hazards and for requisite beneficial levels of chemical species -- e.g., nutrients. "Adequacy" means "certainty" to the layman; if a substance is present (above the specified level of concern) it will surely be detected -- the alarm will go off; if not, there will be no (false) alarm.

- Despite the intrinsic uncertainty (false positives and false negatives) associated with detection, and in fact, with all of measurement, the general public is not schooled to accept such a limitation. Ignorance and suspicion with respect to this issue is reflected also when it comes to our ability to reduce concentrations of "bad" species to zero, or for that matter to detect all concentrations exceeding zero (78). Scientific naivete' regarding newly detected noxious species when detection capabilities improve constitutes another form of ignorance having potentially great political impact.

- Sociopolitical "debates," in both the legislative and judicial arenas, have very different ground rules than scientific debates (3). Advocacy, conflicting societal concerns and perceptions, and even "hidden agenda" drive such debates. They cannot, and probably should not, be conducted like a scientific forum. With patience and honest input from the scientific community in its area of expertise, generally the collective common good is served (79,80). With reference to risk management for "dread risks" affecting large numbers of people, for example, Lave observed that collective decisions are mandatory, but because of the diversity of safety goals, collective decisions are difficult (80).

- Risk perceptions and collective (or delegated) decisions lie behind many of our regulatory limits or hazard "alarm levels," [L_R] which, in turn, drive our measurement Detection Limits. Though certain approaches to decision analysis, especially those incorporating Bayesian strategies, might seem appropriate for simultaneously embracing societal risk and measurement error risk (false positives and negatives), it would seem advisable in practice to decouple the two. Let society (or medicine, or affected industry, etc.) enter the political debate to establish their requisite L_R's. Then, Measurement Science, using the appropriate **scientific** criteria and standards, should attempt to meet these L_R's with scientifically defensible Detection Limits. The late Philip Handler put it well, by stating that "Scientists best serve public policy by living within the ethics of science, not those of politics" (81).

- Societal and scientific perceptions of risk sometimes diverge. Slovic's investigation of ordered risk preferences of laymen vs experts is an interesting case in point (79). Nuclear power, for example, was rated first among representatives of the lay public (League of Women Voters; college students), yet it was 20th in the eyes of experts. Surgery was 5th according to the relevant experts, but it was only 10th in the public view. The nature of our society naturally accords primary weight to that society's public perceptions, when it comes to political decisions. This deserves our respect for many reasons, including the fact that society's judgment is not constrained by a possibly too narrow view or faulty algorithm. In fact, its "basic conceptualization of risk is much richer than that of the experts and reflects legitimate concerns that are typically omitted from expert risk assessments" (79, p. 285).

- Adequacy of detection limits **is** something that society has a right to demand, and support if the cost should be high. Inadequate detection capacity for specified levels of fires, earthquakes, toxic organisms, etc. must be addressed through refined sampling and measurement procedures. Inadequate performance of a Measurement Process not only fails to provide sufficient warning, but it may also produce quite misleading conclusions. Elevated levels of Ni in human serum due to occupational exposure (ca 5 ng/mL), for example, were quite undetectable until an excellent reference analytical method was developed under the auspices of the International Union of Pure and Applied Chemistry [IUPAC]. Prior methods, quite incorrectly implied that **normal** levels of Ni in blood serum were some ten times higher than that occupational exposure level (82).

- The costs of erroneous detection decisions can be quite significant. Disastrous results may follow if irreversible actions are taken. Even the seemingly harmless false positive which can later be shown to be spurious can damage reputations and/or lead to expensive court suits. It is important therefore that scientific detection decisions and detection limits be approached in a quantitative manner, with due attention to the probability of errors of both kinds.

4.2 Technical Issues

- Meaningful detection decisions and detection limits can follow only from rigorous attention to the Measurement Process and an Hypothesis Testing framework for defining detection capability. This is especially appropriate, as hypothesis testing is **the** expression of the Scientific Method. Decision criteria, detection limits, and acceptable false positive and false negative risks must be quantified, and CMP's **designed** to meet their specifications. The scientific expertise required goes deep. This was observed, for example, in the investigation of detection limits for a variety of analytical methods for the International Atomic Energy Agency. As illustrated in Ref. 35, detailed, method specific expertise was essential in order to expose certain subtle, but extremely important factors affecting calibration and the blank [Note 7].

- All is not well in the technical camp. Confusion among scientists between the **design** of the MP to meet requisite levels of performance [L_D], and an experimental **outcome** or detection decision based on a specified criterion [L_C], is at the heart of much our internal disarray. That is, two different (albeit related) issues are under discussion, often unknowingly and with conflicting terminology. Ad hoc rules of thumb, or simplistic consensus ("voted") formulae are proffered -- often in the interest of producing a simple ranking of CMP's according to something labeled as an LOD. This serves no one. In particular, it fails to provide the public with meaningful detection capabilities comprising reliable and adequate false positive and false negative error probabilities. Perhaps the most common extreme is the case where the β-error is unrecognized, such that its de facto value is 50% [Note 8].

o The drive toward facile expressions for limits of detection is partly a matter of attitude and education. Solid training in statistics and drilling with respect to the fundamental concepts of experimental design and hypothesis testing in science is missing from the undergraduate education of many chemists in the U.S. Western Europe fares better; and now that training in Chemometrics is beginning to appear in the American curricula (83), real hope exists for common understanding of these matters by the "ordinary" chemists of the future. An illustration of the present state comes from a survey recently taken by an instrument manufacturer of its users in the nuclear industry. Regarding topical material covered at workshops, comments came back that users would prefer omission of the theory with more time spent on use of the formulas. A personal view is that education related to basic concepts should always have the priority; understanding (and questioning of) formulas is important, but calculators or computers are quite proficient at using them.

o The link between "ordinary" measurements and detection limits needs reinforcing. That is, **both** depend for their validity on all sources of systematic and random error associated with the entire CMP. Thus, for example, detection decisions [tests of significance] and confidence intervals depend on the same assumptions and error components for their validity. If Student's t is appropriate to use with the one, it is equally appropriate for the other.

o Conventions for reporting data, and "black box" algorithms can induce subtle bias into many types of modern chemical/instrumental data, but the problem is exacerbated with the growth of automatic laboratory systems and low level measurements and data bases. The black box may contain mistakes, and all too often its mechanism is unavailable to the user, and on occasion that mechanism (i.e., algorithm) changes for low level observations. Information loss or distortion, whether it occurs within the black box or by the pen of the user, is especially severe for low level data. Its impact on long term storage and data base generation is an issue of some importance (34).

o Quality control at low levels (blank, detection limit) must be addressed both with Standard Reference Materials and Standard Test Data, if we are to certify the accuracy of our detection decisions and detection limits. Since the blank has such a profound influence on the validity of detection decisions, it deserves special attention. The CMP must be designed to incorporate an adequate number of "real" blanks, and it should take advantage of the normalizing tendency of averages from paired comparisons.

- The introduction of Discrimination Limits, such that small non-zero concentrations will rarely produce false positives, should do much to alleviate the public alarm that sometimes follows such "detection." At the same time it could avert the common implicit overcompensation associated with ignoring of the error of the second kind [false negative]. Also, those who decry current usage of detection limits because they are too imprecise, or equivalent to the flipping of a coin, might regard Discrimination Limits as useful, more precise measures of detection capability, still in keeping with the hypothesis testing concept.

- Discrimination Limits and multiple detection decisions lead naturally to univariate and multivariate formulations for identification, an outgrowth of the fundamental concept of hypothesis testing. Methods for treating this link have been developed, so it becomes natural at this point for us to address **together** the **two primary characteristics of Qualitative Analysis**: Detection and Identification.

- Identification differs in one, very critical respect from detection: a consistency test of the null hypothesis is necessary but **not** sufficient for identification. Discrimination limits must be adequate for **all** alternative hypotheses (other substances). At this point scientific intuition or expertise plays a crucial role, for we must somehow discover the universe of all possible alternatives to the substance we wish to identify, in the context of the given measurement process.

4.3 Pre- and Post-History: The Challenge. The concept of the Detection Limit, at least in Analytical Chemistry, was slow to evolve in the early decades of this century from a loose, qualitative idea, to a potentially semi-rigorous numerical attribute for a fully-defined CMP. During the past twenty years or so, important strides have been made in education and in the development of a consistent and practically useful formulation of the Detection Limit, especially in Europe. Unfortunately, diversity in understanding, formulation, and nomenclature among scientists continues. This has been exacerbated by the demand for regulations and simplified rules and formulas, often on relatively short time scales. "Definitions" deriving from polemics or from democratic, consensus tactics are unlikely to meet long term standards for scientific rigor (conceptual rigor, not necessarily uncertainty-free, numerical rigidity).

Although a sound approach to detection has been available for at least two decades, and despite its current successful application to many practical and scientific problems, the current disarray among scientists in the U.S. [cf Fig. 4] can only further mystify the public in an area that seems already inherently mystical. The promise comes from trends in chemical education and from work in progress in reputable international chemical organizations. Statistics and the proper concepts of measurement uncertainty, experimental design, and hypothesis testing are gaining a foothold in the undergraduate chemical curriculum, especially under the stimuli of modern instrumental and computational facilities and Chemometrics (84). Also, at the present time at least two commissions of IUPAC, partly in collaboration with the international chemometrics community, are drafting guidelines and nomenclature documents treating a broad range of chemical measurement issues, including those related to uncertainty, experiment design, reporting of data, and detection.

Cooperation between the two cultures should become increasingly fruitful, as common concern in meeting society's legimitate needs for practical detection capabilities bind us together, and as we each invest our efforts in our respective areas of expertise. Mutual education and inter-cultural communication can only accelerate this process.

Acknowledgment

Special thanks go to the following colleagues, for their important suggestions and care in reading a draft of this chapter: K. R. Eberhardt, M. S. Epstein, R. W. Gerlach, H. M. Kingston, P. A. Pella, C. H. Spiegelman, R. A. Velapoldi, and J. W. Winchester.

Literature Cited

1. Kutschera, W. "Rare Particles"; Nucl. Instrum. Meth. B5, 1984, 233, 420.
2. Cooper, R. M. "Stretching Delaney Till It Breaks"; Regulation, Nov/Dec 1985, 11.
3. Moss, T. "Scientific Measurements and Data in Public Policy Making"; Chapt. 3 in this volume.
4. Science, Risk Assessment Issue, 1987, 236, 267-300.
5. Currie, L.; Klouda, G.; Voorhees, K. "Atmospheric Carbon"; Nucl. Instrum. Meth., B5, 1984, 233, 371.
6. Rogers, L. B. "Interlaboratory Aspects of Detection Limits Used for Regulatory/Control Purposes"; Chapt. 5 in this volume.
7. Rensberger, B. "A Life Is Worth $2 Million, Regulatory Analysis Shows"; Science Notebook, Wash. Post, Mar. 2, 1987. [Science news summarizing highlights of a study to be published in Environmental Science and Technology.]
8. Liteanu, C.; Rica, I. Statistical Theory and Methodology of Trace Analysis. New York: John Wiley & Sons; 1980.
9. Egan, J. P. Signal Detection Theory and ROC Analysis, Academic Press, New York, 1975.
10. Frank, I. E.; Pungor, E.; Veress, G. E. "Statistical Decision Theory Applied to Analytical Chemistry"; Anal. Chim. Acta 133 (1981) 433.
11. Howard, R. A. "Decision Analysis: Perspectives on Inference, Decision, and Experimentation"; Proc. IEEE, 1970, 58, 823.
12. Massart, D. L.; Dijkstra, A.; Kaufman, L. Evaluation and Optimization of Laboratory Methods and Analytical Procedures, New York, Elsevier, 1978.
13. Currie, L. A. The Discovery of Errors in the Detection of Trace Components in Gamma Spectral Analysis, in Modern Trends in Activation Analysis, Vol. II. J. R. DeVoe; P. D. LaFleur, Eds.; Nat. Bur. Stand. (U.S.) Spec. Publ. 312; p. 1215, 1968.

14. Kirchmer, C. J. "The Estimation of Limit of Detection for Environmental Analytical Procedures" - Chapt. 4 in this volume.
15. Boumans, P. W. J. M. A Tutorial Review of Some Elementary Concepts in the Statistical Evaluation of Trace Element Measurements. Spectrochim. Acta 33B: 625; 1978.
16. Keith, L. H.; Crummett, W; Deegan Jr, J; Libby, R. A.; Taylor, J. K.; Wentler, G. "Principles of Environmental Analysis", Analyt. Chem. 1983, 55, 2210.
17. Kateman, G.; Pijpers, F. W. Quality Control in Analytical Chemistry. New York: John Wiley & Sons; 1981.
18. Heydorn, K.; Wanscher, B. Application of Statistical Methods to Activation Analytical Results Near the Limit of Detection. Fresenius' Z. Anal. Chem. 292(1): 34-38; 1978, See also: Heydorn, K. Neutron Activation Analysis for Clinical Trace Element Research, 2 Vol., Boca Raton: CRC Press; 1984.
19. Feigl, F. Tüpfel- und Farbreaktionen als mikrochemische Arbeits-methoden, Mikrochemie 1: 4-11; 1923.
20. Kaiser, H. Z. Anal. Chem. 209: 1; 1965 [Ref. 32].
Kaiser, H. Z. Anal. Chem. 216: 80; 1966.
Kaiser, H. Two Papers on the Limit of Detection of a Complete Analytical Procedure, English translation of the above manuscripts. London: Hilger; 1968.
21. Altshuler, B.; Pasternack, B. Statistical Measures of the Lower Limit of Detection of a Radioactivity Counter. Health Physics 9: 293-298; 1963.
22. Wilson, A. L. The Performance-Characteristics of Analytical Methods. Talanta 17: 21; 1970; 17; 31: 1970; 20: 725; 1973; and 21: 1109; 1974.
23. Currie, L. A. Limits for Qualitative Detection and Quantitative Determination. Anal. Chem. 40(3): 586; 1968.
24. St. John, P. A.; Winefordner, J. D. A Statistical Method for Evaluation of Limiting Detectable Sample Concentrations. Anal. Chem. 39: 1495-1497; 1967.
25. Nicholson, W. L. "What Can Be Detected"; Developments in Applied Spectroscopy, v.6, Plenum Press, p. 101-113, 1968; Nicholson, W. L., Nucleonics, 24 (1966) 118.
26. Grinzaid, E. L.; Zil'bershtein, Kh. I.; Nadezhina, L. S.; Yufa, B. Ya. Terms and Methods of Estimating Detection Limits in Various Analytical Methods. J. Anal. Chem. - USSR 32: 1678; 1977.
27. Ingle, J. D., Jr. Sensitivity and Limit of Detection in Quantitative Spectrometric Methods. J. Chem. Educ. 51(2): 100-5; 1974.
28. Currie, L. A. "Lower Limit of Detection: Definition and Elaboration of a Proposed Position for Radiological Effluent and Environmental Measurements," U S Nuclear Regulatory Commission, NUREG/CR-4007, 1984.
29. IUPAC Comm. on Spectrochem. and other Optical Procedures for Analysis. Nomenclature, symbols, units, and their usage in spectrochemical analysis, Pure Appl. Chem. 45 (1976) 99.
30. IUPAC, Commission V.3, Recommendations for Nomenclature in Evaluation of Analytical Methods, Draft Report, 1986.
31. Kaiser, H. Spectrochim. Acta, 1947, 3, 40.
32. Kaiser, H. Z. Anal. Chem., 1965, 209, 1.
33. W. B. Crummett in Ref. 71.

34. Brossman, M. W., Kahn, H; King; D.; Kleopfer, R.; McKenna; G.; Taylor, J. K. Reporting of Low-Level Data for Computerized Data Bases - Chapt. 17 in this volume.
35. Currie, L. A.; Parr, R. M. "Perspectives on Detection Limits for Nuclear Measurements in Selected National (US) and International (IAEA) Programs" - Chapt. 9 in this volume.
36. Cheeseman, R. V.; Wilson, A. L. Manual on Analytical Quality-Control for the Water Industry - Relating to the Concept of Limit of Detection, WRc Environment, Water Research Center, Medmenham, UK, 1978.
37. Ramos, L. S.; Beebe, K. R.; Carey, W. P.; Sanchez, M. E.; Erickson, B. C.; Wilson, B. E.; Wangen, L. E.; Kowalski, B. R. "Chemometrics"; Anal. Chem., 1986, 58, 294R. [Review and bibliography].
38. Long, G. L.; Winefordner, J. D. "Limit of Detection: A closer look at the IUPAC definition"; Anal. Chem., 1983; 55; 712A.
39. Natrella, M. G. 'The Relation between Confidence Intervals and Tests of Significance'; in Ku, H, Ed. "NBS Spec Publ 300"; 1969.
40. Smit, H. C.; Steigstra, H. "Noise and Detection Limits in Signal Integrating Analytical Methods"; - Chapt. 7 in this volume.
41. Epstein, M. S. "Comparison of Detection Limits in Atomic Spectroscopic Methods of Analysis"; - Chapt. 6 in this volume.
42. Ku, H. H. Edit., Precision Measurement and Calibration, NBS Spec. Public. 300 (1969), p. 315.
43. Saw, J. G.; Yang, M. C. K.; Mo, T. C. 'Chebyshev Inequality with Estimated Mean and Variance'; The Amer. Statistician; 1984; 38; 130.
44. Johnson, J. E.; Johnson J. A. "Radioactivity Analyses and Detection Limit Problems of Environmental Surveillance at a Gas-Cooled Reactor" - Chapt. 14 in this volume.
45. Snedecor, G. W.; Cochran, W. G. Statistical Methods, 6th Edit., Iowa State Univ. Press. (1973).
46. Kingston, H. M.; Greenberg, R. R.; Beary, E. S.; Hardas, B. R.; Moody, J. R.; Rains, T. C.; Liggett, W. S. "The Characterization of the Chesapeake Bay: A Systematic Analysis of Toxic Trace Elements"; National Bureau of Standards, Washington, DC, 1983, NBSIR 83-2698.
47. Scales, B. Anal. Biochem. 5, 489-496 (1963).
48. Currie, L. A. "Model Uncertainty and Bias in the Evaluation of Nuclear Spectra"; J. of Radioanalytical Chemistry 39, 223-237 (1977).
49. Koch, W.; Liggett, W. "Critical Assessment of Detection Limits for Ion Chromatography" - Chapt. 11 in this volume.
50. Watters, R. L.; Wood, L. J. in Ref. 71.
51. Murphy, T. J. The Role of the Analytical Blank in Accurate Trace Analysis, NBS Spec. Publ. 422, Vol. II, U. S. Government Printing Office, Washington, DC, 509 (1976).
52. Kelly, W. R.; Hotes, S. A. "The Importance of Chemical Blanks and Chemical Yields in Accurate Chemical Analysis"; Preprint (1987).

53. Kelly, W. R.; Fassett, J. D.; Hotes, S. A., 'Determining Picogram Quantities of U in Human Urine by Thermal Ionization Mass Spectrometry'; Health Physics, 1987; 52, 331.
54. Currie, L. A. 'Quality of Analytical Results, with Special Reference to Trace Analysis and Sociochemical Problems'; Pure & Appl. Chem., 1982, 54, 715.
55. Currie, L. A. The Limit of Precision in Nuclear and Analytical Chemistry. Nucl. Instr. Meth. 100: 387; 1972.
56. Lub, T. T.; Smit, H. C., Anal. Chim. Acta, 1979, 112, 341.
57. Mellor, R. A.; Harrington, C. L. "Evaluating the Impact of Hypothesis Testing on Radioactivity Measurement Programs at a Nuclear Power Facility" - Chapt. 13 in this volume.
58. Parr, R. M.; Houtermans, H.; Schaerf, K. 'The IAEA Intercomparison of Methods for Processing Ge(Li) Gamma-Ray Spectra'; in "Computers in Activation Analysis and Gamma-Ray Spectroscopy"; U.S. Dept. of Energy, Sympos. Ser. 49, 1979, 544.
59. Liggett, W. ASTM Conf. on Quality Assurance for Environmental Measurements, 1984, Boulder, CO.
60. Watters, R. L.; Spiegelman, C. H., and Carroll, R. J. "Heteroscedastic Calibration in Inductively Coupled Plasma Spectrometry"; Anal. Chem., 1987, 59, 1639.
61. Currie, L. A. "The Many Dimensions of Detection in Chemical Analysis"; in Chemometrics in Pesticide/Environmental Residue Analytical Determinations, ACS Sympos. Series (1984).
62. Owens, K. G.; Bauer, C. F.; Grant, C. L. "Effects of Analytical Calibration Models on Detection Limit Estimates";- Chapt. 10 in this volume.
63. Iyengar, G. V.; Tanner, J. T.; Wolf, W. R.; Zeisler, R. 'Preparation of a Mixed Human Diet Material for the Determination of Nutrient Elements, Selected Toxic Elements and Organic Nutrients: a Preliminary Report', submitted to The Science of the Total Environment, 1986.
64. Currie, L. A. Detection and Quantitation in X-ray Fluorescence Spectrometry, Chapter 25, X-ray Fluorescence Analysis of Environmental Samples, T. Dzubay, Ed., Ann Arbor Science Publishers, Inc., p. 289-305 (1977).
65. Horwitz, W.; in Ref. 71.
66. Zweig, M. "Establishing Clinical Detection Limits of Laboratory Tests"; - Chapt. 8 in this volume.
67. Natrella, M. G.; Experimental Statistics, NBS Handbook 91, 1963.
68. Dixon, W. J.; Massey, F. J. Introduction to Statistical Analysis, McGraw-Hill, New York, 1957.
69. Schlain, B., personal communication, 1987.
70. Currie, L. A. 'The Limitations of Models and Measurements as Revealed through Chemometric Intercomparison'; J. Res. NBS, 1985, 90, 409.
71. Kurtz, D.; Taylor, J. K.; Sturdivan, L.; Crummett, W.; Midkiff, C.; Watters, R.; Wood, L.; Hanneman, W.; Horwitz, W. Real-World Limitations to Detection: A Panel Discussion; Chapt. 16 in this volume.
72. Currie, L. A.; Gerlach, R. W.; Lewis, C. W.; Balfour, W. D.; Cooper, J. A.; Dattner, S. L.; DeCesar, R. T.; Gordon, G. E.; Heisler, S. L.; Hopke, R. K.; Shah, J. J.; Thurston, G. D.;

Williamson, H. J. "Interlaboratory Comparison of Source Apportionment Procedures: Results for Simulated Data Sets"; Atmospheric Environment 18 (1984), 1517.
73. Eisenhart, C.; Zelen, M. Ch 12, "Elements of Probability"; in Handbook of Physics, E. U. Condon and H. Odishaw, Ed., McGraw-Hill, New York, 1958.
74. Kendall, M.; Stuart, A.; Ord, J. K. The Advanced Theory of Statistics; Vol. 3, MacMillan; New York,; 1983.
75. Delaney, M. F. 'Multivariate Detection Limits for Selected Ion Monitoring Gas Chromatography - Mass Spectrometry'; Chemometrics and Intelligent Laboratory Systems, 1987.
76. Wold, S.; Albano, C.; Dunn, III, W. J.; Edlund, U.; Esbensen, K.; Geladi, P.; Hellberg, S.; Johansson, E.; Lindberg, W.; Sjöström, M. "Multivariate Data Analysis in Chemistry"; in Chemometrics: Mathematics and Statistics; B. R. Kowalski, Ed., (Reidel Publishing Co.) 1984; pp. 17-96.
77. Draper, N.; Smith, H. Applied Regression Analysis, Wiley New York, 1981.
78. McCormack, M. "Realistic Detection Limits and the Political World" - Chapt. 2 in this volume.
79. Slovic, P. "Perception of Risk"; 280-285 in Ref. 4.
80. Lave, L. "Health and Safety Risk Analysis: Information for Better Decisions"; 291-295 in Ref. 4.
81. Handler, P. Dedication Address, Northwestern Univ. Cancer Center, 1979. (See also S. J. Gould in the New York Times Magazine, 19 April 1987.)
82. Nieboer, E.; Jusys, A. A. "Contamination Control in Routine Ultratrace Analysis of Toxic Metals"; Ch. 1 in Chemical Toxicology and Clinical Chemistry of Metals, S. S. Brown and J. Savory, Eds., Academic Press, London, 1983.
83. Sharaf, M. A.; Illman, D. L.; Kowalski, B. R. Chemometrics, Wiley, New York, 1986.
84. Kowalski, B. R., Ed. Chemometrics: Mathematics and Statistics in Chemistry, (Reidel Publishing Co.) 1984.
85. Hurst, G. S.; Payne, M. G.; Kramer, S. D.; Young, J. P. Rev. Mod. Phys., 1979, 51, 767.
86. Donaldson, W. T., Envir. Sci. Tech., 1977, 11, 348.
87. Guinn, V., personal communication; 1987.
88. Donahue, D., Fourth Intern. Sympos. on Accelerator Mass Spectrometry, Niagara on the Lake, 1987.
89. Heydorn, K.; Christensen, L. H. "Verification Testing of Commercially Available Computer Programs for Photpeak Area Evaluation"; Intern. Conf. on Meth. Appl. Radioanalytical Chemistry, Kona, 1987.
90. Cochran, W. G.; Biometrics, 1964, 20, 191.
91. Andrews, R. M. "Meat Inspector: 'Eat at Own Risk'"; Wash. Post, May 16, 1987. [Science news following issuance of a National Academy of Sciences Report concerning food poisoning from undetected microorganisms.]
92. Lochamy, J. D. The minimum-detectable-activity concept. Nat. Bur. Stand. (U.S.) Spec. Publ. 456; 1976, 169-172.
93. IAEA; Users' Guide on Limit of Detection; in preparation. (See Ref. 35).

94. "Standard Radiological Effluent Technical Specifications for Pressurized Water Reactors," U. S. Nuclear Regulatory Commission, NUREG-0472, Rev. 3, September 1982.
95. Federal Register, 1984, 49, 43431.
96. Federal Register, 1984, 49, 43430; and Glaser, J., Foerst, D., McKee, G., Quave, S., and Budde, W., "Trace Analysis for Wastewaters," Environ. Sci. Tech., 1981, 15, 1426.
97. Forina, M.; Lanteri, S. "Data Analysis in Food Chemistry;" pp. 305-349; in Ref. 84.
98. Feigl, F. Chemistry of Specific, Selective and Sensitive Reactions; New York: Academic Press; 1949.
99. Emich, F. Ber. 1910; 43; 10.
100. Cox, D. R.; Lewis, P. A. W. The Statistical Analysis of Series of Events; London: Methuen; 1966.

Notes

Note 1. Analytical advances have led to the possibility of "single atom detection" (85). At the same time it is recognized that at concentrations of 1 part in 10^{15} (in water) in principle "every known organic compound could be detected" (86). These measurement realities mandate the setting of regulatory levels on bases other than either non-zero concentrations, or the inherent ability to detect.

Note 2. That Feigl's "Identification Limit" referred to the minimum quantity detectable (L_D) as opposed to the decision or critical level (L_C) is clear from his statement defining the "*'Erfassungsgrenze'* [as] *die Kleinste absolute Menge Substanz ... die ... noch nachweisbar und bestimmbar ist* "(Ref. 19, p. 6). In a later, english language publication, this meaning was amplified in a manner that foreshadowed the modern statistical approach to detection. In the volume "Chemistry of Specific, Selective, and Sensitive Reactions", p. 14 (98), Feigl described a test for magnesium which was "always" positive, for 40 repetitions, using a 0.05% Mg solution. With dilution by factors of 10 and 50, however, the test was positive only in 24 and 6 instances, respectively. With this, Feigl embraced the concept of the "region of uncertain reaction" (99), and a condition for the identification limit that the chance of a false negative be negligible.

Note 3. Symbols introduced in this section include the following: y [gross signal], B [null signal = background, baseline, or blank], S [net signal], x [analyte concentration or amount], A ["sensitivity" or calibration factor], pdf [probability density function], cdf [cumulative distribution function], superscript ^ or est( ) [estimated value], E() or μ [expected value], V or σ^2 [population variance], s^2 [estimated variance], σ_o [standard deviation of the estimated net signal, when E(S)=0], CI [confidence interval], df [degrees of freedom], Δ [bias], Δ_D [bias detection limit; discrimination limit]. Subscripts, $_m$, $_M$, denote lower and upper limits, respectively.

Note 4. In a recent paper (43), Saw, Yang, and Mo present an extension of the Chebyshev Inequality that uses estimated values for the mean and variance. As expected, the limits are generally broader (or limiting α's larger) than would be the case with σ-known, but the extended inequality is directly applicable to our test of the null hypothesis in the form: $B - \bar{B}$, where the mean value for the blank $\bar{B}$ is based on n-replicates. An important conclusion that can be derived from the treatment of Saw, et al. is that the smallest limiting value for α that can be obtained with n-replicates is $1/(n+1)$. This value is obtained once k is at least $\sqrt{(n+1)}$. Thus, for 5 replicates, one can do no better than α (limit) = 0.167, but the requisite standard deviation multiplier k need be no larger than $\sqrt{6}$ which equals 2.45. Larger k's would bring no improvement. To attain an upper limit of 0.05 for the false positive risk, n must be at least 19. The corresponding value for k would be $\sqrt{20}$ or 4.47. The critical level S_C for the detection decision would therefore be $4.47 \cdot s_o$, where s_o is the estimated standard deviation of $B - \bar{B}$, based of $n-1 = 18$ degrees of freedom. Since $\sigma(B-\bar{B}) = \sigma_B \sqrt{(n+1)/n}$, $S_C = 4.47 \sqrt{(20/19)}\ s_B$, or $4.59\ s_B$. Had we been able to assume normality, the multiplier 4.47 would have been replaced with 1.73, Student's t for a 1-sided test at the 0.05 significance level.

Note 5. The **relative** standard deviation [RSD] of $\hat{\sigma}_i$ based on observed counts is $1/(2\sqrt{\mu})$ for Poisson data, or about 6% for $\mu=E(N)=60$. Equivalent precision for $\hat{\sigma}_i$ based on replication would require about 2μ or 120 degrees of freedom. The same is true for confidence intervals for μ, hence V_i, based on counts, vs V_i based on replication. For more detail, including the use of χ^2 to derive both types (counts, replication) of CI's see Ref. 68 and the monograph by Cox and Lewis (100). Adequacy of the large count (normal) approximation, and the exact treatment for extreme low-level data ($\mu \approx 10$ or less) are covered in Ref. 28 and the references therein.

Note 6. Combining the inequality constraint [$V_x \geq 0$] with known internal Poisson variance has been used by Guinn for total variance estimation in activation analysis (87), by Donahue (88) for the same purpose in accelerator mass spectrometry, by Heydorn (18,89) for the analysis of precision of gamma ray spectrum analysis, and by Currie (55) for optimal weighting in counting experiments. The variance weighted t approximation derives from work by Cochran (90). Cochran's work, however, applied to a somewhat different model than used here. The statistical properties of this extension of his technique have not been studied.

Note 7. Method-specific **mechanistic** understanding is, in the last analysis, the **only** route to reliable measurements, reliable detection limits, and meaningful societal decisions involving science related public policy. The need for understanding on the mechanistic level becomes even clearer when one considers the inherent limitations of decisions and policies based solely on the empirical record, such as certain aspects of epidemiology. Sampling statistics and control of the system under investigation empirically, constitute severe limitations -- ones that can be overcome only by directing our research priorities toward increased basic scientific knowledge (81).

Note 8. The drawback of policies based on "detection limits" so defined that $L_D = L_C$, where the false negative risk (β) is 50%, and the false positive (α) is 10 to 400 times smaller, is striking in the light of matters affecting large sectors of the public. A case in point: food poisoning affects several million citizens in the US each year "from poultry because federal inspection fails to detect contamination by salmonella and other bacterial microorganisms" (91). In situations like this we must ask how many would be content with a test whose "detection limit" is designed to catch only half the contaminated specimens! In comparison, one wonders how society is better served in such an instance by reducing false positives to a negligible level. (See also sections 2, 3.2.1, 3.3.1, and Fig. 10.)

RECEIVED September 22, 1987

SOCIOPOLITICAL PERSPECTIVES

Chapter 2

Realistic Detection Limits and the Political World

Mike McCormack

McCormack Associates, Inc., 508 A Street, S.E., Washington, DC 20003

My comments in this chapter are not particularly technical, unless one considers the subject of constructive and realistic political activity to be "social engineering". I propose to explore the interface between the realities of detection limits on the one hand; and the political world, and our obligations within it, on the other.

Probably the best starting point for such a discussion is the recognition that we are all part of the political community. Responsible scientists cannot avoid the obligations of being citizen-politicians in any free society. Scientists are a special subset of that greater political community, and we have, over recent decades, attracted ever-increasing attention within it. That attention has constituted a blessing in times of success, with acclaim and generous financial support; and a frustration in times of public confusion, incomprehension and fear.

The people of any society may, of course, always be categorized according to any number of criteria. For the purpose of this discussion, it is fair, and not uncomplimentary, to define two groups within the population of this country: one in which the concepts of very large and very small numbers are comprehended, and operations with them easily understood, and in which the scientific method -- and especially the meaningful interpretation of analytical data is appreciated; and the other group, within which this is not generally the case. For simplicity, we usually refer to these two groups as the scientific, and the non-scientific communities.

0097-6156/88/0361-0064$06.00/0

It frequently appears to me that there is a mutuality of ignorance (and perhaps a "love-hate" relationship) binding -- and at the same time -- dividing these two groups. For instance, analytical chemists frequently seem unable to understand that their fellow-citizens without technical training cannot understand what the chemists, or many other scientists and mathematicians, are talking about when they attempt to convey scientific or mathematical information to the public on matters of societal concern, or why their information and recommendations are not readily accepted. After all, this attitude assumes, scientists know what they're talking about.

Equally confusing, in the minds of many members of the non-scientific lay public, is the generally more objective, less emotional attitude that most scientists take toward presumed threats to human health and safety that -- the public is told -- flow from scientific and technological activity, especially in the commercial, agricultural and industrial world. "Those guys don't care -- they're prostituting their integrity for a paycheck," so that attitude concludes.

The ancient enemies of humankind have always been ignorance and fear, which in turn are the progenitors of superstition, hatred, bigotry and the isolation of one group from another. To a limited but disturbing extent, this phenomenon has developed within this country; and today a gulf of confusion, suspicion, fear, emotionalism and sometimes hostility lies between significant portions of the scientific and non-scientific communities.

Within the legislative arena, and within some portions of the press and electronic media, this has focused on real or presumed risks to human health and safety from the presence, or the possible presence, of "artificial" (that is, deliberately manufactured by humans) chemicals in our food, water, medicines, air, the things we touch, or the ground upon which we walk.

This often leads to the enactment of unrealistic legislation, and to regulations that attempt to override the basic laws of nature and the realities of the physical world.

The challenge that we in the scientific community face arises from the fact that frequently only we understand that such a contradiction exists, let alone what a rational solution to it may be. This situation is further complicated by the fact that the fear we are trying to overcome often gives rise to an emotionalism within which scientists and even scientific fact become suspect in the minds of some non-scientists. It is not uncommon to encounter the argument that scientists in a specific discipline should be disqualified from providing information on issues involving that discipline because "they are prejudiced".

In addition, a significant number of individuals, from the press and electronic media, from certain activist and religious groups, and even among elected public officials, often encourage emotionalism with respect to science in the minds of susceptible individuals. At that point, scientists may throw up their hands, and tend toward open ridicule of, and hostility toward, such individuals.

Of course, this only exacerbates the situation. Since we are generally the ones who understand the basics of science related issues, it is incumbent upon us, as responsible citizens in a free society, to speak out, patiently helping our non-scientific fellow citizens understand; and leading them to responsible, constructive perspectives. This is a never ending obligation.

Before anyone in the scientific community (and especially in this book) becomes over-inflated with self-righteousness, let me emphasize that we, as individuals and as a profession, are far from faultless or free from legitimate criticism with respect to our responsibilities. Certainly more members of the scientific community -- and chemists in particular -- should have recognized at an earlier date the hazards of industrial pollution or the unlimited application of certain pesticides to the numerous environments in which we live and work and play.

Today it seems almost impossible to us that any chemist could have been so unconcerned only several decades ago that he or she would not speak up to protest such practices. In addition, many members of the scientific community have seriously damaged meaningful communication with the non-scientific community by the display of an unbecoming arrogance; failing to appreciate that this increases hostility among those already suspicious.

These comments are not intended as criticism or soul searching, but to help put the situation into honest context. The fact is that there is a broad chasm of ignorance, suspicion and fear that separates the scientific community from much of the public, the press and legislators at the state and federal level.

An increasing volume of legislation, sincerely intended to protect the physical environment, and to insure human health and safety has been, and is being enacted in this atmosphere. Much of the legislation has been proposed as a result of honest concerns because of accidents involving chemicals or from emotional reaction to suggestions that some chemical which humans may contact may be a carcinogen or cause other physical harm. Legislation is often written in ignorance of the actual degree of hazard involved; and allowable concentrations of hazardous substances have frequently been established without an adequate understanding of the resulting impact.

This is a significant problem for all society, and it will probably continue until vigorous initiatives are undertaken from within the scientific community to assist more public officials and more members of the news media appreciate the realities involved and the importance of observing them.

To be fair, the concepts involved in such legislation make good sense to a casual observer from outside the scientific community. It seems obvious and logical to many honest and sincere citizens that if a substance is "toxic," "hazardous" or "carcinogenic" that none of it should be allowed in our food or drink, in the air we breathe or the ground on which we walk; none at all. The concentration, we are told, should be zero. "Why not?"

The suggestion that it is not practical to try to reduce the concentration of a hazardous substance to the minimal level at which it can be detected (even ignoring cost) frequently seems to constitute a "cop-out"; and those who make such a suggestion may be suspected of being in league with some deliberate polluter who would callously endanger the health (or the lives) of our children.

Analytical techniques that have been developed can now, for most substances, detect concentrations far below hazardous levels. The effect of these advances in analytical capability should be cause for satisfaction. They will help make possible future research, both theoretical and applied, including studies relative to human health and safety.

However, there are those who insist on enacting legislation that would provide that either (1) the concentration of substances designated as carcinogens should be "zero", or (2) that the concentration of some contaminants in drinking water should be as low as detectable. Thus, each new accomplishment in analytical technologies that pushes the limit of detection of any suspect chemical to a lower concentration, brings with it the potential for new political problems.

Such unrealistic provisions in the law or regulations are, in reality, self-defeating. However, it is difficult to persuade the non-scientist that they do not protect the public, and frequently cause the waste of a great deal of public and private money.

There are ways to help the average citizen understand this. For instance, at one part per billion, there are still trillions of molecules of a foreign substance in a glass of water. The average citizen is shocked to learn this. Another way is to point out that in the real world of drinking water and food, at parts per billion, there is some amount of almost everything in almost everything. Another way to express this truth is that a "chemically pure" substance (99.9999% pure) still contains one part per million impurities. This is equivalent to 100 different substances at 10 parts per billion; or, if you like, 10 foreign substances at 100 parts per billion. Thus it becomes extremely difficult to analyze accurately for many substances at the level of a few parts per billion because of potential contamination of the system from equipment, and from reagent impurities.

(Most non-scientists do not appreciate that a billion is one thousand million -- and do not comprehend what a trillion -- or one part per trillion means. Here is an illustration that may be of value. Imagine ordinary glass marbles about 1/2 inch in diameter -- similar to the ones we played with when we were kids. An ordinary square card table will hold about 10,000 such marbles, one layer thick, packed as densely as possible. For comparison, it would take about two million marbles to cover the floor of an average size lecture hall.

It would take about 40 billion marbles to cover Central Park in New York City, one layer deep. Thus, one part per billion would be about 40 marbles among a layer of marbles completely covering Central Park.

We can routinely detect 40 parts per billion of many substances in water, but not necessarily in natural systems containing other unknown substances in higher concentrations; and it may not be practical, or even possible, to guarantee the removal of those 40 parts per billion from any natural system -- without exorbitant cost. One trillion marbles -- one thousand billion, or a million million, would cover all of Manhattan Island and all of the Hudson River from the George Washington Bridge down to the Battery. Even in those few cases when we can detect a few parts per trillion -- the equivalent of a handful of marbles scattered somewhere over Manhattan Island and that stretch of the River -- it would probably not make any sense to try to remove them.)

You must help make that point, that detection of a chemical and standards setting are separate activities, the one comprising a strictly technical activity and the other, a societal value judgment. The basic rationale for disengaging these two activities is that the ubiquitous presence of other carcinogenic agents -- all around us -- swamps the potential hazard, even of known carcinogens, at extremely low concentrations.

Unfortunately, there has been too much uncertainty and confusion regarding the real meaning and reliability of detection limits for their optimal use in the world of legislation and regulation. I urge you to work toward a consistent definition and meaningful realization of the limit of detection from a technical point of view, so that it can be accepted by society as a limit above which chemicals will be found (if present) with an acceptable degree of certainty. This means that such limits must fully take into account the real sample characteristics including the effects of interference, contamination, etc. Otherwise, <u>spurious levels</u> for false positives or false negatives may lead to regulations that are too restrictive or to pressures that make them too lenient.

As with all scientific measurements, working at limits of detection necessarily involves analytical uncertainty. However, this creates a special problem in communication where such uncertainties are not generally understood. We must assure that our detection limits and, in fact, all of our analytical measurements are established with an adequate level of "certainty", especially as assurance for policy makers seeking responsible standards. After all, they must assure their colleagues, constituents and the press that standards based on risk assessment make the most sense, and that scientifically valid data have been used in making such assessments.

Finally, detecting a minute trace of a contaminant does not necessarily make its economic removal practical. This is not immediately obvious to the average layman, but it can be easily explained by extrapolating the concept to the point of absurdity. Can we remove one part per million?

If so, is it still possible to remove one part per billion? If so, and we find one part per trillion, is it possible to remove it, etc.?

These realities must be borne in mind by scientists who work in the realm of detection limits when reporting their findings to non-scientists, and especially to lawmakers.

This brings me back to our obligation as citizens. I hope you will take this matter seriously.

We all face a monumental challenge. Fortunately, we have scientific truth and common sense as our most effective tools.

We must add to that a dedication to serving our fellow citizens by helping them understand, and putting the technology and its purposes into proper context.

RECEIVED September 22, 1987

Chapter 3

Scientific Measurements and Data in Public Policy-Making

Thomas H. Moss

Case Western Reserve University, Cleveland, OH 44106

Sound technical measurements and the integrity of data must seem at times to be almost entirely irrelevant to the course of public policy decision-making. Recent experience in the politics of clean air, acid rain, toxic substances control, pesticide regulatory legislation, as well as other debates, indicates a more tempered view, however. On closer analysis the role of scientific measurement is seen as a vital one, but one which requires a sense of timing and perception of the dynamics of human behavior in seeking solutions to challenging problems.

Technical measurements and the resulting data are used in two distinct ways in the contemporary American policy decision-making process. On the one hand is their role in the systematic building of a body of knowledge which ideally becomes the basis for public policy. On the other hand is their use as weapons in a war of words, or contest for public attention and support. At times one or the other of these functions may seem to dominate, but it is my thesis in this essay that it is extremely risky for all concerned to neglect either aspect. Beyond that, my own experience tells me that participants in public policy discussions can, to a considerable degree, create their own reality from the choice between these two extremes. It is the realization of this relatively strong degree of personal control over the nature of the debate which is so important to participants. By exerting that control wisely, participants can create a process in which they have confidence; by using it unwisely, inconsistently, or lazily, they can find themselves in a process in which they will feel victimized, frustrated, and helpless.

Case Histories of Measurement and Scientific Data in Political Decision-Making:

What are examples to justify this interpretation of the regulatory

0097-6156/88/0361-0070$06.00/0

climate? Clearly my own will be based on interpretations of history and motivations with which some may disagree. However in the cases I mention below I could clearly see points at which one path in the course of public policy debate become dominant over several perfectly feasible alternatives, or in which the debate abruptly changed course. Whether my own interpretation is correct or not, my aim is to at least challenge the reader to come up with an alternative interpretation to help rationalize the curious twists of thinking on public policy.

The classic example is perhaps the long-running debate over the flagship of U.S. environmental law: the Clean Air Act. In a pioneering 1970 effort to establish and enforce meaningful limits on air pollution, the framers of the legislation chose human health as the key and absolute criteria for enforcement of standards (1):

> "In the Committee discussions, considerable concern was expressed regarding the use of the concept of technical feasibility as the basis of ambient air standards. The Committee determined that 1) the health of people is more important than the question of whether the early achievement of ambient air quality standards protective of health is technically feasible: and 2) the growth of pollution load in many areas, even with the application of available technology, would still be deleterious to public health.
> Therefore, the Committee determined that existing sources of pollutants either should meet the standards of the law or be closed down...."

There were many understandable reasons for the choice of this rhetoric at that time. Concern for human suffering seemed to dominate public attitudes, and non-human ecological effects seemed much more subtle and distant. Several well publicized urban air pollution episodes in Europe and the U.S. had vividly shown the connection of acute air pollution and human health impacts through the dramatic increases in hospital admissions and deaths from a variety of cardiac and pulmonary causes. On any less absolute scale, however, legislative leaders were well aware that air quality monitoring and epidemiological data were hopelessly inadequate to demonstrate the subtleties of low-level cause and effect, or to establish anything like reasonable dose-response curves, threshholds, or to validate details of individual and synergistic effects of various pollutants.

The implementation of the 1958 "Delaney Clause" (2) of the Food Additive Amendments to the Federal Food, Drug and Cosmetic Act was also widely considered at this time as a positive political precedent. The Delaney Clause was interpreted for many years as banning absolutely, from any food consumed by man, any chemical found to induce cancer in animals, without reference to dose. Its straightforward language made regulatory action simple and prompt, and it was widely credited with adding a very positive level of protection for the U.S. food consumer. Only in the late seventies and eighties did it begin to be apparent that detection limits were expanding so rapidly that chemical traces far below the clear cancer-causing levels could be routinely detected. In this new technical world, there was much less utility to the strategy of assuming a decremental human health impact as a result of any connection whatsoever of the chemical to animal cancer. (3)

Similarly, the strategy of simplistically focusing on human health to the exclusion of other air pollution impacts, and of considering it as an absolute with no acceptable level of risk, at first seemed to work marvelously in the clean air decision-making process. The Clean Air legislation was perennially popular with the public, and enforcement was vigorous in the '70's. Major benefits were also clear: numbers of days of some types of urban pollution episodes declined dramatically, and visible manifestations of air pollution effects sharply decreased. Even in strictly bureaucratic arenas, the "human health only" approach seemed successful. In the late 70's, while several federal agencies feared being swallowed in a Department of Environment and Natural Resources, the Environmental Protection Agency (EPA) kept itself out of that pool by declaring itself to be primarily a "human health agency".

The problem came, of course, when monitoring and air quality measurements of pollutants began to outstrip the capabilities of health data to show dramatic effects. Suddenly it was obvious that some balance of risks, and costs and/or benefits, would have to be made. Moreover, it became apparent that the impacts of pollutants were not independent quantities, with unique dose-response curves. Instead, many were shown to be closely interdependent and could only be considered in the context of all the others. Most significant of all, as the long range and pervasive damage of acid rain and other ecological effects began to be perceived, it was clear that legislative, regulatory, and research programs addressed solely to human health generated a set of priorities in regulatory law that could be far from those really needed for over-all societal benefits.

Despite the trap created by the original decision to tactically use absolute human health data to dominate in a political argument about air pollution control, the long-standing acid rain debate is an example of a process which illustrates both ends of the spectrum of use of "facts". There have been periods when representatives of utilities, coal companies, environmental scientists, midwestern coal states, northeastern forest and sport fishing areas, and regulatory agencies were able to sit together and jointly review data as well as agree on factual needs for future decisions. However, there have also been periods when these same parties have chosen to take the same body of measurements, selectively pull out "facts", package them with dramatic or manipulative adjectives, and use them almost exclusively to attempt to stampede public or political opinion. Industry groups in these parts of the cycle chose to highlight "facts" such as the existence of lake acidification in the absence of fossil fuel conbustion influences, or improvement of certain crop yields in the presence of deposition of sulfur or nitrogen compounds. Environmental interests, in contrast, focused on "facts" such as forest and lake transitions, or crumbling statuary. Both sides published glossy literature with titles like "Facts About

Acid Rain", but presenting very different "facts" in very different ways.

What is also remarkable is that during these latter adversarial periods of "fact" use, both sides generally felt abused by the others' manipulations. Both bemoaned the deterioration of the honesty of the exchange, and both wrung hands about the gullibility of the public or vulnerability of the political system to the distortions or manipulations of the other side. Both sides regretted also that their own level of discussion was "forced" to a level of simplistic slogans or selective use of "facts", but explained that this was the only way that they could counter the tactics of their opponents.

Despite the sense of being "trapped" or dragged into the fact manipulation arena, however, there is ample evidence that it is possible to break out of that entrapment. Before passage was finally achieved, the Toxic Substance Control Act (TOSCA) had languished in several Congresses, with coalitions of industry groups on one side, and environmental health groups on the other, trading "fact sheets" as they jockeyed for public and political support. Not surprisingly, when confronted by two sets of experts citing "facts" as proof of opposite positions, the political bodies typically tend to avoid decisive action or even any action at all. TOSCA is not the only example of a major piece of legislation which was pushed enthusiastically by proponents, and bitterly opposed by adversaries, but given determined neglect by a much larger group rather than voted up or down. In such situations, endless procedural steps seem enevitably to drag out because there is no collective will to shorten them, and somehow deadlines of bringing decisive votes are never met.

However, in TOSCA's case, both sides finally acknowledged the risk of using measurements and facts primarily as manipulative tools, and eventually switched to use them as a rational basis for action. The risk of stalemate for TOSCA opponents was that their success in delay of any action could have led to a sudden acute political frustration, perhaps triggered by an external event, which might lead to enactment of an extreme measure much stronger than that which was achievable by working constructively with facts to build a mutually tolerable solution. The risk for proponents was just the converse: they might, indeed, by entering into compromise problem-solving with their opponents, have missed a chance for achieving something closer to their ideal goals.

Naturally, I can't know in detail how these risks were balanced in the minds of all participants in the political struggle over TOSCA, acid rain, or other issues. Whatever these internal considerations, however, the results usually show a definite pattern of swinging from one approach to facts and measurements to the other. A not unfamiliar pattern in the political system is that of initial fact and measurement manipulation and exploitation to stake out position, and later intelligent use to build reasonable positions after the fatigue and futility of adversary "fact" games have set in. A common variation is that of initial cooperative work with a common base of fact and measurement among technical colleagues, which switches to the adversary position when professional "adversary coaches" (lobbying and public relations firms, certain kinds of general counsel offices, etc) become involved. And though this pattern may eventually evolve, due to fatigue or other reasons, back to the cooperative fact-use mode, the adversary mode in many debates (like the continuing inability to reach a up-dated version of the Clean Air Act), may simply continue indefinitely with a few minor accomodations or actions, but continual underlying bitter "fact" manipulation back and forth.

Another particularly classic example of this cycle unfolded concurrently with this writing. After a fourteen year stalemate of posturing and adversary confrontations, Congress appeared to be ready to substantially revise federal pesticide regulatory law (the Federal Insecticide, Fungicide, and Rodenticide Act (FIFRA)) with the support of both major industry and environmental groups. "Facts" and "measurements" had been, for seven Congresses, merely pawns in a continuing struggle in which each side was convinced of the others' ignorance at best, but more typically of their narrow greed and self-interest. Studies and test results had been cited endlessly over the course of the fourteen years, and yet these "facts" and measurements at least superficially had provided little help in forming national policy. They were designed or used to upstage or negate, rather than build on each other. However, by October of 1986 both the House and Senate had voted overwhelmingly (329-4 in the House) to bring forth a compromise version of the legislation.

What caused the change in the style of the FIFRA debate? As usual in most major controversies, the dynamics of the periods of tension, or of positive attitudes, are hard to trace. In this case, as is often true, there was an external factor which drove the parties toward compromise, and in so doing motivated them to use their data and measurements constructively rather than in a gaming pattern. Industry groups in this case were very much interested in legislative extension of their patent rights to compensate for long regulatory delays; environmental group support was needed for this and thus there was a potential for barter between the elements of pesticide regulatory and patent legislation. In a broader sense, however, as the New York Times reported (4):

> "The changes in the law reflect growing scientific knowledge and public awareness about the dangers of pesticides. In recent years increasingly advanced testing has found that some pesticides can cause cancer, birth defects and mutations in humans."

The important observation is probably to note that during the fourteen years of pitting "fact" against "fact", measurement against measurement, the two sides did not <u>exactly</u> balance to zero knowledge gained. However frustrating in appearance to the scientific community or public on both sides, retrospectively one can see that slow progress was being made in establishing a base of knowledge and a body of "accepted fact" as opposed to the isolated anecdotal "facts" used only for debating purposes.

The FIFRA revisions eventually failed in the rush for Congressional adjournment, because of House-Senate Conference disagreement on technicalities of the patent provisions. Only the coming Congress will tell us whether a stable consensus has been achieved, which can be resurrected and legislated again, or whether another cycle of "fact" manipulation will be necessary to the process.

The Risks of Misunderstanding the Debate Cycles:

I've anecdotally documented the cycle of measurement and fact use in public policy debate, but need to add a further observation on the need for awareness of this cycle, and risks of naivete concerning it. In an ideal world we might hope that all concerned would be willing to plunge immediately into a constructive and purely objective mode of using a body of measurement data to optimize policy courses. Various segments of the scientific community periodically express the hope for this ideal world, with frustration at the true state of affairs. There certainly is no fault in being disappointed at the adversary and distorted state of public debate at the low points in its cycles. However, it is crucial to understand that adversary posturing is often a natural and even inevitable human group (and thus political) behavior, and not an indication that a more rational or constructive approach is hopeless. One risk of naivete is to find oneself in a state of shock in reaction to the adversary phase, leading to drop-out from the process. This can result in non-participation at the critical time in the cycle when participants are ready, for whatever reason, to swing back to the constructive problem-solving mode. The converse risk is to find oneself spending an inordinate amount of time speaking in objective scientific language in a discussion which is proceeding with an entirely different approach. For the technical person not tuned to the adversary style, it may be best to simply wait out the adversary periods, like times of stormy weather, but in a state of readiness for the moment when the storm blows over.

On the other side are those so committed to a fundamentally adversary approach to politics that they may completely fail to recognize an opportunity for true win-win compromise with their opponents. Though the hardened adversary attitude may seem often to be "realistic", its proponents risk being totally disenfranchised from participation in eventual compromise when the course of decision-making turns in that direction. In all of the debates I've mentioned; clean air, acid rain, pesticide regulation, and toxic chemicals, I've seen both industry and environmental groups which had held too long to rigid adversary roles be simply left out of the final discussions which led to compromise action. The lesson is simply that though one must guard against naivete in expectation for objectivity and constructive attitudes, it is equally naive not to remain vigilant and ready for opportunities for gain through cooperation and the honest mode of seriously integrating scientific measurement and fact into decision-making.

Can Participants Control the Debate Reality in Which They Operate?

The hardest part to argue of my initial thesis is that public policy debate participants can to an appreciable extent "create their own reality" in the tone of the political debate. The negative side of this seems most plausible: in early acid rain discussions it was particularly obvious when many industry groups began to total up potential control costs and switch to a strategy of embarrassing rather than working with control proponents, and when environmental groups realized that acid rain concerns were likely to be a major political force in maintaining or strengthening other aspects of the Clean Air Act and turned away from an approach of compromise or incremental pilot approaches. The actions of either side were sufficient to create a reality of a bitter debate in which measurements and "facts" were weapons rather than tools for finding optimal solutions.

On the positive side, it is not easy to argue convincingly that any person or groups, no matter how well-motivated, has the power to single-handedly convert an adversary confrontation to a constructive discussion. What my own experience does tell me, however, is that in the political world there are always some very real benefits to conserving the energy and resources otherwise dissipated in adversary battles, and that there is always a steady though unpredictable appearance of external factors which may increase the self interest of those involved in creating a constructive problem-solving process.

Thus, the participant in such a debate who decides to move from the adversary course to that of using measurement and facts in a constructive manner is usually not without allies. The self-interest of the other side, and powerful external influences may at any moment push in the same direction. Creating the reality of a constructive debate may thus take a persistence and patience, but nonetheless the opportunity is very likely to come along if it is seriously sought. As mentioned above, the final House vote this fall on the 14-year stalemated pesticide legislation was 329 to 4. It is a vivid example of the kind of consensus on the meaning of measurements and data that can be eventually found if commitment to do so is sufficiently strong and enduring.

The lesson of these actual cases is that we in the scientific community must realize that public policy debates have cycles of objectivity and distortion, in which the latter is not a sign of the disintegration of the political process but only of its human-centered nature. Measurement and data that may seem wasted or prostituted at a low point in the debate process may, however, be quietly moving a center of consensus toward a point and credibility which will later be the core of sound decisions. This is the enduring reason for the vital need for measurement experts and other scientists as participants in the political process. Though sometimes ignored, distorted, or abused, they are the ultimate key to progress in reaching sound public policy.

Literature Cited:

1. U.S. Senate Report 91-1196, 91st Congress, 2nd Session (1970)
2. Food Additive Amendment, 1958, Federal Food, Drug and Cosmetics Act (Sec. 409(c)(3)(A))
3. "Stretching Delaney Till It Breaks", Richard M. Cooper, Regulation, November/December 1985, pg. 11-17,41
4. New York Times, September 20, 1986

RECEIVED September 28, 1987

FUNDAMENTAL AND COMPARATIVE ASPECTS OF DETECTION

Chapter 4

Estimation of Detection Limits for Environmental Analytical Procedures

A Tutorial

Cliff J. Kirchmer[1]

Ecology & Environment, Inc., Suite 404, Cloverleaf Building 3, Overland Park, KS 66202

A review of theory indicates that the variability of blank responses is the preferred basis for defining the lower limits of measurement. The variability of blank responses has been used in England to estimate the limits of detection for official methods of water analysis. In the United States, the variability of sample or standard responses has been more often used to estimate limits of detection. Practices also differ with respect to whether or not blank correction is done. These practices are compared and recommendations made regarding the most appropriate procedures for estimating lower limits of measurement for several types of environmental analyses.

During the last 20 or so years there have been significant advances in the theory for defining the lower limits of measurement. Unfortunately, in practice we have not always applied what theory has told us. Theory tells us that the variability of blank responses is the preferred basis for determining when the sample response indicates that the determinand* concentration is greater than zero. Heinrich Kaiser was one of the first scientists to recognize that fact, and he likened the situation to that of "searching for a ship in a stormy sea. Is it a ship or a higher wave than usual?"(3) In this analogy, the height of the waves in comparison with the

*"Determinand" means that which is to be determined (1,2).

[1]Current address: Manchester Laboratory, State of Washington Department of Ecology, P.O. Box 346, Manchester, WA 98353

0097-6156/88/0361-0078$06.00/0

height of the ship limit the ability of one to detect the presence of the ship, just as the variability in the background (i.e. blank) response limits the ability to detect the presence of the determinand. Note that it is not the depth of the ocean that limits the ability to detect the presence of the ship, just as it is not the absolute blank response that limits the ability to detect the presence of determinand. According to Kaiser, "the cause of the uncertainty in the analytical value is not due to the size itself of the blank measure, but to the size of the fluctuations in it. A constant blank measure of whatever size can always be compensated."(3) Thus, in determining the criterion of detection one must measure the variability of blank responses, and in determining whether a sample response indicates the presence of determinand one must first correct for blank response. Then, the blank corrected response can be compared with the calculated criterion of detection to decide whether the determinand has been detected. The expression criterion of detection has been used here in a general sense, but will be more precisely defined in the next section.

Variability of the Blank Responses

Several authors have published papers regarding the calculation of the limit of detection based on the variability of the blank responses. Prominent among these have been H. Kaiser(3) in Germany, A.L. Wilson(4) in England, and L. Currie(5) in the U.S.A. The following theoretical treatment is based on the work of A.L. Wilson. The conclusions of Wilson are similar to those of Kaiser and Currie, the primary differences being the terminology and the choices made for errors of the 1st and 2nd kinds.

Cheeseman and Wilson have stated that "it is a well-established concept that each method has a lower concentration limit below which the determinand cannot be detected. However, there is a great deal of variability in how this limit is chosen and in how results are reported when the determinand was not detected."(6) Wilson proposed determining the limit of detection based on the variability of the blank as a means of contributing uniformity as well as accuracy to the reporting of results at low concentrations. Wilson, following Kaiser on this matter, emphasized that "**the blank should generally be analyzed by exactly the same procedure as that used for samples.** This simple and obvious conclusion is worth stating because it appears often to be ignored. Of course, situations arise where it is impractical or not essential to analyze blanks and samples identically, but such situations generally require experimental confirmation."(6)

Criterion of Detection. According to Wilson, an analytical result, R, is equal to the sample response minus the blank response (i.e. R = S - B). This is the 'paired-comparison' situation, in which we compare individual sample and blank responses. If we were to determine the distribution of results when sample and blank are identical (i.e. sample does not contain the determinand), we would obtain the distribution represented in Figure 1. In this distribution, the mean result is zero with equal numbers of positive and negative results distributed around the mean. A normal distribution is assumed. In order to conclude that the sample contains the determinand we must choose the level of risk we are prepared to take in committing an error of the 1st kind (that is, a false positive, concluding that the determinand has been detected when in fact none is present). Wilson chose a level of 5% (α=0.05) for this error. As illustrated in Figure 1, this choice results in the definition of a criterion of detection equal to $1.645(\sqrt{2})\sigma_B$ or $2.33\sigma_B$ (where σ_B is the within-batch standard deviation of the blank response), meaning that values greater than 2.33 are considered to indicate that the determinand has been detected, with 1 chance in 20 of being wrong.

Limit of Detection. One must also consider the possibility of errors of the second kind (that is, false negatives or the probability of falsely concluding that the sample does not contain determinand, when in fact it is present). For a sample whose true concentration is equal to the criterion of detection, that probability is equal to 50%. Wilson chose to reduce that value to 5%, as illustrated in Figure 2. The limit of detection is defined as being twice that of the criterion of detection, or $4.65\sigma_B$. Thus, the limit of detection is the smallest sample concentration that can be detected with 95% probability.

Discussion. The relationship of the criterion of detection to the limit of detection is illustrated in Figure 3. In applying these concepts to sample results, the determinand is considered detected if the sample result is equal to or greater than the criterion of detection. However, if a result is less than the criterion of detection, it is reported as less than the limit of detection to take into account the possibility of false negatives. Suppose, for example, that the criterion of detection, C_D, is equal to 5 and the limit of detection, L_D, is equal to 10. Then a result of 3 would be reported as <10, while a result of 7 would be reported as such (i.e. 7). While consistent with the definitions, the results have rarely been reported in this manner. In order to avoid the loss of information, two alternative ways of reporting results at small concentrations have been suggested. Currie(7) has suggested that results below the criteria of detection be reported as N.D. (not detected), but that

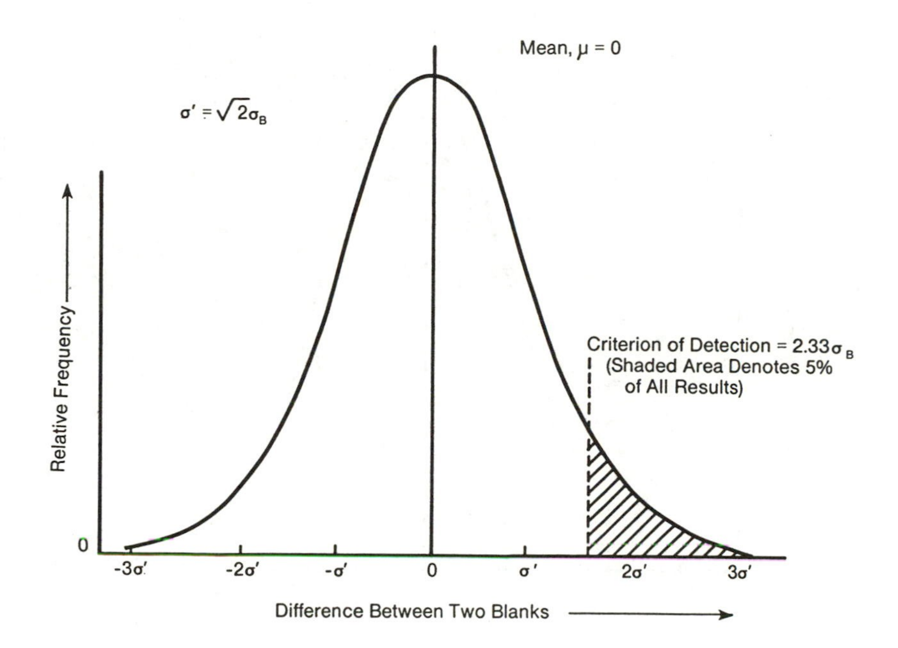

Figure 1 Distribution of the Differences Between Analytical Responses for Two Blanks - Definition of the Criterion of Detection (Adapted with permission from Ref. 6. Copyright 1978 Water Research Centre).

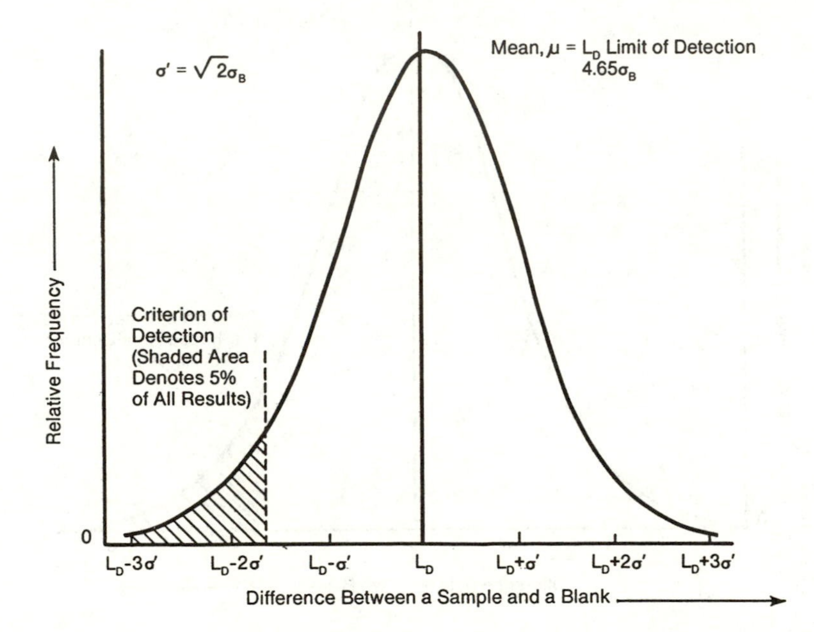

Figure 2 Distribution of the Differences Between Analytical Responses for a Sample and a Blank When the Sample Concentration is Equal to The Defined Limit of Detection (Adapted with permission from Ref. 6. Copyright 1978 Water Research Centre).

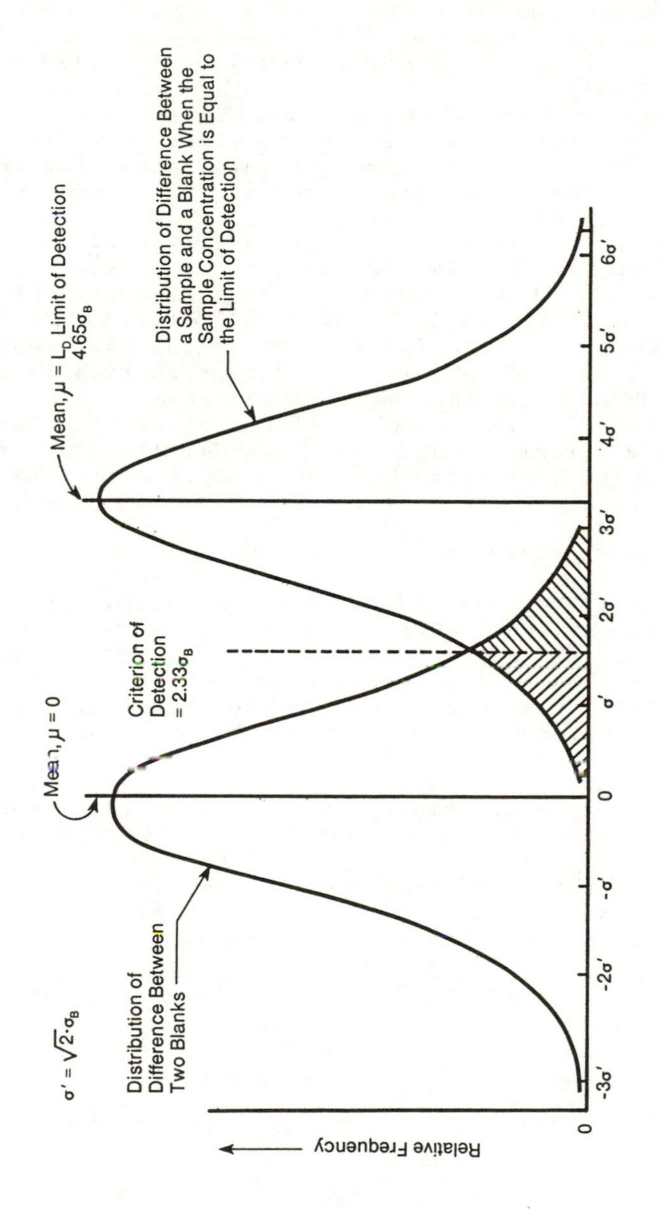

Figure 3. Illustration of the relation between the criterion of detection and the limit of detection. (Adapted with permission from Ref. 6. Copyright 1978 Water Research Centre.)

the estimated value also be reported. (i.e. in the example above, the estimated value of 3 would be reported, along with a not detected decision). Hunt and Wilson(8) have suggested that all results at low concentrations simply be reported as the result plus or minus the confidence limits at a stated level of confidence ($R \pm t_{\alpha}s$). This radical suggestion would completly eliminate the use of the criterion of detection and the limit of detection in reporting results. However, they could be provided separately as reference information for the user to aid in the interpretation of the reported results and associated confidence limits.

Wilson used "paired observations" in defining C_D and L_D. That is, each sample response is considered to be individually blank corrected. It is also possible to define a detection limit based on a "well-known" blank. In this case, the variability of the difference between sample and blank response is reduced by a factor of $\sqrt{2}$. (5) Hunt and Wilson have presented a general expression for C_D based on n replicate blank responses and m replicate sample responses, in which the standard deviation of the difference between sample and blank responses (σ_{S-B}) is given by the following equation:

$$\sigma_{S-B} = \sigma_B[(n+m)/nm]^{1/2}$$

Only the more common situations of a "well-known" blank or "paired-comparisons" will be further considered here.

In summary, in defining the detection limit based on the variability of blank responses, one must make two choices. First, one must choose whether blank correction is to be based on a "well-known" blank or on "paired-comparisons". Second, one must choose the values for α and β, corresponding to the risks of errors of the first and second kinds. Table I summarizes some of the definitions that have been proposed by Currie.(5)

Table I. "Working" Expressions For Critical Level, Detection Limit, and Determination Limit Proposed by Lloyd Currie

	Critical Level L_C	Detection Limit L_D	Determination Limit L_Q
Paired Observations	2.33 σ_B	4.65 σ_B	14.1 σ_B
"Well-known" Blank	1.64 σ_B	3.29 σ_B	10 σ_B

Wilson's definitions generally agree with those of Currie except that he uses the terms criterion of detection and limit of detection rather than critical level and detection limit, and restricts himself to a consideration of paired observations. Currie also introduces a new term, the determination limit, L_Q, which is defined as that concentration for which the relative standard deviation is 10%. Results above the determination limit may therefore be considered satisfactory for quantitative analysis. Table II compares the definitions by Currie with those of Kaiser for "paired observations".

Table II. A Comparison of Definitions by L. Currie and H. Kaiser Based on "Paired Comparisons" of Blank Responses

	Kaiser	Currie
Definitions Based On Error of the First Kind	Limit of Detection (Nachweisgrenze) $=4.24\ \sigma_B$ $\alpha = 0.0014$	Critical Level $=2.33\sigma_B$ $\alpha = 0.05$
Definitions Based On Errors of the First and Second Kinds	Limit of Guarantee (Garantiegrenze) $=8.49\sigma_B$ $\alpha = \beta = 0.0014$	Detection Limit $=4.65\sigma_B$ $\alpha = \beta = 0.05$

Note that the differences lie in the choice of α and β values. The important thing is not how these definitions differ, but rather that which they have in common. All are based on a consideration of the variability of the blank and an assumption that sample variability is the same as blank variability at low concentrations. All require some sort of blank correction, based on "paired observations" or a "well-known" blank. All are based on determining the limit of detection of a "complete analytical procedure". And finally, all are based on some choice of acceptable errors of the 1st and 2nd kinds.

As Figure 1 shows, when the sample does not contain the determinand, the result is negative half of the time. But these negative results are not reported since below the C_D all results are indistinguishable from the blank due to random error. However, an analytical result is not equal to the sample response, but rather to the difference between sample response and blank response. This result, which may be positive, negative, or zero is compared with C_D and the appropriate decision regarding detection made.

From the discussion to this point, it would appear that we only need apply theory to practice in order to arrive at our desired criterion or limit of detection. Unfortunately, the theory is based on the following assumptions, which may not always hold:

1) That the within-batch standard deviations of both the blank and samples containing very small concentrations of the determinand are the same

2) That the analytical response is not zero for finite concentrations of the determinand

3) That the sample and blank are not biased with respect to each other (that is, there are no interfering substances in the sample or the blank)

If any one of the above assumptions is not true, then the limit of detection cannot be calculated using the equations given previously. In some cases, however, adjustments can be made in the equations. Currie, for example, has presented an analysis which allows for corrections when assumptions 1) and 3) above are not met.(9) For example, as illustrated in Figure 4, adjustments can be made to allow for differences in the standard deviation for blank and sample responses ($\sigma_B \neq \sigma_S$) and for different values for errors of the 1st and 2nd kinds. Also, when systematic error cannot be assumed negligible, the limit of detection must be increased by an amount, $2\Delta_m$, where Δ_m is the assumed upper bound for the bias

i.e. $C_D' = C_D + \Delta_m$ and $L_D' = 2\ C_D'$

Existing Practice in Water Analysis

After this brief review of theory, let us turn our attention to existing practice, as exemplified in environmental methods of analysis. Environmental methods of analysis employ many of the common analytical instruments in analyzing a wide spectrum of chemicals in a variety of matrices. Instruments commonly used include spectrophotometers (atomic absorption, visible, inductively coupled plasma), gas chromatographs (with a variety of detectors, including the mass spectrometer), and automatic analyzers.

Variability of Blank Responses. In order to limit the discussion, let us focus on water analyses as representative of environmental analyses. In the United Kingdom, the Standing Committee of Analysts of the Department of the Environment issues analytical methods in a series of booklets. Included among these are the 'Methods for the Examination of Waters and Associated Materials.'(10-14) Several of these methods have been evaluated by individual laboratories to determine the limit of detection based on the variability of the blank and using 'paired comparisons' for blank correction. Published values for the limit of detection for several of these methods are listed in Table III. The Standing Committee of Analysts has adopted a policy of including an estimate of the limit of detection (or the within-batch standard deviation of the blank, which is used to calculate the limit of detection) as one of the 'Performance Characteristics of the Method'.

Table III. Estimated Limits of Detection for Selected Methods taken from "Methods for the Examination of Waters and Associated Materials"

Element	Estimated Limit of Detection**	Degrees of Freedom Used in Estimate(s)
Copper (10)*	1.7-2.9 ug/L	5
Chromium (11)*	3.2-7/4 ug/L	7-9
Phosphorus (12)*	0.003-0.006 mg/L	35
Silicon (13)*	0.03 mg/L	10
Aluminum (14)*	0.013 mg/L	10

*Note--These are reference numbers
**Note--Range for some elements is due to outcomes in in different laboratories.

In the United States there are several published methods for the analysis of waters. These include methods published by the Environmental Protection Agency and the 'Standard Methods for the Analysis of Water and Wastewater'.(15) In Standard Methods for the Examination of Water and Wastewater there is surprisingly little guidance on how to determine detection limits. For flame atomic absorption spectrophotometry, the detection limit is defined as the concentration that produces absorption equivalent to twice the magnitude of the background fluctuation. No mention is made of the blank or blank correction. This definition implies an instrument detection limit rather than a detection limit of a 'complete analytical procedure.' Finally, no mention is made of the need to determine the variability of responses.

Variability of Sample Responses. Under the Clean Water Act the U.S. Environmental Protection Agency has published guidelines establishing test procedures for the analysis of pollutants. A review of these methods can give us a reasonable picture of existing practice in determining the limit of detection. In Appendix B to Part 136 of the Federal Register a definition and procedure for the determination of the method detection limit is given.(16). The method detection limit (MDL) is defined as "the minimum concentration of a substance that can be identified, measured and reported with 99% confidence that the analyte concentration is greater than zero and is determined from analysis of a sample in a given matrix containing the analyte". The term method detection limit is a misnomer, since the value will depend on the instrument sensitivity, the nature of the samples, and the skill of the analyst, as well as the method. The equation given for calculating the MDL is:

$$MDL = ts$$

Where t = the Student's t value appropriate for a 99% confidence level (one-sided) and a standard deviation estimate with n-1 degrees of freedom and s = the standard deviation of the replicate analyses of standards or samples with low concentration of analyte. Figure 5 depicts the MDL as an error distribution. As originally proposed, the procedure did not allow for blank correction. The latest version of the procedure does allow for blank correction, although the standard deviation of analyte response rather than the standard deviation of the blank is used to calculate the detection limit. The distribution shown in Figure 5 is that of the result (presumably after blank correction if necessary).

This definition is mathematically very similar to that presented earlier based on the variability of the differences between sample and blank when the sample concentration is equal to the criteria of detection. The 99% confidence level corresponds mathematically to the case in which the error of the 1st kind is chosen to be 1% (and the error of the 2nd kind is equal to 50 %). However, the MDL is based on the variability of analyte response rather than blank response. The MDL was first defined for application to the analysis of trace organic compounds and apparently was based on the conclusion that the first assumption listed above was not met. (i.e. the within-batch standard deviations of both the blank and samples containing very small concentrations of determinand are judged not to be the same). In fact, the implication is that there often may be no significant response for the blank. Hence, no need for blank correction and theory therefore emphasizes sample response.

Blank Correction Omitted. There are two common situations in water analysis where blank correction is not made, or apparently not made, even though the detection of traces of determinands is important.(8) The first corresponds to those procedures in which blank determinations are made but their responses are adjusted to zero by instrumental subtraction of a constant from all responses. This procedure is equivalent to making separate measurements of blank and sample responses,

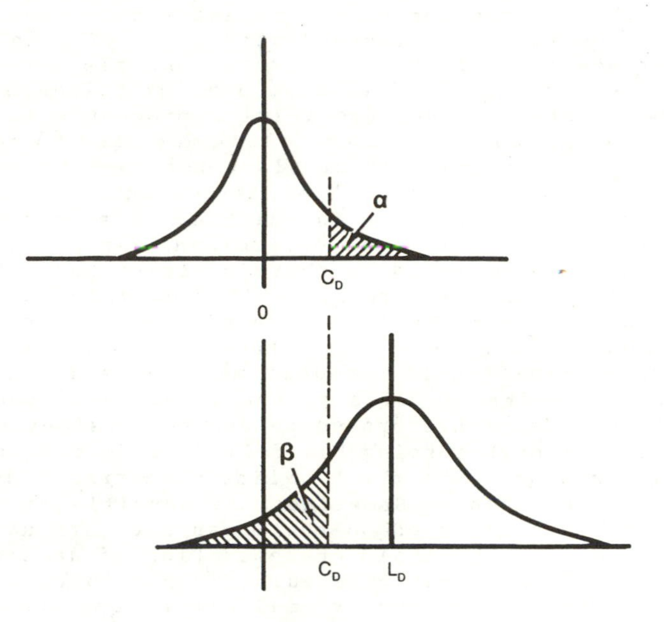

Figure 4. Illustration of the case in which the standard deviation for sample and blank responses differ and in which the values chosen for errors of the first kind (α) and the second kind (β) also differ. (Adapted with permission from Ref. 9. Copyright 1978 Wiley.)

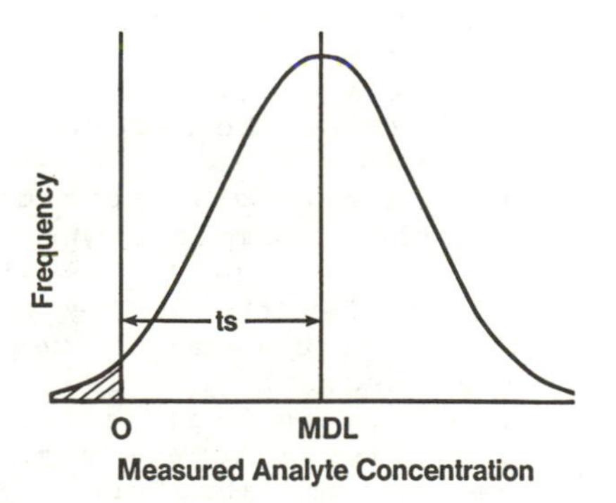

Figure 5. Method detection limit depicted as an error distribution. (Adapted from Ref. 19. Copyright 1981 American Chemical Society.)

since the subtraction of a constant from a measurement does not affect its standard deviation. The only major difference is that this procedure conceals any systematic, temporal variations of the blank response from batch to batch. Also, the initial procedure to determine the variability of the blank may need to be altered, since no independent blank response values are obtained with instrumental blank correction.

The other procedure in which no blank correction is used is that in which the determinand's presence is revealed by a peak rising above a background; chromatographic techniques are a familiar example, and analyses using an AutoAnalyzer have also been reported to fall into this category. In these types of procedures, a blank-correction is commonly made, in effect, by measuring the height or area of the determinand peak above the base-line. This procedure can be considered valid only if thorough testing has failed to detect an analytical response above the baseline when true blank determinations are made. Sometimes the sensitivity of the instrument is adjusted so that responses are not obtained until a threshold concentration of determinand is reached. This situation should, if possible, be avoided whenever detection of very small concentrations is of primary interest by adjusting the sensitivity so that the discrimination has negligible effect on the apparent size of random errors. As discussed previously, the solution for chromatographic methods proposed by the U.S. Environmental Protection Agency is to determine the MDL based on the standard deviation of low concentration sample or standard responses.(16) Hunt and Wilson have suggested that a practical limit of detection for cases where no blank correction is necessary would be that concentration producing an analytical response greater than some value d, which can be determined from the analysis of standards.(8)

The EPA Contract Laboratory Program under Superfund (CERCLA) provides another example of how blanks and detection limits are treated.(17)(18) With respect to blanks, the statements of work specify that the laboratory should not blank correct sample responses. In the case of organics analyses, the EPA evaluator and/or data auditor has the authority to blank correct sample responses. In practice this is never done. For both organics and inorganics, the absolute blank level is primarily used as a control to determine if samples need to be reanalyzed. Detection limits are based on replicate analyses of a standard at 3-5 times the required detection limit concentration. The instrument detection limit is calculated as being equal to 3 times the standard deviation of the measured value. Since blank correction is not permitted or not done, sample results will all be biased high by an amount equal to the blank response. The absolute blank value (actually usually a multiple of 5 or 10 times the blank value) rather than the variability of the blank determines the lower reporting limit.

Summary and Conclusions

It is apparent from the above descriptions of official methods of analysis in the United States that none of them specify that detection limits should be determined based on the variability of the blank for a 'complete analytical procedure.' This contrasts sharply with the policy of the Standing Committee of Analysts in the United Kingdom, which is to preferably determine the limit of detection based on the variability of the blank. Why is this? One reason is the difference in emphasis on the blank itself. While in the U.K. they have required that blank correction be a part of the procedure, in the U.S.A. we have not only not required blank correction but have even prohibited blank correction in some of our official methods.

In the case of the U.S. Environmental Protection Agency, its Method Detection Limit was developed initially for organic methods of analysis using chromatographic techniques.(19) Later its use was suggested for other methods for which blank correction and a detection limit based on variability of the blank is the correct procedure to follow. Winefordner and Long have, for example, recognized the importance of the blank in spectrometric determinations.(20)

Another obstacle is the widely held belief that bias due to interference is so large that the random variability of the blank is negligible in comparison. This may be true for some analyses, but should not be assumed to hold in all cases. If anything, modern instruments are becoming more selective and more interference free. And bias can always be incorporated into the estimation of the detection limit.

Finally, there is the attitude that equates the instrument response to the analytical result. In this context, blank correction and the calculation of the variability of the blank is not understood to be a necessary additional step. **Too much emphasis has been placed on following certain rules**, rather than doing what is necessary to obtain the best possible estimate of the true value.

Detection limit theory has not been adequately applied to practice for many common officially approved or standard methods of analysis. To improve this situation, the following recommendations are offered:

1. In all cases, blank correction should be done before comparing the analytical result with the criterion of detection to decide if the determinand has been detected.

2. The preferred procedure for determining detection limit is that based on the variability of blank responses for a complete analytical procedure. This

should always be attempted first for a method before resorting to alternatives. It is usually readily applied in the case of spectrometric methods of analysis, since these methods do not generally exhibit significant bias due to interference.

3. When an analysis involves a peak rising above a background and no true blank response is obtained, an alternative procedure must be used. This might be done by determining the MDL from the standard deviation of low level standard responses, or by establishing some concentration, d, that must be exceeded, based on responses obtained for standards. Always work at the highest practical level of sensitivity in order to avoid the situation where response is zero for finite concentrations of determinand.

4. Instrumental detection limits should be avoided, since they do not include the influences of sample preparation, cleanup, etc. on the detection limit. In this regard, detection limits based on signal to noise ratio should also be avoided. If used, they should be obtained from a 'complete analytical procedure' and not just the instrument.

5. When an estimate of standard deviation is required (either for blanks or low concentration samples) sufficient replicates should be done to obtain a good estimate of the population value.

6. Estimates of systematic errors, such as interference, should be included in the estimate of detection limit.

Literature Cited

1. Wilson, A. L. Talanta 1965, 12, 701.
2. International Organization for Standardization, ISO 6107/2-1981, ISO, Geneva, 1981 (BS 6068: Part 1: Section 1.2: 1982).
3. Kaiser, H. Two Papers on the Limit of Detection of a Complete Analytical Procedure; Adam Hilger, Ltd.: London, 1968.
4. Wilson, A. L. Talanta 1973, 20, 725.
5. Currie, L. A. Anal. Chem. 1968, 40, 586-93.
6. Cheeseman, R. V.; Wilson, A. L. Manual On Analytical Quality Control For the Water Industry; Technical Report TR66: Water Research Centre: England, 1978.
7. Currie, L. A. In X-Ray Fluorescence Analysis of Environmental Samples; Dzubay, T. G., Ed.; Ann Arbor Science Publishers, Inc., 1977; Chapter 25.
8. Hunt, D.T.E.; Wilson, A. L. The Chemical Analysis of Water; 2nd edition; The Royal Society of Chemistry: London, 1986.

9. Currie, L. A. In Treatise On Analytical Chemistry, Part 1, Volume 1, 2nd edition; Kolthoff, I.M. and Elving, P. E. Eds.; John Wiley & Sons: New York, 1978; Chapter 4.
10. Standing Committee of Analysts Copper in Potable Waters by Atomic Absorption Spectro photometry ,1980; Her Majesty's Stationery Office, London, 1981.
11. Idem Chromium in Raw and Potable Waters and Sewage Effluents, 1980; H.M.S.O., London, 1981
12. Idem Phosphorus in Waters, Effluents and Sewages, 1980; H.M.S.O., London, 1981.
13. Idem Silicon in Waters and Effluents, 1980; H.M.S.O., London, 1981.
14. Idem Acid-Soluble Aluminum in Raw and Potable Waters by Spectrophotometry, 1979; H.M.S.O., London, 1980.
15. Standard Methods For the Examination of Water and Wastewater; American Public Health Association: Washington, D.C., 1985.
16. Methods for Organic Chemical Analysis of Municipal and Industrial Wastewater; Environmental Protection Agency Publication EPA-600 14-82-057, July 1982.
17. Statement of Work for Organics Analysis, Multi-Media, Multi-Concentration, U.S. Environmental Protection Agency, Contract Laboratory Program, January, 1985.
18. Statement of Work for Inorganic Analysis, Multi-Media, Multi-Concentration, U.S. Environmental Protection Agency, Contract Laboratory Program, July, 1985.
19. Glaser, J.A.; Foerst, D. L.; McKee, G. D.; Quave, S. A.; Budde, W. L. Environ. Sci. Technol. 1981, 15, 1426-1435.
20. Long, G. L.; Winefordner, J. D. Anal. Chem. 1983, 55, 712A.

RECEIVED May 19, 1987

Chapter 5

Interlaboratory Aspects of Detection Limits Used for Regulatory and Control Purposes

L. B. Rogers

Department of Chemistry, University of Georgia, Athens, GA 30602

The factors that influence the relative standard deviation for a determination and, hence, the detection limit that should be used for interlaboratory measurements are briefly reviewed. In addition, the need for a more general approach to the estimating and reporting of selectivity is pointed out. A possible basis for a suitable scheme is suggested that should be applicable to environmental and clinical samples that usually contain a large number of completely unknown species, one or more of which may interfere in the measurement of the sought-for species.

A more nearly complete picture of the problems encountered in interlaboratory measurements, especially those at the trace level, may be found in two recent publications (1,2) and in their references. In contrast, the present report emphasizes the variables found in interlaboratory studies that are not adequately accounted for in intralaboratory studies. If the detection limit is calculated by taking a multiple of the relative standard deviation (RSD), and if the value for the pool of interlaboratory measurements is (as is usually the case) larger than that for intralaboratory measurements, then the former RSD is the appropriate one for calculating the detection limit for regulatory purposes.

This report also discusses the question of selectivity, especially with respect to determinations of amounts that are near the lower limit of detection. Classically, two different, and as nearly independent procedures as possible, were recommended. In that case, the minimum detectable amount (MDA) would be set by the RSD for the less sensitive procedure. On that basis, one can conclude that, for regulatory/control measurements involving interlaboratory measurements of trace amounts, the proposed method (discussed later) for arriving at the Acceptable Minimum Detectable Amount (AMDA) of the American National Standards Institute (ANSI) can be made more explicit. Once that has been done, one can proceed

0097-6156/88/0361-0094$06.00/0

to calculate the "regions of detectability" and "regions of quantitation" as recommended by a committee of the American Chemical Society (3).

However, the original goal of using two procedures was to increase one's confidence in the reliability of the result through an independent check on the selectivity. In trace analyses, the usual tests of selectivity, which examine the relative responses from known interferences that produce the same type of signal (e.g., in mass spectrometry, the same mass-to-charge ratio) as the sought-for substance, appear to be of limited value for environmental and clinical analysis. There, one is faced with the possibility of "direct" interference from generation of an identical signal from one or more of a very large number of truly unknown species. This situation is in contrast to the more common case where the signal of the sought-for analyte is only decreased or increased as a result of the presence of the interfering species. Hence, a new approach has been suggested for the estimation and expression of selectivity in cases of direct interference.

VARIABLES IN INTERLABORATORY MEASUREMENTS

All of the changes that can occur within one laboratory can, of course, occur between laboratories. However, they are often much more visible when they occur in interlaboratory studies, as for example, the effects of contamination from dust, reagents, or walls of containers. Two beautiful examples from the Manhattan Project were reported to me in conversations in 1947. In one case, C. J. Rodden reported that the National Bureau of Standards had great difficulty in measuring low parts per million of cadmium in uranium until they removed all of the cadmium-plated ironware from their laboratories. Similarly, Charles Metz of Los Alamos Scientific Laboratory reported that they did not permit containers of macroscopic amounts of cadmium or boron compounds to be carried into the facilities where parts-per-million traces of those elements were being analyzed in uranium. More recently, Nordberg (4) reported that his attempts to make measurements of drugs in body fluids led to erratic, meaningless data until he isolated the location and the equipment with which the concentrated doses of drugs were administered from the equipment and facilities where the measurements were made.

When a problem of contamination or interference is encountered in an interlaboratory study, it is easier to detect. Figure 1 shows the results of a very old study done by Taft Sanitary Engineering Center, a forerunner of the U. S. Environmental Protection Agency, when they sent portions of a sample of aluminum ion to a number of different laboratories. Each laboratory calibrated its procedure for aluminum ion against a primary standard and did the analysis in triplicate. The fact that a variety of methods was used suggests that the high result for the mean was not due to an interference in a particular method but, instead, due to contamination of the entire lot or the presence of an interference from the walls of the sample containers.

Another major problem arises from losses, usually due to adsorption, but sometimes also from decomposition or volatility of the analyte. For example, if micropipettes are used for taking

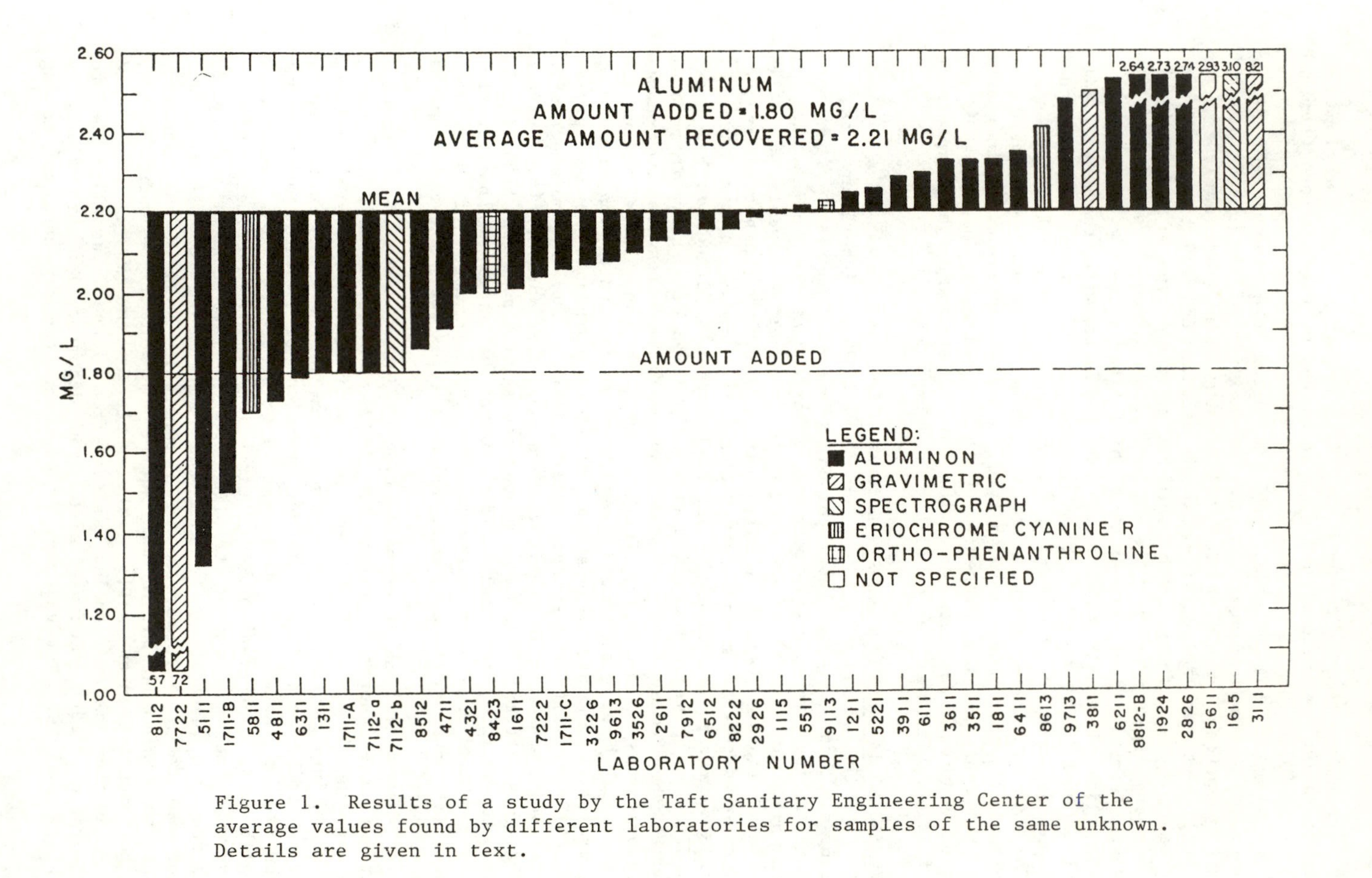

Figure 1. Results of a study by the Taft Sanitary Engineering Center of the average values found by different laboratories for samples of the same unknown. Details are given in text.

samples, the large surface-to-volume ratio may result in relatively large losses of the sought-for species. One of the earliest examples of this was a report that a substantial fraction of radioactive silver ion was lost to the walls of a micropipette even after it had been rinsed three times (5). Similarly, another interlaboratory comparison performed by Taft Sanitary Engineering Laboratory showed that the mean found for Cr(III) was substantially less than the amount originally added to the sample (6). When one considers that competing factors of contamination and loss are present in every determination, one can readily understand why relative standard deviations are much larger at the trace level, especially when carried out in different laboratories under different environmental conditions.

The next variable reflects the participation of more than one person in doing a particular determination. One of the best examples is found in analyses of dioxin samples that were performed by a number of different laboratories in Italy (7). For a given amount in a particular sample, one person doing the entire analysis was found to have a RSD of slightly under 5%. If two people in that laboratory combined their efforts and did different parts of the determination, the RSD jumped to somewhat greater than 10%. However, when two or more people from different laboratories were involved, the RSD value was greater than 26%. This example clearly illustrates the trend that has been confirmed in other studies to be mentioned later.

Another variable that leads to highly unpredictable results is derived from a universal sin of which we are all guilty - subjective use of statistics, especially with respect to discarding data. For example, in the previous dioxin study, samples were sent to two laboratories and the results compared. If they were too widely discrepant, the sample was sent to a third laboratory, and the two results that agreed the closest were reported. Furthermore, of the results that remained, 5% were judged to be outliers and were discarded before calculating the 99% confidence level of the mean. In contrast, a committee of the American Chemical Society recommended that, in general, outliers should usually only be discarded if reasons are known why the results are bad (2). Certainly, the dangers of discarding data are easier to see in other people's studies than in one's own.

The next variable is concerned with the experience a particular analyst has had with a given procedure. There is no disagreement with the conclusion that, in general, an experienced analyst should have a lower RSD than one who is inexperienced. Furthermore, one would expect that the duration of the contact of an experienced analyst with a particular procedure might have a noticeable influence on the resulting RSD. However, there are two surprising results in the data presented by Horwitz et al. (8) which were obtained from their examination of data from interlaboratory studies conducted by the Association of Official Analytical Chemists (AOAC) (9). First, there was a steady decrease in the RSD to roughly 45% of its initial value, over a period of five years! Second, at the end of that time, although the value leveled off, it was usually between 1.5 and 3 times the value that a single analyst usually obtained for that same procedure. It is important to note not only that this type of result was observed for a wide variety of procedures conducted by the AOAC but also that it is consistent with the trends in the data for dioxin analyses discussed earlier.

The final major contribution to the RSD, the lack of truly blind quality assurance samples, was documented in an interlaboratory study but, undoubtedly, holds as well for a quality assurance program within one laboratory. The need for quality assurance samples that are "completely blind" was clearly demonstrated by Liddle (10) in the following way. A group of laboratories participated in a study involving the determination of lead in blood serum using unknowns distributed by Liddle. Although it took some time for the laboratories to meet the hoped-for overall performance level as measured by the RSD of the pooled results, when those same samples were distributed later as complete unknowns that were not suspected of being associated with the quality assurance program, the RSD doubled! This result does not mean that the analysts were dishonest. Instead, it simply reflects the fact that all of us take special care when performing analyses that we know are important, and such extra care results in a significant improvement in the resulting data.

The relative contributions of the known variables that contribute to increases in the RSD for interlaboratory measurements are summarized in Table I. The important thing to note is that, in general, one would expect the RSD for quality assurance samples measured under truly blind conditions to be between 5 and 10 times larger than those measured on known standards by a single person in a laboratory. The resulting larger value for the RSD has an important bearing on the limit of detection for a procedure that is to be used in interlaboratory measurements.

Table I. Variables That Increase the Standard Deviation

VARIABLE	MULTIPLY BY
EXPERIENCE (time) using stated procedure (AOAC Horwitz)	>1
TWO OR MORE PEOPLE	
Same Lab. (Seveso Study)	2
Different Labs. (Seveso; AOAC)	2-4
TRULY BLIND QUALITY ASSURANCE (Liddle-CDC)	2+
LOWER CONCS. (AMTS.) that require added indep. confirmatory measurements	>1

Finally, it is important to note that the limit of detection will be larger under some conditions in which a measurement involves consumption of a sample. For example, a mass spectrometer equipped with a single detector is sometimes used to measure more than one mass-to-charge ratio in a given sample so as to increase one's confidence in the identification of the analyte. If the spectrometer spends equal time on each of three masses (and the switching time is negligible), the quantity of sample must be three times larger in order to attain the same detection limit as that for the similar procedure that measures only one mass for the same total time. We shall see later that such a trade-off of improved selectivity and reliability for a higher detection limit will usually be desirable.

SELECTIVITY ASPECTS

Selectivity is generally inversely related to the relative amount of interference one can expect from a particular species (above a given level) in attempting to measure another sought-for species. Before going farther, one should recall that the extent of an interference is usually expressed in terms of its concentration or amount that will produce the same signal as the unit amount of the sought-for species. It is important to note that use of the standard addition or internal standard method to estimate the amount of a sought-for species provides no means of compensating for an interference that contributes directly to the measurements. The extent of signal from the interference itself can often be taken into account and subtracted by measuring a blank that contains a known amount of the interfering species in the presence of all other species except the sought-for species. However, as will be pointed out below, this possibility is ruled out if completely unknown substances are involved because one is unable to prepare an appropriate blank.

It is worth noting that in classical quantitative analysis, which usually did not involve analyses of trace amounts, the problem of unknown interferences was attacked by using two methods that were as nearly independent as possible. In that way, the chance that an interference would give the same response for each method was minimal. In contrast, there is an example in clinical chemistry which clearly suffers from a lack of selectivity in the accepted method. Figure 2 (11) shows that the use of the determination of sugar in serum as a method for diagnosing diabetes is clearly unsatisfactory. The distribution for sugar content in sera of diabetics distinctly overlaps the distribution for those who are not. Hence, significant fractions of both false positives and false negatives will be obtained over a relatively wide range of sugar concentration. A similar conclusion would have been reached if sugar were indeed a highly selective basis for diagnosing diabetes but there was a second unknown substance present in some sera that contributed an interfering "sugar signal". Hence, the use of two "independent" procedures should enhance one's confidence in the results.

The situation involving the completely unidentified source of interference becomes increasingly important as the concentration or amount of sought-for substance decreases (2). Donaldson (12) pointed out this principle, which can be illustrated using 99.999% water that contains 10 parts per million of total impurity. If we make the simplifying assumption that all impurities are present at the same level, we calculate that 10 impurities can be present at the 1 ppm level, 10^4 at the 1 part per billion level or 10^7 at 1 part per trillion. The last figure corresponds quite closely to the estimated total of known chemical species. Hence, when one is faced with the trace analysis of a complete unknown, such

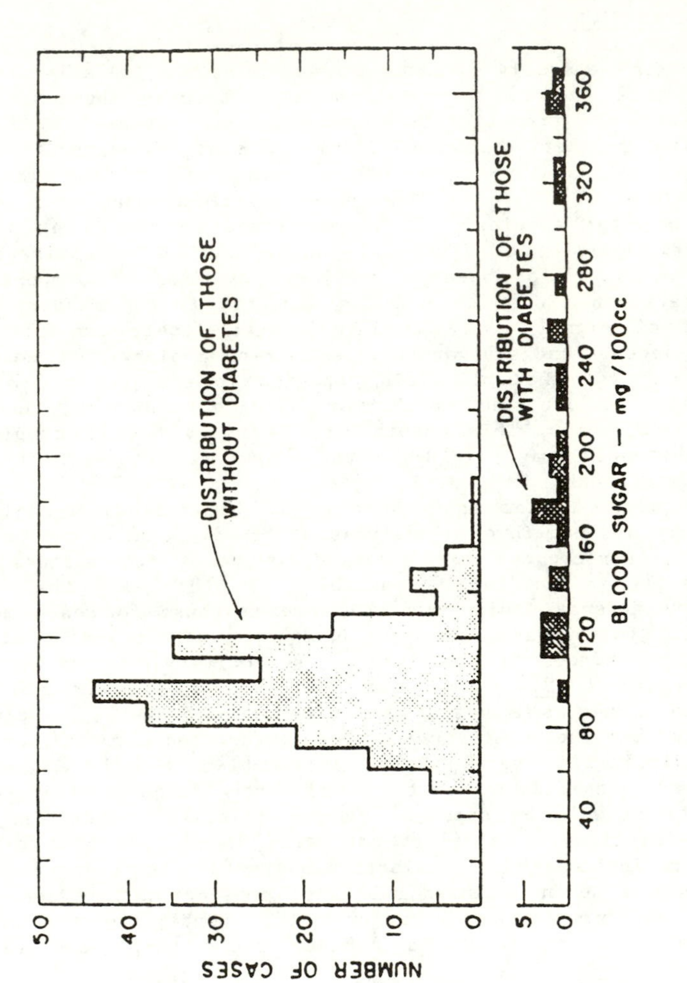

Figure 2. Distribution of blood sugar in diabetics and nondiabetics.

as the water in a drainage ditch from an industrial plant, the approach to selectivity of estimating the extent of interference based on a few known species is totally inadequate.

One can get a feeling for the reality of such interferences from unusual examples that came to light that relate to my service with a team of consultants to advise the Dow Chemical Company concerning the adequacy of their procedures for analyzing dioxins. The overall procedure used by Dow involved multiple high-resolution steps (13): an initial liquid-liquid extraction followed successively by liquid chromatography, gas chromatography, and, finally, a few selected ions in mass spectrometry. One of their chemists remarked that they found, quite by accident, when using highly purified nitrogen to evaporate a solvent from a sample prior to mass spectrometric measurement of the dioxin, high erratic values were obtained unless they first passed the nitrogen through a column packed with silica gel. That seemingly unnecessary step removed the unidentified interference. The second example came to my attention only recently. When the original sample was a ground-up fish that had been extracted, and "cleaned up" by liquid chromatography before being analyzed by mass spectrometry using three different mass-to-charge ratios, unexpectedly high results were obtained. By recording the entire spectrum over a wide range of masses, the selected masses were found to be sitting on top of a very broad high background which the chemist speculated might have been a large glyceride, unanticipated and unidentified, that broke up into a broad continuum of different masses in the range of interest. (Telephone conversation with Lewis Shadoff in July 1986)

A third example from the laboratory of Veillon has recently been described, together with its historical background (14). In brief the graphite furnace measurements of chromium using atomic absorption spectrometry, were confounded with an unidentified species in urine samples that was contributing a background signal which was inadequately corrected for when using a deuterium lamp source. Veillon showed a linear plot relating the magnitude of the background signal to the apparent chromium concentration in the urine. He estimated that erroneous values had been reported in the literature for more than 10 years!

Earlier, mention was made of the classical way of insuring higher selectivity by doing two unrelated determinations in parallel. The Dow determinations of dioxins cited above are examples in which selectivity was increased by performing multiple, relatively high-resolution steps in one quantitative procedure, but they still encountered unexpected interferences. There are also other examples of procedures consisting of multiple high-resolution steps such as those in mass spectrometry-mass spectrometry, mass spectrometry-infrared, and other "hyphenated" techniques which also increase one's confidence in the results. However, it is clear that we need a way to estimate the selectivity of such techniques when a large number of species, most of them unknown, might be present. This aspect will be addressed more fully later.

REGULATORY/CONTROL LIMITS FOR TRACE AMOUNTS

The term "regulatory limit" usually implies that a governmental body has set the limit whereas the term "control limit" is usually associated with process specifications for private or non-governmental purposes. It is the writer's impression that, in arriving at such limits, the usual practice is to: (a) use the most sensitive procedure, (b) test the procedure with known or likely interferences, (c) use data from intralaboratory studies to estimate the RSD for use in calculating the detection limit for the procedure, (d) use a quality assurance procedure that employs samples of unknown concentrations which, however, are known to the analyst to be standards, and (e) apply guidelines of the American Chemical Society in selecting the "regions of detection" and "regions of quantitation" (3).

In contrast, a general protocol recommended by the AOAC, that has long been used by the Food and Drug Administration, may differ from the above approach in the following ways. First, the procedure selected for use may or may not be the most sensitive one available. Second, the RSD obtained from the pooled results has been recognized as usually being approximately double that of the value obtained in a single laboratory. The fact that a long time period may be required to reach that minimum multiple of an intralaboratory value has also been recognized.

The American National Standards Institute (ANSI), after much deliberation, has added another concept, that of an Acceptable Minimum Detection Amount (AMDA) for use in interlaboratory studies. Although it has been discussed in more detail by (Brodsky, A., Nuclear Regulatory Commission, Washington, D.C., unpublished data.), the AMDA, in brief, is the MDA that is obtainable in practice by competent analytical scientists using the best state-of-the-art at economically affordable costs to the customer. Where possible, the AMDA has been established at least several times higher than the MDA, if the higher value is adequate for protection purposes. In essence, the AMDA permits the flexibility for different laboratories to use cheaper and more rapid procedures when they are available and meet the limit. In the writer's opinion, this concept has been implicit, if not explicit, in the selection of procedures by the AOAC and in EPA regulations.

The AMDA concept of ANSI is attractive, and the writer would like to expand upon it. First, if the RSD and the MDA have been based on intralaboratory studies, a procedure for estimating the interlaboratory RSD and the AMDA could be done by applying an appropriate series of approximate corrections shown in Table 1. Such corrections could be applied to either the original procedure for which an AMDA was to be stated or to the alternate less sensitive procedure which was to be used for the AMDA. Of course, if an AOAC-type interlaboratory study were conducted to determine a pooled RSD, only the corrections for variables not included in the AOAC-type study would be used.

To apply the AMDA concept to critical regulatory decisions, it would be easy to adopt the classical concept of requiring analyses by two as nearly independent methods as possible. This would be an important start in the direction of addressing the questions of possible contributions to the measured signals by unsuspected and unrecognized interferences. However, since the second procedure will rarely be as sensitive as the first, adoption of the second procedure will provide a basis for deciding more explicitly the minimum multiple that should be applied by ANSI in calculating the detection limit on going from an MDA to the AMDA (or to a second, confirmatory AMDA procedure).

The use of such a double-check will also contribute to the reliability through having another test of the selectivity. Remember that the usual way of discussing selectivity and interferences is to work with species that are known or suspected to be present. However desirable that exercise may be, it is, unfortunately, unrealistic in dealing with most environmental samples and, possibly, many clinical samples. If one recalls the large number of possible interferences that might be present when making an analysis at the low part per billion level, a new approach to thinking about possible interferences and estimating selectivity of a procedure appears to be needed. An approach for attacking this seemingly hopeless, undefined problem is proposed below based on an idea of Kaiser (15).

Let us assume that one can do a computerized search of physical properties of all known organic and inorganic substances so that one can, for example, assemble lists of their molecular weights, boiling points, mass-to-charge ratios of positive (or negative) molecular ions, major bands in the infrared, and so forth. Then, it should be possible to plot a distribution of the frequencies with which different values are found along the axis of each property. Taking the boiling point at atmospheric pressure in degrees Kelvin as an example of one property, it would then be possible to estimate first the number of compounds that had boiling points between 300°K and 350°K (or any other desired range) and, second, what fraction of the total possible number of species to which that corresponded. Hence, if a distillation were performed over that range, one would be making an estimate of the number of possible compounds that might be present. If a second property were employed in the overall analysis, such as a positive molecular ion spectrum in mass spectrometry, a similar estimate could be made of that distribution. (For example, Harvan et al. (16) reported that there were 700 elemental compositions within 0.2 dalton of the mass for tetrachlorodioxin that satisfied the criteria for valence, contained four chlorine atoms and, in addition, only C,H,N,O and S atoms.) It should then be possible to calculate for the n properties involved in a given overall procedure a "volume" in n-dimensional space which would provide an estimate of the number of possible interferences present. That number, when viewed in terms of the total number of compounds considered could provide a basis for estimating the selectivity of the overall procedure.

To cite a specific example, a computerized search of the current Heilbron compilation, published by Chapman and Hall, was made using the Lockheed Dialog system. That file currently includes 150,000 organic compounds. First, a boiling-point range between 200 and 220°C was examined, and after less than 5 seconds of search time, 1321 compounds were reported as having been found. Second, molecular weight was searched between 300 and 305. Again, after less than 5 seconds of search, 1492 compounds were found. Finally, a search for the compounds found in both lists took less than 2 seconds to come up with a total of 6. Hence, when suitable properties are available and the software permits, appropriate searches can be made very quickly.

It is important to note that implicit assumptions are made in doing such a calculation. The first is that the compounds in the collection are representative of the entire body of organic compounds. If so, one can make the following extrapolation. If there are 1.5×10^6 compounds estimated to exist, one can extrapolate to 60 possible "hits"; if 1.5×10^7 compounds are assumed, 600 "hits" are estimated. If further assumptions are made that (a) the interval is small relative to the width of the overall distribution and (b) the distribution of boiling points is uniform across the 20° interval, then, if one can justifiably narrow the range of boiling points to 10°C instead of 20°, the number of estimated "hits" could be reduced to 300 out of the 1.5×10^7 compounds. Clearly, assumptions (a) and (b) are weak when the total number of compounds is small (17). Finally, although each property parameter will have its own characterisitic ability to impart selectivity, especially when combined with one or more other parameters, the use of additional parameters should produce significant reductions in the estimated number of "hits". Unfortunately, the number of compounds that can be searched for specific infrared bands, mass-to-charge ratios of ions, gas chromatographic retention indices, and liquid chromatographic indices is much smaller than 150,000 species. However, one can expect the data bases for all types of compounds - organic, inorganic, and metallorganic - to grow quickly during the next few years. Although one can foresee disagreements about the total to be used for the number of possible species, this should not be a major deterrent to the use of this approach for estimating relative selectivities of different properties and the gains to be made from an improved precision of measurements. (For example, many oligomeric series, e. g., those of polystyrene terminated with butyl, as well as the corresponding series terminated with other alkyl groups, will quickly add many species to the estimated total. Improvements in precision and accuracy with which properties can be measured will usually conteract that increase by reducing the percentage of "hits". However, in special cases, such as in high resolution mass spectrometry, of large organic compounds, where isotopic species can easily double or treble the number of major molecular ions that should be detectable within 5 millimass units, greater resolution does not necessarily insure a better qualitative or quantitative result. (17A)) In any case, the approximate approach to the estimation of overall selectivity of the type outlined here should be useful to the analyst for comparing two overall procedures.

In a lighter vein, it would be of interest for a chemist to find out, for example, the mean value for the boiling point of all chemicals (ruling out those that decompose as well as a few extremes like liquid helium and magnesium oxide). Similarly, it would be interesting to speculate about the mean value in the distribution of vibrational frequencies or of negative molecular ions.

Before concluding this discussion of selectivity, one must realize that there are situations in which the analyst might like to have two methods that exhibited as nearly the same selectivity as possible for a wide variety of compounds. That selectivity goal, which represents the opposite extreme to the earlier one, has been represented schematically (19) in Figure 3 which shows the scopes of different procedures for determining the organic carbon content of water. Instead, to a greater or lesser extent that depends on the selections made, the reported carbon content is a function of the types of compounds present *and* the procedure employed for the determination.

CONCLUSIONS

First, it is clear that any stated minimum detection limit for a procedure used for interlaboratory measurements should incorporate the RSD that includes all known sources of interlaboratory uncertainty. Second, because, as the concentration or amount of the sought-for substance goes down, the number of criteria needed for minimizing the effects of unknown interferences gets larger, it would be highly desirable that, in critical regulatory decisions, a minimum of two as nearly independent procedures as possible be used. In that case, the less sensitive procedure would be the one that would have to be used to establish the AMDA. At that point, one would then apply the recommended criteria of the American Chemical Society Committee to calculate values for the "region of detection" and "region of quantitation" (3).

Third, a basis has been presented for a general approach for estimating and reporting selectivity for a sample in which a large number of unsuspected and unknown interferences may occur. The approach depends upon *n*-dimensional screens of the properties of the sample itself (e.g., solubility) and of those involved in the isolation and measurement procedures.

Finally, it is important to note that this discussion has focussed almost entirely on uncertainties arising largely from chemical sources. These always assume prior calibration against a primary chemical standard and, frequently, additional internal references, except in clinical chemistry where the rule has not yet been universally adopted (20). However, failures to control the laboratory environment (21) and to calibrate instruments properly (22 and correspondence from G. N. Bowers, Jr. on May 15, 1984.) (e.g., wavelength, detector response, instrument and room temperatures, flow rate, resistance between electrodes) and computer algorithms (23-27) (e.g., peak deconvolution, curve-fitting, multivariate analysis) are also major sources of uncertainty and error. It seems clear that one must not assume that the use of costly complex instrumentation, including those incorporating on-line computers for instrument control, data acquisition and data analysis, eliminates the need for careful environmental controls, frequent calibration of the critical components of the instrumentation against primary standards, and calibration (validation) of the algorithms for analyses of data.

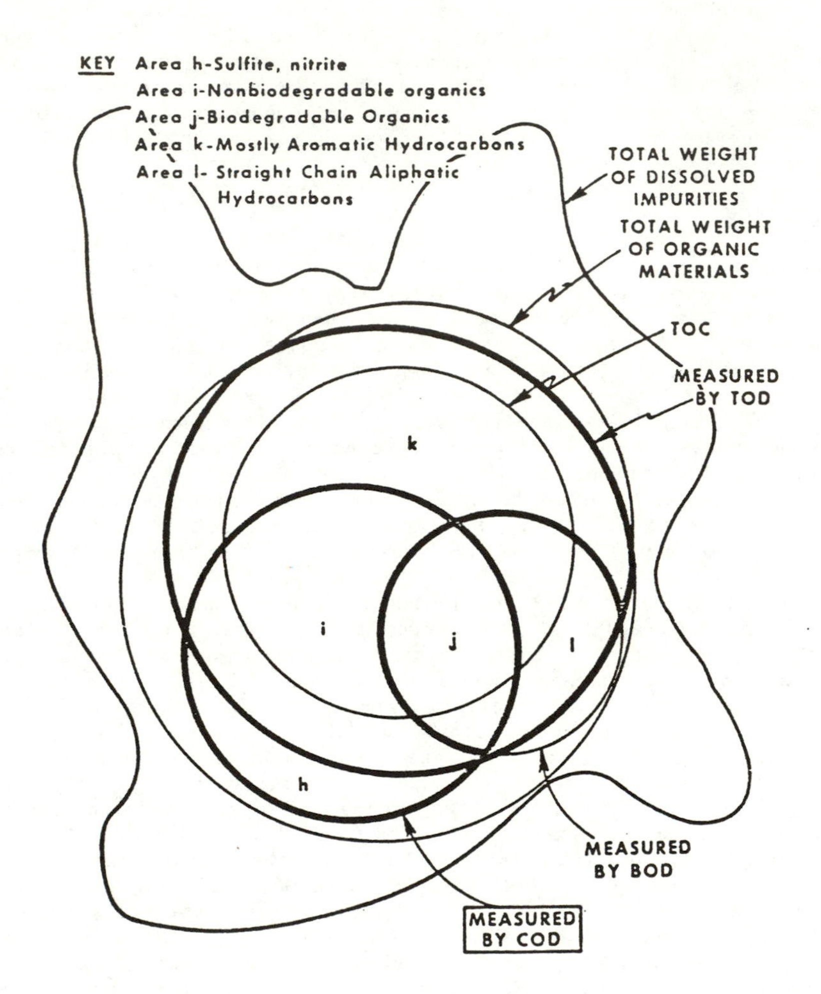

Figure 3. Schematic diagram of materials measured by chemical oxygen demand (COD) as related to total organic materials. (Reproduced with permission from Ref. 19. Copyright 1973 American Water Works Association.)

ACKNOWLEDGMENT

The writer wishes to thank the U. S. Department of Energy, Department of Basic Energies, Contract No. DE-AS09-76ER00854, for long-term support of his research which has led to awareness of the problems discussed in this paper.

LITERATURE CITED

1. Rogers, L. B. J. Chem. Ed. 1986, 63, 3.
2. Ad Hoc Subcommittee Dealing with the Scientific Aspects of Regulatory Measurements, ACS Joint Board/Council Committee on Science, Improving the Reliability and Acceptability of Analytical Chemical Data used for Public Purposes May 10, 1982; see also Chem. Eng. News 1982, 60(23), 44.
3. ACS Committee on Environmental Improvement, Principles of Environmental Analysis Anal. Chem. 1983, 55, 2210-18; MacDougall, D.; Crummett, W. B. Anal. Chem. 1980, 52, 2242.
4. Nordberg, R. The Sampling Situation and the Analytical Result Symposium on Trace Analysis of Drugs and Related Compounds in Complex Mixtures, Stockholm, Sweden, November 18, 1981; Abstracted in Acta Pharm. Suecica 1982, 19(1), 51; see also: Borgå, O., **Idem.** 52.
5. Hershenson, H. M.; and Rogers, L. B. Anal. Chem. 1952, 24, 219; see also, Rogers, L. B. J. Chem. Ed. 1952, 29, 612.
6. Data shown in Figure 4 of reference 1.
7. di Domencio, A.; Merli, F.; Boniforti, L.; Camoni, I.; Di Muccio, A.; Taggi, F.; Vergori, L.; Colli, G.; Elli, G.; Gorni, A.; Grassi, P.; Invernizzi, G.; Jemma, A.; Luciani, L.; Cattabeni, F.; De Angelis, L.; Galli, G.; Chiabrando, C.; Faneli, R. Anal. Chem. 1979, 51, 735..
8. Horwitz, W.; Kamps, L. R.; Boyer, K. W. J. Assoc. Off. Anal. Chem. 1980, 63, 1344.
9. Garfield, F. M. Quality Assurance Principles for Analytical Laboratories, Association of Official Analytical Chemists, Arlington, VA, 1984.
10. Liddle, J. A.; through Maugh, T. H., II Science, 1982, 215, 490.
11. Weinstein, M. C. Med. Decis. Making 1981, 1, 309 through Lusted, L. B. Recent Advances in Analytical Methodology in the Life Sciences, L. A. Beaver, Ed.; U. S. Food and Drug Administration, Office of Science Coordination, Washington, D.C., 1982; p. 13.
12. Donaldson, W. T. Environ. Sci. Technol. 1977, 11, 348-351.
13. Nestrick, T. J.; Lamparski, L. L.; Stehl, R. H. Anal. Chem. 1979, 51, 1453, 2273.; see also Hummell, R. A.; Shadoff, L. A. Anal. Chem. 1980, 52, 191.
14. Veillon, C. Anal. Chem. 1986, 58, 851A; see also Guthrie, B. E.; Wolf, W. R.; Veillon, C. Anal. Chem. 1978, 50, 1900.
15. Kaiser, H. Spectrochim. Acta 1978, Part B, 33B, 551; see also Trehy, M. L.; Yost, R. A.; Dorsey, J. G. Anal. Chem. 1986, 58, 14.
16. Harvan, D. J.; Hass, J. R.; Schroeder, J. L.; Corbett, B. J. Anal. Chem., 1981, 53, 1755..

17. Kateman, K; Pijpers, F. W. In Chemical Analysis; Elving, P.J.; Winefordner, J. D.: Wiley, New York, 1981, Vol 60, p.155.
17A. Mahle, N. H.; Shadoff, L. H. Biomed. Mass Spectrom. 1982, 9, 45; see Table 9 and the related discussion.
18. Analytical Reference Service Training Program, Report of Water Metals, No. 2, U. S. Dept. of Health, Education and Welfare, Robert A. Taft Sanitary Engineering Center, Cincinnati, OH, 1962.
19. Stevens, A. A; Symons, J. M. Proc. Amer. Water Works Assoc. Water Quality Technology Conference, 1973, pp. xxiii-1 to xxiii-26.
20. Bowers, G. N., Jr.; McComb, R. B. Clin. Chem. 1984, 30, 1128.
21. de Haseth, J. A. Appl. Spectrosc. 1982, 36, 544.
22. Horwitz, W. J. Assoc. Off. Anal. Chem. 1984, 67, 1053.
23. Meglen, R. R. Practical Consideration for Acceptance of Pattern Recognition and Other Chemometric Methods 100th AOAC Annual Meeting, Scottsdale, AZ, September 16, 1986.
24. Sprouse, J. F. Spectroscopy, 1986, 1(6), 14.
25. Gritton, V. Calibration Curves for Flame AAS - A Study No. 123, Div. Anal. Chem., Natl. Mtg. Amer. Chem. Soc., Anaheim, CA, Sept. 11, 1986.
26. Marshall, A. G.; Chen, L.; Cottrell, C. E. Limits to Precision in Measurement of Peak Height, Position, and Width in Fourier Transform Spectrometry No. 67, Div. Anal. Chem., Natl. Mtg. Am. Chem. Soc., Anaheim, CA, Sept. 10, 1986.
27. Rothman, L. D. Chromatogr. Forum 1986, 1(2), 13.

RECEIVED November 13, 1986

Chapter 6

Comparison of Detection Limits in Atomic Spectroscopic Methods of Analysis

Michael S. Epstein

Inorganic Analytical Research Division, National Bureau of Standards, Gaithersburg, MD 20899

The comparison of detection limits is a fundamental part of many decision-making processes for the analytical chemist. Despite numerous efforts to standardize methodology for the calculation and reporting of detection limits, there is still a wide divergence in the way they appear in the literature. This paper discusses valid and invalid methods to calculate, report, and compare detection limits using atomic spectroscopic techniques. Noises which limit detection are discussed for analytical methods such as plasma emission spectroscopy, atomic absorption spectroscopy and laser excited atomic fluorescence spectroscopy.

The comparison of detection limits is a fundamental part of most decision-making processes for the analytical chemist. Whether the decision involves the purchase of a new instrument or the design of a trace analysis protocol, the figure-of-merit [1] which influences the choice will most likely be the detection limit. Since one or more of the techniques being compared is often unfamiliar, the decision will be based on information that can be retrieved from the literature, both from manufacturer advertising and the open scientific literature. Unfortunately, despite the efforts of organizations such as the International Union of Pure and Applied Chemistry (IUPAC) to standardize methodology to calculate and report detection limits [2], there is still a wide divergence in the way that detection limits appear in print. While there is hopefully no deliberate attempt on the part of authors and manufacturers to bias detection limits towards a particular technique, the manner of calculating and reporting can lead to a misinterpretation of detection limits by the careless or unfamiliar reader. If the detection limit methodology is not well documented, a comparison can be biased by several orders of magnitude.

It is impossible to completely eliminate bias in detection limit comparisons, particularly when comparing detection capabilities in real sample matrices. However, if the basic principles behind the techniques to be compared are understood and we are aware of the common ways in which detection limit comparisons can be misinterpreted, reasonably valid conclusions can be drawn. Thus, this discussion will concentrate in general on valid and invalid ways to compare detection limits and in specific detail about limiting noises which determine those detection limits using several of the most common atomic spectroscopic techniques, including flame and plasma emission spectroscopy, atomic absorption spectroscopy, and laser-exited atomic fluorescence spectroscopy.

Before detection limits are discussed in any detail, it is necessary to define the scope of the process to which the detection limit applies. For example, the detection limit determined for an element in the absence of concomitants (i.e., in pure water solution) is likely to be significantly less than the detection limit determined for a complete analytical protocol which includes sampling, sample preparation, and analysis. The former, which is the type of detection limit most often reported in the literature, may be referred to as **fundamental** in that it reflects only the instrumental noise sources which are inherent in the analytical instrument used. Fundamental detection limits are often of limited value to the practicing analytical chemist who must determine that element in real and often very complex matrices. The latter type of detection limit, reflecting the entire analytical protocol, may be referred to as **methodological**. Methodological detection limits are also of limited value since they include many variables which cannot be easily reproduced. The detection limits to be discussed here will be called **instrumental** and will be defined as falling between fundamental and methodological in that they will consider variations induced by the instrument alone **and** by the interaction of the sample with the instrument, but will not consider the entire analytical scheme which includes blunders and contamination in the sampling and sample preparation process. It is noteworthy that instrumental detection limits will approach fundamental detection limits when the sample matrix is simple or when noise reduction methods specific to sample-matrix-induced noises are applied.

While the discussion will deal with atomic rather than molecular spectroscopic methods, many of the points to be made will apply to both atomic and molecular methods. The major difference between the noise characteristics of the two methods is usually the dynamic or flowing state of an atomic system, such as a high temperature flame or plasma, compared to the static state of a molecular system in which the sample usually is placed in a small transparent cuvette. The dynamic state of the atomic system generates an analyte signal-carried noise which is proportional to analyte signal magnitude and thus becomes limiting at high analyte concentrations. (A signal-carried noise is one whose magnitude is a constant percentage of the amplitude of a signal, which may be due to background or to the analyte. Thus, an analyte signal-carried noise is a fluctuation in the phenomenon caused by the analyte, where the phenomenon is used as a measure of the analyte concentration, such as absorption or emission of electromagnetic radiation). The static state of the molecular system limits the magnitude of analyte signal-carried noises, except where the static state is disturbed (i.e., vibration, cell position changes, etc.) or where radiation source fluctuations are significant at high analyte concentrations (i.e., molecular fluorescence spectrophotometry).

Bias in Detection Limit Comparisons

There are several ways that detection limit information can be presented in order to bias the observer. Again, it must be emphasized that in most cases sufficient information will be presented in a figure or in the accompanying text to allow the knowledgeable reader to properly interpret detection limit comparisons.

Real or Artificial Detection Limits. Certainly, one of the most common ways to report detection limits is in "pure aqueous solution." Whether the analytical conditions or the instrumentation used is capable of those detection limits when real samples are analyzed is another question. This source of bias is most often encountered when a new analytical technique is developed. An example is the early development of flame atomic fluorescence spectroscopy (FAFS), where cadmium was "detected" with a total consumption burner in an oxy-hydrogen flame at 1 pg/mL [3]. Certainly no one would attempt real sample analysis in such a flame because of its turbulent flow and poor dissociation characteristics. More realistic detection limits are on the order of 200 pg/mL in an air-acetylene flame [4]. In flame atomic absorption spectroscopy (FAAS), tin detection limits are significantly better (≈4x) in a cool air-hydrogen flame than in hotter flames as a result of increased sensitivity and lower flame background emission [5]. The use of the sampling boat [6] in FAAS also improves detection limits for many elements by an order of magnitude because of increased sample transport efficiency. However, neither of these techniques is widely used in FAAS, since both exhibit significant chemical interferences with real samples. Recently developed techniques, such as inductively-coupled plasma mass spectrometry (ICP-MS) [7] and laser enhanced ionization spectroscopy (LEIS) [8] exhibit similar sample-related degradation of detection limits. Certainly, most technique detection limits suffer somewhat when real samples are analyzed and noises induced by the sample matrix become limiting. The extent of this effect will vary, however, from technique to technique, and will usually diminish as the method reaches maturity.

Detection Limit Criterion. The criterion used to define the detection limit, or perhaps as important, the protocol used to measure it, can be critical in establishing a valid detection limit. Currie [9] has described the wide variation in detection limit definitions for radiochemical measurements reported in the literature. IUPAC [2] recommends the detection limit, c_L, be defined as the concentration of an analyte equal to a background-corrected signal, $x_L - x_B$, three times the estimated standard deviation of a single determination using 20 measurements of the blank.

$$x_L = x_B + ks_B \tag{1}$$

$$c_L = ks_B/m \tag{2}$$

where

x_L = uncorrected signal
x_B = blank measure
s_B = estimated standard deviation of the blank measure
c_L = detection limit, which is the concentration derived from the smallest measure (x_L) that can be detected with reasonable confidence.
k = numerical factor chosen in accordance with the confidence level desired.
m = analytical sensitivity

As pointed out by Long and Winefordner [10], the use of k=3 allows a confidence level of 99.86% for a normal distribution of x_B, or an 89% confidence level for a non-normal distribution. While x_B will often be normally distributed when instrumental noise limits detection, the presence of analyte contamination in the blank, either in the sample preparation process or as a series of discrete events (i.e., Na or Fe airborne particulates) in the instrumental measurement process, will result in a non-normal distribution. Such a distribution may be bimodal or skewed depending on the source and characteristics of the contaminant. Long and Winefordner [10] have also presented several examples of the influence of measurement protocol on c_L. The use of values of $k < 3$ or the use of the standard deviation of the mean or pooled standard deviation rather than the standard deviation of a single measurement, can lead to c_L values which deviate by an order of magnitude from the IUPAC model. Measurement protocols which include the error in the analytical sensitivity as well as the error in the blank can also cause c_L to deviate significantly from the IUPAC model, which assumes a well-defined sensitivity. Finally, the presence of very low frequency noise or drift may not be incorporated into the IUPAC definition of detection limit [11]. The calibration scheme used for real samples may be spread out over a longer time period than was used for the determination of the detection limit and thus noises which were insignificant during the detection limit measurement may be encountered. Ideally, a technique detection limit should be determined using the measurement protocol employed for real sample analysis.

Analytical Blank. If the detection limit is not measured from the true analytical blank, a critical part of the detection limit determination has been ignored. Since the emphasis in this discussion is on "instrumental" rather than "methodological" detection limits, only blanks resulting from the instrumentation will be considered. Although method blanks can certainly be limiting, particularly for elements such as Fe, Na, and Ca which are common in the laboratory environment, they are not as predictable as instrumental contamination blanks, since variations in laboratory procedures and design will be much greater than variations in instrument design. For example, in FAFS one can significantly improve a detection limit in a situation limited by flame scatter of source radiation by making measurements with no water being introduced into the flame. By eliminating the scattering species, unvaporized water droplets, the detection limit is improved. Similarly, one can measure a graphite furnace atomic absorption spectroscopy (GFAAS) detection limit without actually atomizing a blank sample, assuming the noise to be independent of the atomizer. This is certainly an invalid assumption when determining an element whose most sensitive absorption line lies in the visible region of the spectrum, such as barium, where thermal emission from the graphite tube is significant, or where contamination in the tube is limiting, such as when zinc is determined. For these elements, published detection limits may be invalid, unless they were measured under actual analysis conditions. **The moral is thus to measure the blank under conditions as similar as possible to the analysis conditions used.**

Instrument Noise Characteristics. Depending on the frequency domain spectrum of the signal from the analytical instrument, that is, if the noise is white (shot) or 1/f (flicker) in nature [1], the integration time or time constant used for the detection limit determination may have a significant effect on the detection limit. In cases where shot noise is most often limiting, such as in FAFS at all wavelengths or FAAS above 230 nm, or ICP emission spectroscopy (ICP-ES) below 250 nm, the detection limit can be improved as the square-root of the integration time. In flicker noise limited cases, there may be little or no improvement in the detection limit with an increase in the integration time. [12,13] The improvement in detection limit for an increase of integration time will be ultimately limited by the significance of very low frequency drift and the availability of large volumes of sample solution.

Measurement Units. Perhaps the most obvious yet confusing aspect of many detection limit comparisons is the use of "relative" versus "absolute" units. Relative units reflect a mass per unit volume, such as micrograms per milliliter, while absolute units reflect a mass only, such as micrograms. Obviously, "relative" and "absolute" units should not be directly compared. However, absolute units can be converted into relative units and vice versa, employing the volume of solution utilized by a particular technique. Nonetheless, how that conversion is done or how it is documented can significantly bias the observer. Table I illustrates several examples, taken from the scientific literature, of the use of detection limit values in a table for comparison purposes. In each case the author provides adequate information for the informed reader to make an accurate comparison. Nevertheless, the conclusions drawn by the careless or uninformed reader who does not read or understand the footnotes or the text which describes the table, can be biased by several orders of magnitude.

Table Ia, presents a comparison of FAAS and GFAAS detection limits [14]. Without reading the text which refers to the table, one is impressed by the significant, 3 to 4 orders of magnitude, improvement using the graphite furnace. However, the text clearly indicates that both flame and graphite furnace detection limits assume a 1 mL sample volume which, while certainly valid in a flame, is not valid for a graphite furnace since the maximum sample volume is usually about 50 to 100 μL. Some systems, like the carbon rod atomizer [15], can only accommodate 1 to 2 μL of solution. Thus, for a valid comparison, the graphite furnace detection limits must be degraded by one to two orders of magnitude.

Table Ia. Detection Limits Reported for Atomic Absorption

Element	Detection Limit (μg/mL) Flame	Nonflame
Ba	0.02	6×10^{-6}
Ca	0.002	4×10^{-7}
Fe	0.004	1×10^{-5}
Mn	0.0008	2×10^{-7}

"...the limits are based on a signal-to-noise ratio=2 criterion and the assumption that a volume of 1 mL is the minimum required for a determination. For example, if an absolute detection limit is given (e.g., nonflame atomizer) as 10^{-9} g, this is expressed as a concentrational detection limit of 0.001 μg/mL. One must bear in mind that most current nonflame atomizers cannot handle samples larger than, say, 0.1 mL, and that most flame atomization systems cannot handle samples (for a reliable reading) of much less than 1 mL. The 1 mL criterion...is thus more for the purpose of direct comparison than for the very lowest possible detection limits..." Reprinted from [14] by permission of John Wiley and Sons, copyright 1976.

Source: Reproduced with permission from Ref. 14. Copyright 1976 Wiley.

In Table Ib [16] absolute rather than relative detection limits are compared for several techniques. Unless one looks at the footnotes however, it is not obvious that the detection limit for one method is based on a 1 μL sample size, another on a 5 μL sample size and another on a 1 mL sample size, thus biasing the careless observer of this table by 2 to 3 orders of magnitude.

Table Ib. Absolute Detection Limits Using Atomic Fluorescence Spectrometry and Several Other Methods

Element	Detection Limits (pg) AFS	AAS	AEICP
Ag	0.4	0.2	200
Cd	0.0015	0.1	70
Mg	1	0.06	3
Ni	5	10	200

AFS = Atomic fluorescence spectrometry - 1 μL sample size
AAS = Atomic absorption spectrometry - 5 μL sample size
AEICP = Plasma emission using the ICP - 1 mL sample size [16]
Reprinted from [16] by permission of Pergamon Journals Ltd.

Source: Reproduced with permission from Ref. 16. Copyright 1979 Pergamon Press.

Finally, in Table Ic [17] a comparison is made of detection limits for carbon furnace atomic emission spectroscopy (CFAES), flame emission spectroscopy (FES), and CFAAS. Note that sensitivities are used as pseudo-detection limits for CFAAS. These are not really sensitivities as defined by IUPAC [18], but are characteristic concentrations, since they represent a concentration equivalent to an absorbance of 0.0044. Furthermore, noise measurements are not made in the calculation of this parameter, so the true detection limit will likely be much smaller, particularly in the case of CFAAS, where transmission flicker noise is negligible.

Table Ic. Detection Limits Using Carbon Furnace Atomic Emission Spectrometry and Other Techniques

Element	Detection Limits (μg/mL)		
	CFAES	Flame emission	CFAAS
Mo	0.016	0.03	0.005
Si	0.088	10	0.01
Be	0.46	10	0.0002

CFAES = Carbon furnace atomic emission spectrometry - 20 μL aliquot of solution

CFAAS = Carbon furnace atomic absorption spectrometry - Sensitivity in μg/mL/0.0044 A - based on a 20 μL aliquot of solution

Reprinted from [17] by permission of Elsevier Science Publishers

Source: Reproduced with permission from Ref. 17. Copyright 1978 Elsevier Scientific.

Now, let us summarize the questions that should be considered when comparing detection limits.

First, what is the noise bandwidth (defined by the integration time or time constant for each measurement) of each instrument? Were the measurements made under similar conditions and, when using a method such as ICP-ES [12,13], does it make any difference?

Second, are we dealing with absolute or relative units and have the units been correctly converted to allow a valid comparison?

Third, does sample-induced noise, that is noise resulting from components in the sample matrix, significantly degrade detection limits? This may be more significant for some techniques than others. For example, scatter or molecular absorption in FAAS, when compensated for by a background correction method such as Zeeman splitting or a continuum source, will usually result in only a small increase in shot noise due to attenuation of primary source intensity and no significant change in detection limits will occur. The same matrix components in an ICP-ES system, which is flicker noise limited, may show a far more significant degradation of detection limit when flicker in the sample matrix emission becomes the limiting noise.

Fourth, does the sensitivity of the technique decrease in the presence of the sample matrix? Often conditions which favor the best detection limits, such as low background emission or high sample introduction rates also result in reduced sample dissociation and thus decreased analyte sensitivity when a complex sample matrix is present. Are detection limits determined under unrealistic conditions or with apparatus unsuitable for real sample analysis?

Fifth, are we dealing with conditions optimized for a single element or multielement analysis? Compromise conditions degrade detection limits but improve the informing power of the method (i.e., the total amount of information about a sample that can be obtained from an analytical method).

Sixth and finally, what criteria were used to define the detection limit and how was it calculated?

Noises Which Limit Detection

Let us now look briefly into the three major classes of analytical spectrometric methods: emission, absorption, and fluorescence. Noises will be defined, and examples of how and when they limit detection will be given. Table II lists the major noises which limit detection for the three atomic spectroscopic techniques to be discussed. Detailed definitions of these noises may be found in the paper by Epstein and Winefordner [1].

TABLE II. Noises Which Limit Detection in Atomic Spectroscopic Methods

EMISSION

PMT shot noise induced by dark current, atomizer background emission, or sample matrix emission.
Electronics noise (including RF)
Atomizer background intensity fluctuations induced by atomizer gases, sample matrix components, or contamination.

ABSORPTION

PMT shot noise induced by the radiation source, atomizer background emission, or sample matrix emission.
Electronics noise
Radiation source intensity fluctuations
Atomizer transmission fluctuations induced by flame or furnace gases, sample matrix components, or contamination.

FLUORESCENCE

PMT shot noise induced by dark current, atomizer background emission, sample matrix emission, or scattered radiation source intensity.
Electronics noise (including RF)
Radiation source intensity fluctuations carried by scatter, contamination fluorescence, or broadband fluorescence from flame and furnace gases or from sample matrix components.
Atomizer background intensity fluctuations induced by flame or furnace gases, sample matrix components, or contamination.

Every spectrometric system consists of four of the components shown in Figure 1: (a) a source of atoms; (b) a spectrometer to isolate the radiation whose intensity and frequency contains information about the analyte; (c) a photodetector to convert photons to electric current; and (d) a signal processing scheme to

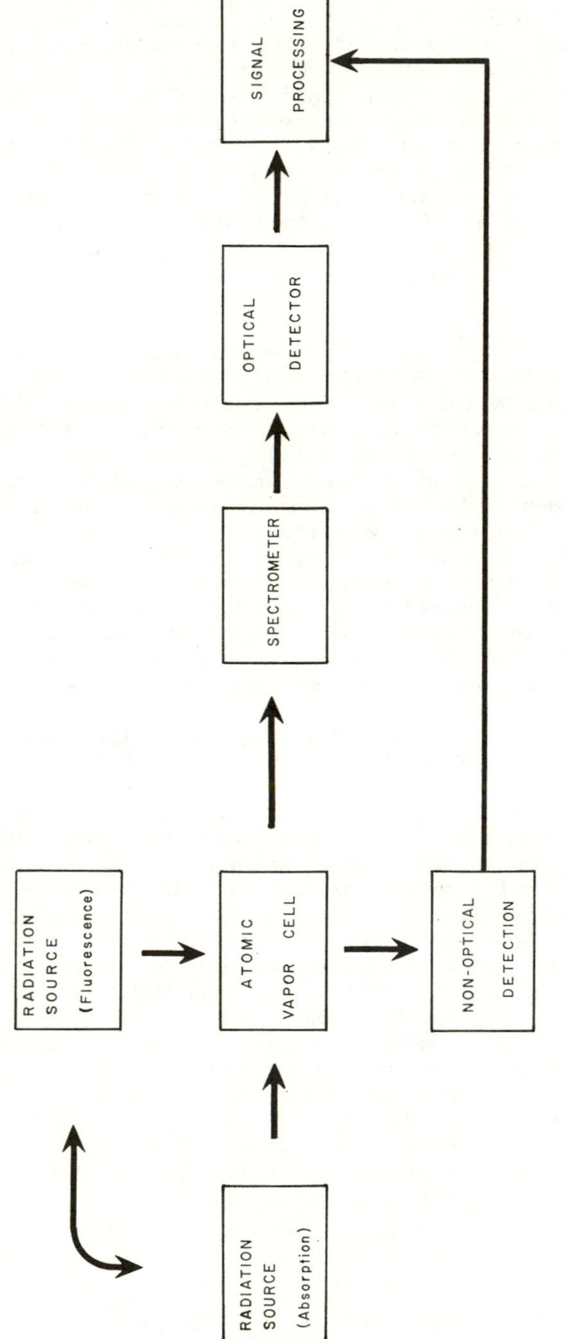

Figure 1. Components of an analytical spectrometric system.

decode the information stored in the radiation. In emission methods, the high temperature of the atom source provides excitation energy to promote electron transitions, while in absorption and fluorescence methods, external radiation sources are used to induce the electron transitions of the analyte atoms. All of these instrument components can result in noise which limits detection. Spectroscopic techniques which use non-optical detection, such as LEIS or photoacoustic spectroscopy, are characterized by noise sources similar to fluorescence, since the information-carrying phenomenon is energy release following absorption. Rather than radiational deactivation of the excited state, in LEIS the energy release mechanism is flame ion current generation, and in photoacoustic spectroscopy, it is thermal or collisional deactivation.

Emission Noise Sources. Noises in the emission technique are the simplest to understand. The inductively coupled plasma (ICP) and direct current plasma (DCP) are both emission sources which have become popular in recent years. The noises which limit detection using these emission sources are easily characterized. With very low optical throughput, such as when narrow slit widths are used in the far UV, photomultiplier dark current noise may be significant. However, in most cases, shot noise induced by the source background radiation, or flicker noise carried by the source background are limiting. The background intensity may result from argon emission in the source or may be induced by interaction of the source with the sample matrix. In the case of flicker noise, that is, the fluctuation in the background intensity, the noise usually results from temporal variations in the sample transport system or the external gas flows.

The major question when comparing detection limits using emission techniques is whether the signal-to-background ratio (SBR) or the signal-to-noise ratio (SNR) was used as the measure of detection limit. The SBR requires a measure of the concentration corresponding to a multiple of the background intensity, rather than the noise, and thus requires only one measurement of background. The measurement of background is usually made in the presence of the analyte-containing sample by measuring at a wavelength slightly offset from the wavelength of the analyte intensity maximum. In a multielement system, it is thus much simpler to monitor instrument performance by measuring the SBR for each channel, rather than the SNR, which would require multiple measurements. The detection limit is then calculated by assuming that the background-carried flicker noise is limiting and that there is a constant relative standard deviation of the background emission, usually about one percent. The limitation to this procedure has been clearly pointed out by Boumans [19], who describes the relationship of the relative standard deviation of the background to the flicker noise and shot noise components by the following equation:

$$(RSD)_B = (\alpha_B{}^2 + g\beta/x_B)^{1/2} \qquad (3)$$

where

$(RSD)_B$ = observed relative standard deviation of the background emission.

α_B = flicker factor induced by variations in instrumental components such as nebulizer or gas flow controls.

x_B = measured background signal in units of anode current.

g = photomultiplier gain

β = constant coefficient which includes components due to the effective system noise bandwidth, the electronic charge, and gain fluctuations due to secondary electron emission.

The validity of assuming a constant standard deviation of the background emission depends on the dominance of background flicker noise. With that noise, which Boumans points out is limiting at wavelengths greater than 300 nm, the SNR and thus the detection limit can be characterized by the SBR divided by a flicker factor, α_B, the first term in equation 3. The flicker factor will be a function of a particular instrument and will have a magnitude which depends on the stability of various instrument components. Thus, as long as flicker noise is limiting and the flicker factor does not change, the approximation is valid. Deviations from the assumption occur at shorter wavelengths, where the spectrometer optical throughput and plasma background intensity decrease. The background shot noise intensity, represented by the second term in Boumans' equation, $(g\beta/x_B)^{\frac{1}{2}}$, will make a significant contribution to the variation of the background intensity, and the simple relationship of flicker factor to SNR mentioned previously breaks down. Thus, detection limits calculated from SBR's without consideration of shot noise may be in error.

Absorption Noise Sources. Noises in atomic absorption spectroscopy are more complex than in emission. When a source of radiation is introduced, whose attenuation carries the analyte information, several new limiting noise sources are introduced. Flicker noise due to emission from the high temperature atomic vapor cell is not as significant as it is in emission techniques, because atomic absorption uses source modulation to discriminate against such noise by encoding the analyte information signal at a high frequency. Shot noise is still observed as a result of background emission from the flame or from sample matrix components, but no significant flicker noise is measured. However, new noises are shot and flicker from the radiation source, flame transmission flicker noise which becomes limiting at wavelengths less than 230 nm, and molecular absorption or scatter noise from sample matrix components.

All of the flicker noises can be effectively eliminated by the use of double-beam optics in conjunction with a background correction system such as Zeeman splitting or a well-aligned (or wavelength-modulated) continuum source. Thus the ultimate limiting noise in atomic absorption is source shot noise, which can be reduced (relative to total source intensity or I_o) by increasing the source intensity, up to the point of optical saturation.

Table III presents some examples of limiting noises in different atomic absorption determinations. These measurements are a compilation of information from several sources, but primarily from the work of Ingle [20,21] using a very simple, single-beam atomic absorption spectrometer. Aluminum in a nitrous oxide-acetylene flame is limited by flame transmission flicker and source flicker. The flame transmission flicker results from absorption by molecular species (e.g., OH). Barium in a similar flame, but with a double-beam instrument, is limited only by source induced shot noise. The double-beam system reduces the source flicker noise component. Calcium in an air-acetylene flame is limited by both source flicker and shot noise, while in the hotter nitrous oxide-acetylene flame, flame emission shot noise becomes greater than the source shot noise. The flame emission shot noise results from the intense molecular emissions of CN and CH in the higher temperature flame. Copper is limited by both source flicker and shot noise at a one second integration time, but limited by only flicker noise at a ten second integration time, a result of the reduction of the shot noise component. An increase in integration time will improve the detection limit in a shot noise limited case.

Table III. Dominant Noises in Atomic Absorption Spectroscopy

Element	Wavelength (nm)	Flame type	Limiting noise (Absorbance<0.2)[a,b]
Al	309.3	N_2O/C_2H_2	Flame transmission flicker Radiation source flicker
Ba[c]	553.6	N_2O/C_2H_2	Radiation source-induced shot noise
Ca	422.6	Air/C_2H_2	Radiation source flicker Radiation source-induced shot noise
Ca	422.6	N_2O/C_2H_2	Radiation source flicker Flame background emission noise
Cu (1s)[d]	324.7	Air/C_2H_2	Radiation source flicker Radiation source-induced shot noise
Cu (10s)[d]	324.7	Air/C_2H_2	Radiation source flicker
Zn	213.8	Air/C_2H_2	Flame transmission flicker

[a]Determined with a single-beam AAS instrument [20,21], except as noted.
[b]Noises with a variance of at least 33% of the most significant noise.
[c]Determined with a double-beam AAS instrument.
[d]Integration time

Finally, zinc is an example of an element whose absorption wavelength is lower than 230 nm, the range where flame transmission noise dominates. Let us look a bit more closely at the case of zinc. The instrument performance can be characterized using a precision plot, shown in Figure 2, where the relative standard deviation of concentration is plotted on the vertical axis and concentration on the horizontal axis. The detection limit is defined by the intersection of the precision curve with the RSD which represents the criterion used to define detection, about 30% RSD for k = 3. Note that flame transmission flicker noise limits detection.

If the burner head is rotated to reduce sensitivity, we find that the limiting noise is no longer flame transmission flicker, but source shot noise, since the absorption path has been reduced by a factor of 20. Although the sensitivity is decreased by a factor of 20, the detection limit is decreased by only a factor of 10, since the flame transmission noise is no longer limiting. Thus, referring back to a statement made earlier, sensitivity, or more correctly characteristic concentration [18] cannot be used as an accurate measure of detection limit in AAS. Unlike the case of SBR in emission, because of the complexity of noises in atomic absorption, a general and simple relationship cannot be derived to relate characteristic concentration and detection limit.

<u>Laser Fluorescence Noise Sources</u>. Finally, let us examine a technique with very complex noise characteristics, laser excited flame atomic fluorescence spectrometry (LEAFS). In this technique, not only are we dealing with a radiation source as well as an atomic vapor cell, as in atomic absorption, but the source is pulsed with pulse widths of nanoseconds to microseconds, so that we must deal with very large incident source photon fluxes which may result in optical saturation, and very small average signals from the atomic vapor cell at the detection limit [22]. Detection schemes involve gated amplifiers, which are synchronized to the laser pulse incident on the flame and which average the analyte fluorescence pulses [23].

The limiting noises can vary significantly depending on the configuration of the optical system, the type of flame, and the type, temporal pulse width, and intensity of the laser used. Electronic noise, dark current noise, and flame background shot and flicker noise are temporally continuous noises that tend to limit laser-based spectroscopic systems whose detection system gatewidths are in the microsecond range and wider. At smaller detector gatewidths, the temporally continuous noises will be exceeded by pulse-type noises which occur only during the laser pulse and thus during the open gate of the detection system. These are laser-induced noises such as laser scatter flicker and shot noise, laser induced nonanalyte fluorescence flicker and shot noise, and radiofrequency (RF) noise.

If we look at a few cases in the literature, we see a wide variation in the reported limiting noises [24-27]. Table IV lists noises and detection limits for several laser systems. A flashlamp-pumped system, because of the relatively low intensity and wide temporal pulse width, about 1 μs, is limited by flame emission shot and flicker, which are temporally continuous noises. The

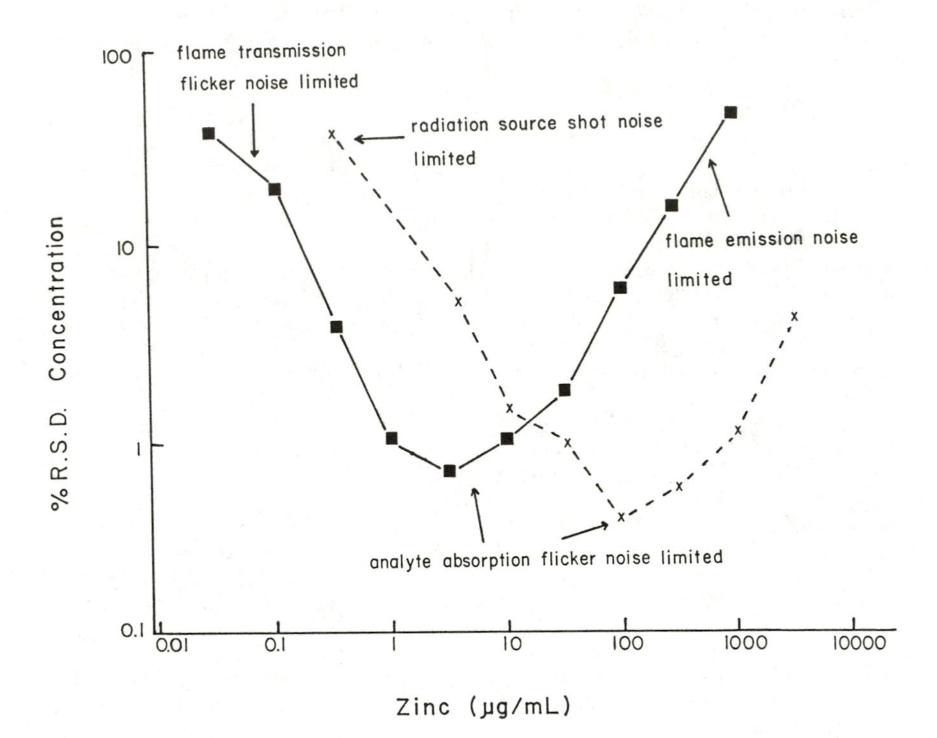

Figure 2. Precision plot for the determination of zinc by FAAS using a measurement cell (burner head) whose long axis (10 cm) is parallel (-▪-) or perpendicular (-X-) to the optical path of the spectrometer.

higher intensity lasers, with very narrow temporal widths, on the order of 5-to-20 ns, are limited by pulse type noises such as RF or scatter shot and flicker. Detection limits for these systems can be correlated quite well with the intensity, temporal and spectral pulse width, repetition rate, and excitation area of the laser [27]. The laser most likely to provide improved detection limits is the copper vapor laser, since it is not limited by a fundamental noise, such as shot noise, but rather by RF, which can be reduced by proper shielding.

TABLE IV. Limiting Noises and Detection Limits in Laser-Excited Atomic Fluorescence Spectrometry using the 296.7/373.5 nm Transition of Iron

Dye Laser Pump	Major Limiting Noises	Detection Limit (ng/mL)[a]
Flashlamp	Flame emission shot/flicker	0.6 [25]
N_2 laser	RF, boxcar	30 [24]
Excimer laser	Contamination, boxcar, RF,	0.2 [26]
Nd:YAG laser		
Focused	Scatter shot/flicker	7 [27]
Beam expanded	Flame emission shot	36 [27]
Cu vapor laser	RF	0.8 [27]

[a]Detection limits assume a 1 s time constant and SNR = 3.

Conclusions

Finally, a few suggestions should be made for determining detection limits.

Whatever criterion and protocol are used should be reported in detail. When sufficient information is provided, the reader can normalize the reported values to other detection limit methodologies, particularly if he is familiar with the noise characteristics of the methods compared.

Determine the analyte signal or response (i.e., the sensitivity) with an analyte concentration close enough to the blank noise level to assure that linearity of response exists down to the noise level. Do not assume that variation of analyte response equates with that of the blank response.

Measure the variation of the blank under instrumental conditions and with the measurement protocol typically used for analysis of real samples. Establish the effect of real sample matrices on the sensitivity and noise level of the instrument when operated under these conditions.

It is certain that despite all precautions, the validity of

detection limit comparisons will still remain in question. Nevertheless, by carefully evaluating the information presented, and with a fundamental knowledge of noise sources and the effect of sample-induced noises on an analytical technique, the reader can reach an intelligent decision based on the information available.

Literature Cited

1. Epstein, M. S.; Winefordner, J. D. Prog. Analyt. Atomic Spectrosc. 1984, 7, 67-137.
2. IUPAC Commission on Spectrochemical and Other Optical Procedures for Analysis, Anal. Chem. 1976, 48, 2294-2296.
3. Zacha, K. E.; Bratzel, M. P.; Winefordner, J. D.; Mansfield, J. M. Anal. Chem. 1968, 40, 1733-1736.
4. Larkins, P. L. Spectrochim. Acta 1971, 26B, 477-489.
5. Capacho-Delgado, L.; Manning, D. C. Spectrochim. Acta 1966, 22B, 1505-1513.
6. Kahn, H. L.; Peterson, G. E.; Schallis, J. E. At. Abs. Newslett. 1968, 7, 35.
7. Gray, A. L. Spectrochim. Acta 1986, 41B, 151-167.
8. Turk, G. C. Anal. Chem. 1981, 53, 1187-1190.
9. Currie, L. A. Anal. Chem. 1968, 40, 586-593.
10. Long, G. L.; Winefordner, J. D. Anal. Chem. 1983, 55, 712A-718A.
11. Winefordner, J. D.; Ward, J. L. Anal. Lett. 1980, 13, 1293-1297.
12. Belchamber, R. M.; Horlick, G. Spectrochim. Acta 1982, 37B, 71-74.
13. Belchamber, R. M.; Horlick, G. Spectrochim. Acta 1981, 36B, 581.
14. Veillon, C. In Trace Analysis: Spectroscopic Methods for Elements; Winefordner, J. D., Ed.; Wiley; New York, 1976; Chapter 6, pp 164-166.
15. West, T. S.; Williams, X. K. Anal. Chim. Acta 1969, 45, 27.
16. Omenetto, N.; Winefordner, J. D. Prog. Analyt. Atom. Spectrosc. 1979, 2, 1-183.
17. Littlejohn, D.; Ottaway, J. M. Anal. Chim. Acta 1978, 98, 279-290.
18. IUPAC Commission on Spectrochemical and Other Optical Procedures for Analysis, Appl. Spectrosc. 1977, 31, 348-364.
19. Boumans, P. W. J. M.; McKenna, R. J.; Bosveld, M. Spectrochim. Acta 1981, 36B, 1031-1058.
20. Bower, N. W.; Ingle, J. D. Anal. Chem. 1977, 49, 574.
21. Bower, N. W.; Ingle, J. D. Anal. Chem. 1979, 51, 73.
22. Omenetto, N.; Winefordner, J. D. In Analytical Laser Spectroscopy; Omenetto, N., Ed.; Wiley; New York, 1979; Chapter 4, pp 167-217.
23. O'Haver, T. C. In Trace Analysis: Spectroscopic Methods for Elements; Winefordner, J. D., Ed.; Wiley; New York, 1976; Chapter 2, pp 56-58.
24. Weeks, S. J.; Haraguchi, H.; Winefordner, J. D. Anal. Chem. 1978, 50, 360-368.
25. Epstein, M. S.; Bayer, S.; Bradshaw, J.; Voigtman, E.; Winefordner, J. D. Spectrochim. Acta 1980, 35B, 233-237.

26. Seltzer, M. D.; Hendrick, M. S.; Michel, R. G. Anal. Chem. 1985, 57, 1096-1100.
27. Epstein, M. S.; Travis, J.; Turk, G. C., in preparation.

RECEIVED December 24, 1986

Chapter 7

Noise and Detection Limits in Signal-Integrating Analytical Methods

H. C. Smit[1] and H. Steigstra[2]

[1]Laboratory for Analytical Chemistry, University of Amsterdam, Nieuwe Achtergracht 166, 1018 WV Amsterdam, Netherlands
[2]Faculty of Medicine, Radboud Hospital, Nijmegen, Netherlands

The uncertainty in the estimation of signal parameters is discussed, particularly the 0th moment (area) of a peak determined via an integration procedure. An overview is given of the derivations of the error variance due to integrated noise, both in the frequency domain and in the time domain. As an example the uncertainty in case of some typical kinds of noise is calculated, using the derived expressions. The theory is extended with the derivation of the optimum integration interval on basis of known peak shapes and known noise characteristics, assuming stationary noise without a deterministic drift component. Finally, the influence of uncorrected linear drift on the integration variance is determined, while an expression for the variance after a frequently applied drift correction is derived, using correction intervals.

One of the basic problems in analytical chemistry is how to calculate the uncertainty in the determination of the parameters of a noisy analytical signal. Although this uncertainty is important, it is not the only factor influencing the detection limit. It must be emphasized that errors and uncertainties originating from sample preprocessing, sample introduction, lack of standardization of the measurement conditions, etc., may be just as important as noise perturbing the signal. However, it is certainly useful to calculate the contribution of that noise to the total uncertainty, determining the detection limit, in the analytical result.

If only <u>one</u> measurement (data)point is considered, then the problem reduces to a simple comparison of the measured amplitude with the standard deviation of the noise, determined by repeated measurements. Ordinary statistics can be applied to calculate the uncertainty. However, often dynamic signals, like peaks in chromatography, are produced and signal parameters like the 0th moment (peak area) or higher moments are representative for the desired analytical information.

Determining these parameters always includes an integration

0097-6156/88/0361-0126$06.75/0

procedure, where of course, together with the signal, the noise is integrated as well. We might formulate the problem as follows: What is the uncertainty in the determination of the analytical signal parameters due to the influence of the integrated noise?

In this paper we emphasize the determination of the peak areas. Particularly in quantitative chromatography the uncertainty in the area determination is directly related to the detection limit. To elucidate the problem formulation, a (Gaussian) peak and the time integral of the peak is shown in Figure 1. The relevant information is the height I of the integral with respect to baseline, assuming a constant (flat) noiseless baseline. If noise is added and the peak is integrated again, then the final value will probably differ from the true value. Repeating the same procedure with a similar peak with noise with the same statistical properties yields a number of statistically distributed data points. If the noise is assumed to be stationary, i.e. if the statistical properties like mean and variance are not changing with time, the mean of the mentioned data points is an estimate of the true area. The standard deviation σ_I or error variance σ_I^2 determines the uncertainty in the peak area determination.

The problem of determining the variance of integrated noise is not restricted to peak parameter determination. Measurement and calculation of the average intensity of a spectroscopic line means integrating too, however, the final result including the standard deviation of the integrated noise has to be divided by the integration time.

Altogether, this brings us to the desirability to derive an expression for σ_I or σ_I^2, respectively, containing all factors influencing the error variance. Of course, this expression can be used to calculate the detection limit in, for instance, chromatography as far as determined by the baseline noise. However, it is also usable to make an optimum choice of parameters and conditions. Besides, some rules of thumb can be given, usable in daily practice.

One has to keep in mind that such a derivation always implies some assumptions concerning the stationarity of the analytical system and particularly the stationarity of the noise. In general, stationarity and the absence of a deterministic drifting baseline is assumed, although some derived expressions in the general form are valid for non-stationary noise. However, the derived theory can be used as a basis for the calculation of the remaining uncertainty in the case of a correction procedure for deterministic (for instance linear) baseline drift.

Basic Theory

The derivation of the error variance requires some theory from different fields. For the convenience of the reader a very short overview will be given, including some basic principles and definitions of the required theory. Another reason to give some textbook theory is that the definition of several quantities can differ; literature is not very consistent in that respect. A detailed description of signal theory, system theory, stochastic processes and of course mathematics can be found in several textbooks (1-5).

It is necessary to solve the problem of deriving σ_I^2 both in the

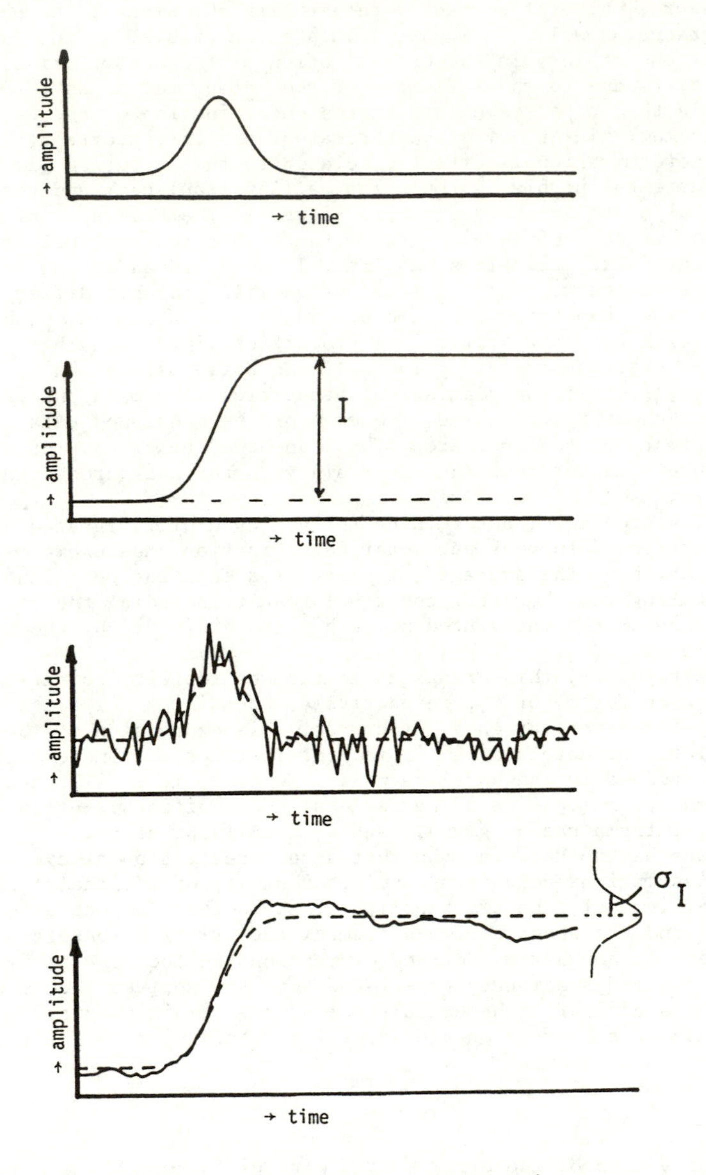

Figure 1. Integrated peak and noise.

time domain and in the frequency domain by Fourier transforming the time signals or functions derived from the time signals. The Fourier transform (FT) of a function of the time f(t) is defined in the usual way:

$$F(j\omega) = \int_{-\infty}^{+\infty} f(t)\, e^{-j\omega t}\, dt \qquad (1)$$

$$j^2 = -1$$

We wish to consider random variables in a continuous domain. This can be done by using different descriptive functions. For our derivations we need the concepts of probability density function (PDF), auto-correlation function (ACF) and power spectral density (PSD). Moreover, the following system functions are used: the weighting function h(t) and the complex frequency response $H(j\omega)$.

The well known probability density function can be considered as the limiting value of the probability that an amplitude of noise n(t) lies in an interval around a certain value, divided by the width of that interval. The shape, which is often Gaussian, and the width of the PDF, expressed in the standard deviation σ_n, are used for statistical calculations of detection limit etc. A random signal, or in general a family of functions of time (random process) of which the values vary randomly even if it is stationary, is not uniquely specified by a PDF, as is demonstrated in Figure 2. Both random signals have the same PDF, but they are obviously different.

An important quantity, summarizing much information about a random process, is the ACF. To illustrate the concept of the ACF, Figure 3 shows a family of stochastic signals (signals evolving in time according to probability laws), a random (stochastic) process or an ensemble. An example of an ensemble is a set of possible noise records from a chromatographic detector, each recorded during a certain time interval. Now we have to distinguish ensemble statistics and time statistics. For instance, the mean value at the time t_1 (ensemble statistics) is defined:

$$\mu_n(t_1) = \lim_{N\to\infty} \frac{1}{N} \sum_{k=1}^{N} n_k(t_1) \qquad (2)$$

k refers to signal k.
The (ensemble) ACF is defined:

$$R_n(t_1,t_1+\tau) = R(t_1,t_2) = \lim_{N\to\infty} \frac{1}{N} \sum_{k=1}^{N} n_k(t_1)n_k(t_1+\tau) \qquad (3)$$

being the average product of the values of the stochastic process at time t_1 and t_2. If there is no relation or better correlation between the values at time t_1 and t_2, then the average will tend to the product of the mean value at t_1 and t_2. In case of fast fluctuating noise, no correlation will exist even after a relatively short time, where slowly fluctuating noise still shows an average relation between the amplitudes.

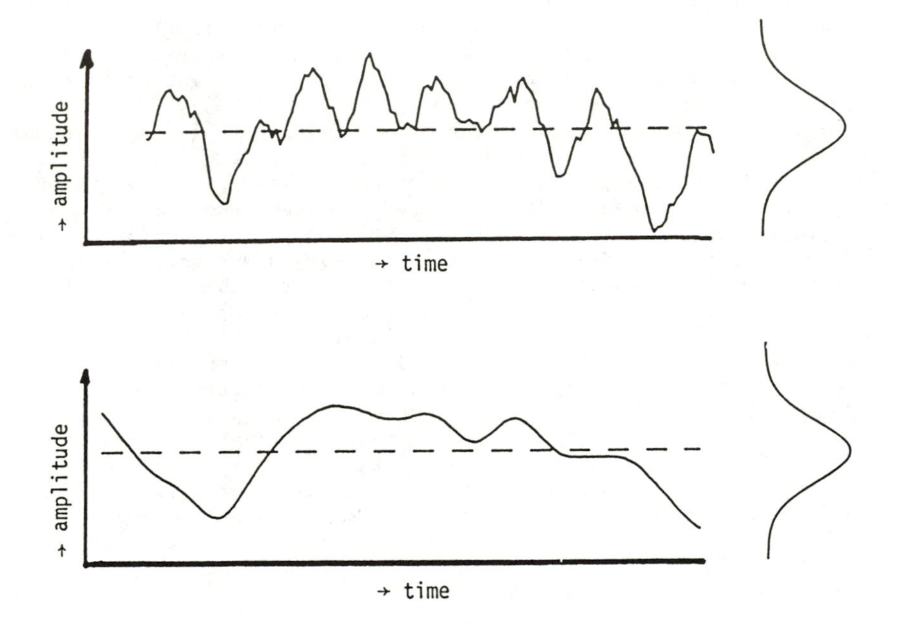

Figure 2. Fast and slowly fluctuating noise with similar PDF.

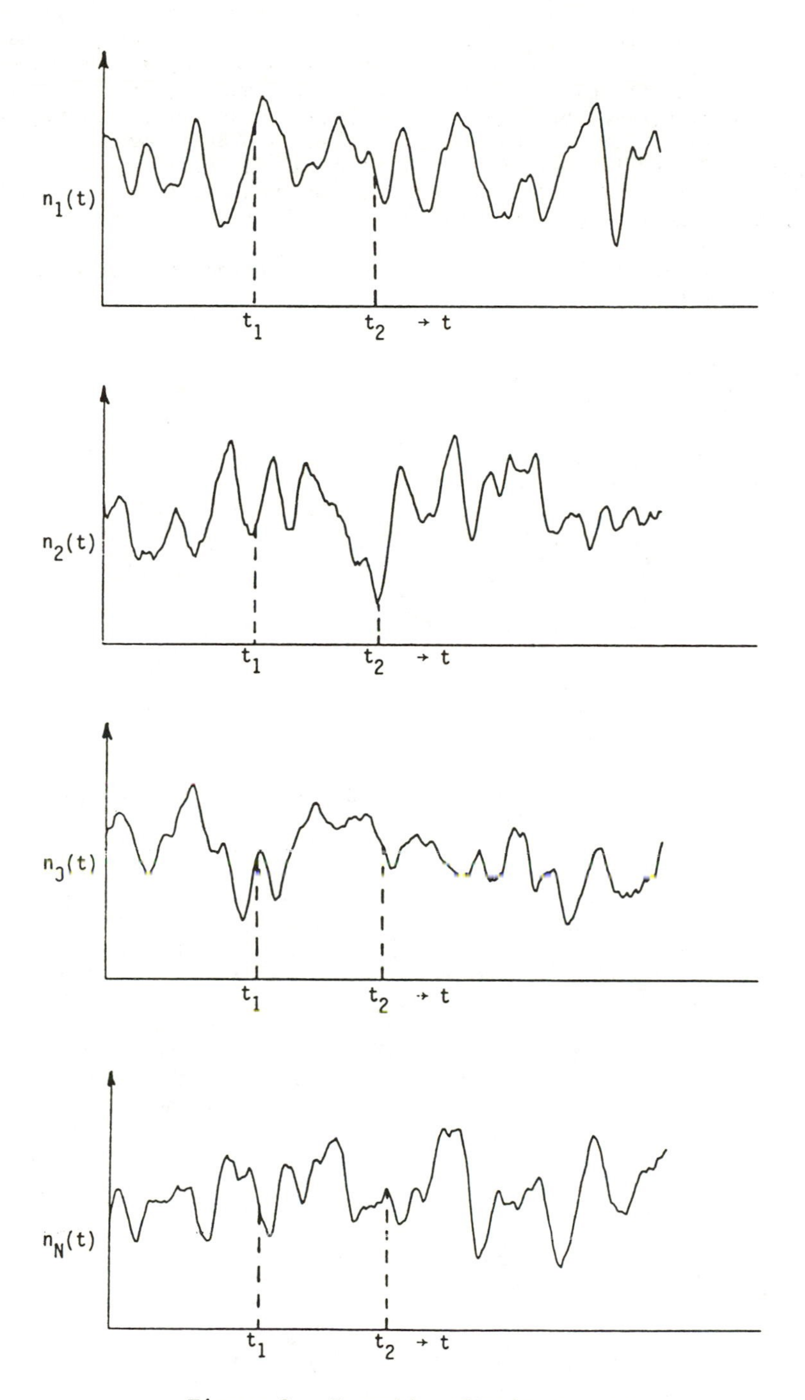

Figure 3. Ensemble of noise records.

A stochastic process $x_k(t)$ is stationary if the two processes $x(t)$ and $x(t+\varepsilon)$ have the same statistics for any ε. In other words, the values of t_1 and t_2 do not influence the ensemble statistics, only the difference $t_1-t_2 = \tau$ is important, $R(t_1,t_2)$ can be replaced by $R(t_1-t_2)$. $x_k(t)$ can be a set of noise records n_k or a set of other random variables. If the expected value $E[x(t)] = \mu =$ constant, and $E[x(t+\tau)x(t)] = R(\tau)$ is only dependent on τ and does not vary with the time (t_1-t_2 is replaced by τ), then the process is weakly stationary.
Ergodicity means: all statistics can be determined from a single function $x_k(t)$:

$$\mu_x(k) = \lim_{T\to\infty} \frac{1}{T} \int_0^T x_k(t)dt \tag{4}$$

and

$$R_{xx}(\tau,k) = \lim_{T\to\infty} \frac{1}{T} \int_0^T x_k(t)\, x_k(t+\tau)\, dt \tag{5}$$

If $\tau = 0$, Equation 5 reduces to:

$$R(0) = \lim_{T\to\infty} \frac{1}{T} \int_0^T x^2(t)dt = E[x^2(t)] \tag{6}$$

being the mean square value of $x(t)$, determining the average power and thus the energy of the signal.

Of course, fast and slowly fluctuating noise can also be distinguished in the frequency domain. However, noise usually goes on indefinitely in time and, actually, its energy is unbounded; the Fourier transform does not exist. Nevertheless, FT techniques can be applied if the average power is bounded. The introduction of the power spectral density function (PSD), not suitable for a simple representation of the (stochastic) signal, allows the introduction of expressions concerning signal energy. The PSD gives the average power per unit of frequency. Of course, the ACF and the PSD are not independent; strictly speaking, both are representing the same properties of the signal (energy, fast or slow fluctuation). The formal definition of the PSD is:

$$S(\omega) = FT\left\{R_{xx}(\tau)\right\} \tag{7}$$

Because $R_{xx}(\tau)$ is real and even for real processes, $S(\omega)$ is also real and even, and:

$$S(\omega) = \int_{-\infty}^{+\infty} R_{xx}(\tau)\cos\omega\tau\, d\tau = 2\int_0^{\infty} R_{xx}(\tau)\cos\omega\tau\, d\tau \tag{8}$$

Finally, the physical realizable one-sided PSD function $G(\omega)$ is defined as:

$$G(\omega) = 4 \int_0^{\infty} R_{xx}(\tau) \cos \omega\tau \, d\tau \tag{9}$$

The energy of the signal can be calculated from $G(\omega)$:

$$E[x^2(t)] = \int_0^{\infty} G(\omega) \, d\omega = R_{xx}(0) \tag{10}$$

$E[x^2(t)]$ is the mean square value ψ_x^2 of the signal $x(t)$. If the mean of $x(t)$ is not zero, for instance if $x(t)$ is a noise with a Direct Current (DC) component, then the mean (DC) value contributes to the total energy. However, we are only interested in the stochastic variations of the signal and not in the mean value which can be estimated and corrected. Therefore, in the following we assume a mean value of zero, in which case $E[x^2(t)]$ becomes the variance σ_x^2.

Some definitions from linear system theory are required. The weighting function $h(t)$ is the response (output) of a linear system applied to an impulse signal, theoretically a Dirac-delta function $\delta(t)$, or more precisely, a δ-distribution with the properties:

$$\int_{-\infty}^{+\infty} \delta(t) \, dt = 1$$

$$\delta(t) = 0 \qquad (t \neq 0) \tag{11}$$

The FT of the impulse response is the complex frequency response $H(j\omega)$. Suppose $x(t)$ with a PSD = $G_x(\omega)$ is the input signal of a linear system with a complex frequency response $H(j\omega)$ and suppose $y(t)$ is the resulting output. Then it can be proved (1) that the PSD of $y(t)$ is given by:

$$G_y(\omega) = |H(j\omega)|^2 G_x(\omega) \tag{12}$$

Variance of Integrated Noise

The system functions, mentioned in the previous paragraph, can be determined for an integrator. The response of an integrator applied to an impulse $\delta(t)$ is the value 1 $(t \geq 0)$, as follows from the definition of the δ-distribution (Equation 11). The FT can easily be calculated, resulting in:

$$H(j\omega) = FT\left\{h(t)\right\} = \int_{-\infty}^{+\infty} h(t) \, e^{-j\omega t} \, dt = \int_0^{\infty} e^{-j\omega t} \, dt = \frac{1}{j\omega} \tag{13}$$

According to Equation 12 we can calculate the PSD of the output, assuming a stochastic input signal with known PSD:

$$G_y(\omega) = |H(j\omega)|^2 G_x(\omega) = \frac{1}{\omega^2} G_x(\omega) \tag{14}$$

At first sight, it appears possible to calculate, without any problem, the variance of this output signal by integrating the calculated output PSD over the ω-range from zero to ∞ (see Equation 10):

$$k_{\sigma_I^2} \stackrel{?}{=} \int_0^\infty \frac{1}{\omega^2} G_x(\omega) d\omega \tag{15}$$

However, this is not correct, as already is shown in previous papers (6,7). A signal is never integrated from $-\infty$ to $+\infty$, but always during a limited time interval.
In reality, the impulse response h(t) of an integrator is not 1, but is given by:

$$\begin{aligned} h(t) &= 1 \qquad (0 \leq h(t) < T) \\ &= 0 \qquad (\text{else}) \end{aligned} \tag{16}$$

T is the integration interval.
Fourier transforming Equation 16 gives a different expression for $H(j\omega)$:

$$H(j\omega) = \int_{-\infty}^{+\infty} h(t)\, e^{-j\omega t}\, dt = \frac{1-e^{-j\omega T}}{j\omega} \tag{17}$$

The PSD of the output signal is now:

$$G_y(\omega) = |H(j\omega)|^2 G_x(\omega) = \frac{\sin^2 \omega T/2}{(\omega/2)^2} G_x(\omega) \tag{18}$$

The error variance is:

$$k_{\sigma_I^2} = \int_0^\infty \frac{\sin^2(\omega T/2)}{(\omega/2)^2} G_x(\omega)\, d\omega = 2T \int_0^\infty G_x(\omega) \frac{\sin^2 \omega T/2}{(\omega T/2)^2}\, d\,\frac{\omega T}{2} \tag{19}$$

One has to keep in mind that a signal with a PSD given by Equation 18 is not present at the output of a single integrator. It is necessary to consider a set of possible outcomes of k similar integration procedures, in other words, to use the already mentioned concept of an ensemble. The k in Equation 19 denotes an ensemble representation. As a simple application of Equation 19, baseline noise with a rectangular PSD, i.e. white noise with energy distributed uniformly over all frequencies bandlimited by an ideal low pass filter with rectangular passband, will be treated. In formula:

$$G(\omega) = K \qquad (0 \leq \omega \leq \omega_0)$$
$$G(\omega) = 0 \qquad (\omega > \omega_0) \tag{20}$$

K = constant.

This kind of noise is not very realistic, a true ideal filter is "real-time" impossible. However, such a spectrum can be approximated with a higher order filter with sharp cut-off characteristics.
The variance of the non-integrated baseline noise can be calculated, using Equation 10:

$$\sigma_n^2 = \int_0^\infty G(\omega)d\omega = \int_0^{\omega_0} G(\omega)d\omega \tag{21}$$

resulting in:

$$K = \frac{\sigma_n^2}{\omega_0} \tag{22}$$

Substituting Equation 22 and Equation 20 into Equation 19 we obtain:

$$^k\sigma_I^2 = \sigma_n^2 \cdot \frac{2T}{\omega_0} \int_0^{\omega_0 T/2} \frac{\sin^2(\omega T/2)}{(\omega T/2)^2} \, d(\omega T/2) \tag{23}$$

giving a quantitative expression for the variance of integrated noise.
Equation 23 does not look very attractive for routine use, but often it can be simplified. For example, $\omega_0 T/2$ has a relatively large value in chromatography, as will be proved. Let us assume chromatographic peaks with a Gaussian peak shape and standard deviation σ_p, determining the peak width. The minimum integration interval with an acceptable systematic error (< 1‰) in the area determination is about 7 σ_p. The frequency spectrum of the peak can be determined by Fourier transforming the Gaussian peak function. The result is also a Gaussian function in the frequency domain, however, with a standard deviation $\sigma_\omega = 1/\sigma_p$. To prevent unacceptable peak distortion, the minimum cut-off frequency ω_0 of a low pass filter has to be about $3\frac{1}{2}.\sigma_\omega$; higher frequencies can be neglected in case of a Gaussian PSD.
Hence:

$$\frac{\omega_0 T}{2} \approx \frac{\frac{7}{2} \cdot \frac{1}{\sigma_p} \cdot 7\sigma_p}{2} \approx 12 \tag{24}$$

In practice, ω_0 is determined by the peak with the smallest peak width. All the other peaks require a larger integration time and $\omega_0 T/2 > 12$. The contribution to the integral resulting from the interval between $\omega_0 T/2$ (≥ 12) and infinite is negligible. The value of the integral with integration limits 0 and ∞ is $\pi/2$. The final result is:

$$^k\sigma_I^2 = \sigma_n^2 \frac{\pi T}{\omega_0} \tag{25}$$

In this particular case the variance of the integrated noise is proportional to the integration interval and to the variance σ_n^2 of the original baseline noise. We note that σ_n^2 and ω_0 are not independent, reducing ω_0 means reducing σ_n^2. Noise with a strong $1/f$ character is much more realistic (7), particularly in chromatography. The PSD of $1/f$ (or $1/\omega$) noise is proportional to $1/\omega$. Because of the singularity in $\omega = 0$, a slightly modified model is more realistic. Such a PSD might be:

$$\begin{aligned} G(\omega) &= K/\omega_\ell \qquad 0 < \omega < \omega_\ell \\ G(\omega) &= K/\omega \qquad \omega_\ell < \omega < \infty \end{aligned} \tag{26}$$

ω_ℓ is a fixed (low) frequency.

Substitution in Equation 19 gives:

$$^k\sigma_I^2 = 2T \frac{K}{\omega_\ell} \int_0^{\frac{\omega_\ell T}{2}} \frac{\sin^2(\omega T/2)}{(\omega T/2)^2} d\omega T/2 + 2KT^2 \int_{\frac{\omega_\ell T}{2}}^{\infty} \frac{\sin^2(\omega T/2)}{(\omega T/2)^3} d(\omega T/2) \tag{27}$$

For low values of ω_ℓ the first term is approximately KT^2, the integral in the second term has to be calculated numerically. An important conclusion is that in case of $1/f$ (flicker) noise the variance σ_I^2 is proportional to T^2.

A treatment in the frequency domain is not always optimal. For instance, calculations with non-stationary stochastic processes are difficult. Moreover, the PSD is mostly determined by Fourier transforming the ACF, therefore an expression using directly the ACF avoids Fourier transforming. The derivation happens to be not difficult. Assuming a random signal $n(t)$ with mean value μ, we write the random variable:

$$I = \int_a^b n(t)dt \qquad E[n(t)] = \mu \qquad E[I] = \int_a^b \mu(t)dt \tag{28}$$

Now we write:

$$I^2 = \int_a^b n(t_1)dt_1 \int_a^b n(t_2)dt_2 \tag{29}$$

using two dummy (time) variables t_1 and t_2.
The integration limits are independent and the iterated integral can be written as a double integral:

$$I^2 = \int_a^b \int_a^b n(t_1)n(t_2)dt_1dt_2 \quad (30)$$

Taking the expected value and interchanging the expected value procedure and the integration gives:

$$E[I^2] = \int_a^b \int_a^b E[n(t_1)n(t_2)]dt_1dt_2 \quad (31)$$

$$= \int_a^b \int_a^b R(t_1,t_2)dt_1dt_2 = \sigma_I^2 \quad (32)$$

More or less naturally the ACF $R(t_1,t_2)$ appears, giving a generally usable expression. If, however, stationarity or ergodicity is assumed, then we get:

$$E[I^2] = \int_{-\frac{T}{2}}^{\frac{T}{2}} \int_{-\frac{T}{2}}^{\frac{T}{2}} R(t_1 - t_2)dt_1dt_2 \quad (33)$$

assuming an integration interval from $-\frac{T}{2}$ to $\frac{T}{2}$.
Equation 33 can be simplified to:

$$E[I^2] = 2\int_0^T (T - \tau)R(\tau)d\tau \quad (34)$$

$\tau = t_1 - t_2$.

The proof, being purely mathematical, is omitted (6).

As a final result we have two expressions (Equations 19 and 34) with certain restrictions quite usable in practice. One can prove that the expressions essentially are the same and they can be derived from one another.

Figures 4 and 5 show some typical types of noise: first order noise, i.e. white noise bandlimited by a simple first order filter with time constant T_1, and noise with a strong 1/f component, originating from an Inductively Coupled Plasma - Atomic Emission Spectrograph (ICP-AES).

The ACF of first order noise is:

$$R(\tau) = \sigma_n^2 \exp \frac{-|\tau|}{T_1} \quad (35)$$

Substitution in Equation 34 results in:

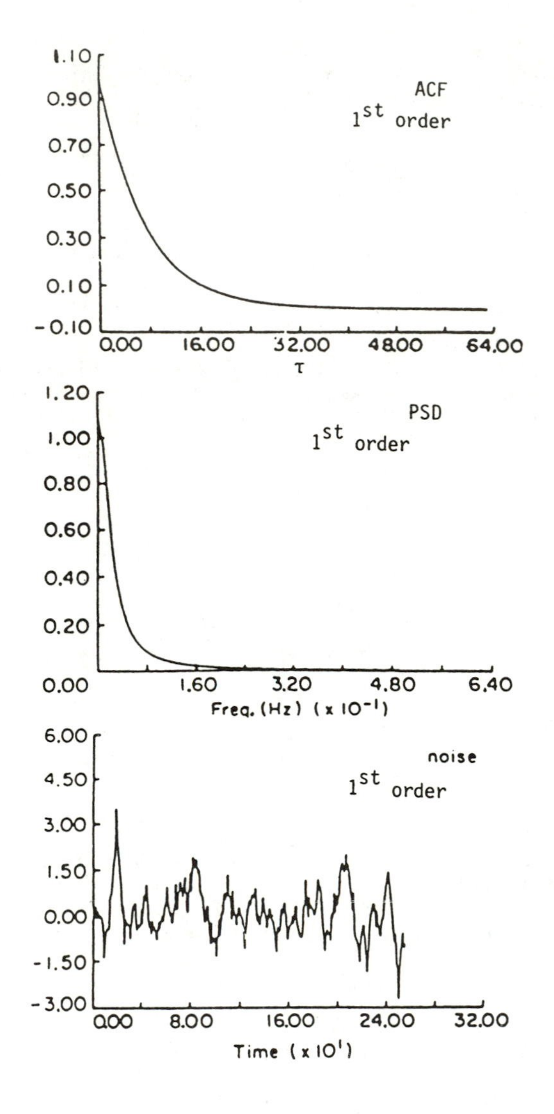

Figure 4. ACF, PSD and record of first order noise.

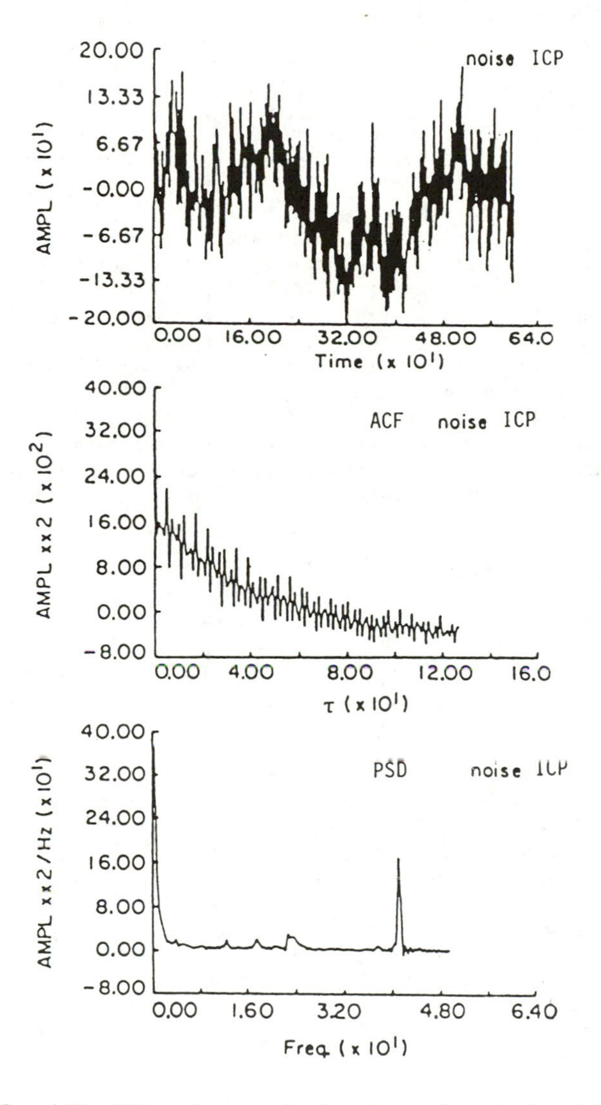

Figure 5. ACF, PSD and record of noise of an Inductively Coupled Plasma - Atomic Emission Spectrograph.

$$\sigma_I^2 = \sigma_n^2 \left[2TT_1 + 2T_1^2 \left\{ \exp\left(-\frac{T}{T_1}\right) - 1 \right\} \right] \tag{36}$$

In chromatography $T \gg T_1$, and the expression can be simplified to:

$$\sigma_I^2 \approx \sigma_n^2 \cdot 2TT_1 \tag{37}$$

Again, the variance is proportional to T, σ_n^2 and T_1 are not independent. Figure 6 shows the result of the computer calculation of the error variance as a function of the integration time. The origin of the noise is a Flame Ionisation Detector (FID). The ACF is estimated from a limited number of baseline noise data points, resulting in a confidence interval derived with the Bartlett formula (4,8).

An interesting question is: Is filtering prior to integration useful, particularly in case of peak-shaped signals? Of course, a decreasing cut-off frequency of a filter means a decreasing variance of the baseline noise and that is a favorable effect. However, a decreasing cut-off frequency results in a distorted peak; the peak width is increasing and implying that an increasing integration time is needed to avoid systematic errors; as is shown, this is not favorable. A study carried out with first order and second order filters has shown that filtering prior to integration is not advisable, as the second effect dominates (6).

Optimum Integration Limits

Another interesting question is: Is it possible to determine optimal peak integration intervals on the basis of known or even unknown peak shapes and known noise characteristics? And if an optimum integration interval can be estimated, what is the error variance?

As is extensively shown in a previous paper (9), it happens to be possible in some cases to determine optimum integration limits. As an example let us consider a symmetric peak with, for instance, a Gaussian peak shape with known peak maximum. Decreasing the integration interval means decreasing the random error in the peak area estimate, as is shown. But the systematic error is increasing; the peak is not completely integrated and the resulting area will be biased.

Figure 7 shows a signal x(t) composed of a peak s(t) (dashed line) and noise n(t). I_∞ is the true peak area and $\hat{I}_\infty$ is the peak area estimate, taken as the area of the noisy peak within the integration interval divided by the normalized integrated peak fraction $I_{norm}(T) = I_s(T)/I_\infty$, being the fraction of the peak area in the interval T divided by the true peak area. This is equal to the shaded part of the small peak with unit area.

We want to minimize $\mu_2(T)$, i.e. the expected value of the squared difference between the true area and the estimate:

$$\mu_2(T) \equiv E\left[I_\infty - \hat{I}_\infty\right]^2 = \frac{\sigma_{I_n}^2(T)}{I_{norm}(T)^2} = \text{error variance} \tag{38}$$

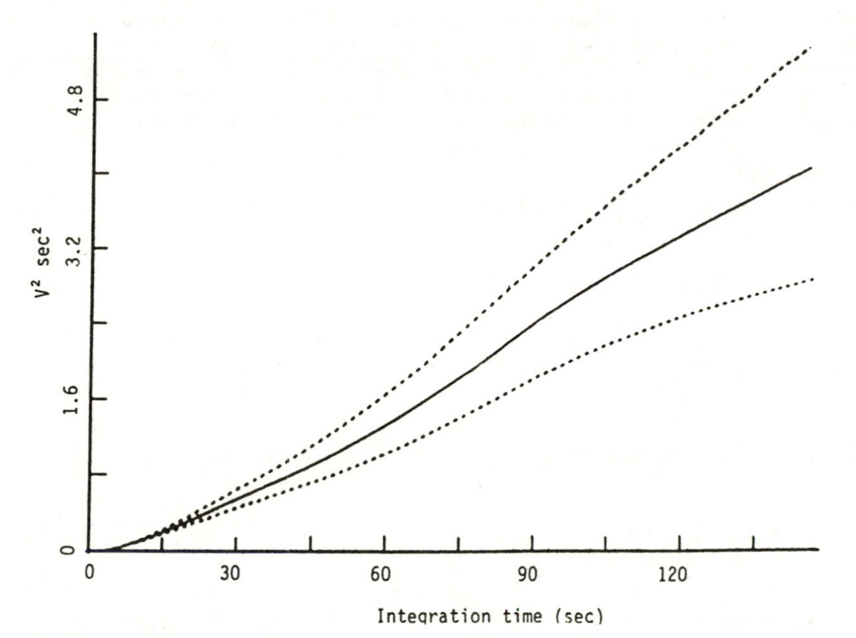

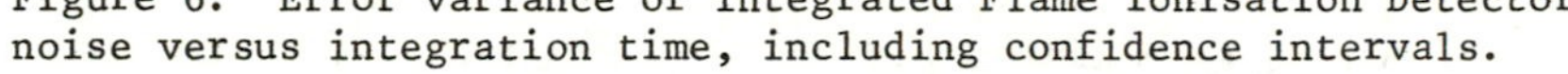
Figure 6. Error variance of integrated Flame Ionisation Detector noise versus integration time, including confidence intervals.

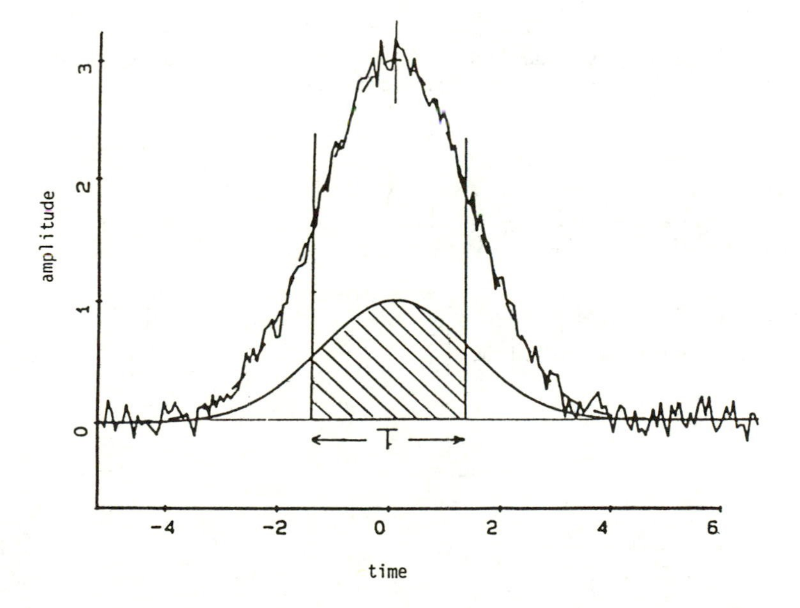

Figure 7. Noisy peak and normalized peak with optimum integration interval.

$\mu_2(T)$ has to be minimized with respect to T. The normalized integrated peak area in case of a known peak shape is known, and we have derived an expression for the integrated noise variance. As usual, we can determine the derivative, setting it equal to zero and the desired result can be calculated. This procedure leads to:

$$I'(T)\,\frac{d\sigma^2_{I_n}(T)}{dT} = 2\,\sigma^2_{I_n}(T)\,\frac{dI'(T)}{dT} \qquad (39)$$

In case of symmetric peaks:

$$I'(T) = \frac{I(T)}{I(\infty)} = \frac{1}{I_\infty}\int_{u-\frac{1}{2}T}^{u+\frac{1}{2}T} s(t)d(t) \qquad (40)$$

An expression for the variance is known (Equation 34):

$$\sigma^2_{I_n}(T) = 2\int_0^T (T-\tau)\,R_{nn}(\tau)d\tau \qquad (41)$$

Hence, using Equations 40 and 41 Equation 39 can be written as:

$$\int_{u-\frac{1}{2}T}^{u+\frac{1}{2}T} s(t)dt\,.\int_0^T R_{nn}(\tau)d\tau = 2\int_0^T (T-\tau)R_{nn}(\tau)d\tau\,.\,[s(u+\tfrac{1}{2}T)] \qquad (42)$$

Evaluation generally leads to a minimum for some value of T. Similar equations can be derived for asymmetric peaks (9).

The final result in case of a Gaussian peak with first order noise is:

$$\mathrm{erf}\left(\frac{T}{2\sigma_p\sqrt{2}}\right) = \sqrt{\frac{2}{\pi}}\,.\,\frac{T}{\sigma_p}\,\exp\left(-\frac{T^2}{8\sigma_p^2}\right) \qquad (43)$$

σ_p = standard deviation of the peak.

This relation is satisfied if:

$$T_{opt} \approx 2.8\,\sigma_p \qquad (44)$$

The corresponding error variance is:

$$\mu_2(T_{opt}) = \frac{5.6\,\sigma_n^2\,T_x\,\sigma_p}{\mathrm{erf}^2(0.99)} \approx 8\,\sigma_n^2\,T_x\,\sigma_p \qquad (45)$$

T_x = time constant of the first order noise.

A remarkable result is that in this particular case the optimum integration interval is independent of the time constant and of the variance of the noise. Of course, the resulting estimation error depends on both parameters. The theory can be extended to skewed peaks with known or unknown shape, other kinds of noise, etc. A detailed treatment is given in (9).

Integration Variance after Baseline Correction

The expressions derived so far are only valid if the noise is assumed to be stationary. However, it is unfortunate that this is not always the case. Particularly in a technique like chromatography, a non-stationary baseline drift is often present due to, for instance, stripping of the column, contamination of detectors, etc. The non-stationary drift, not to be confused with stationary low frequency noise with properties defined in probalistic terms, can often be considered as a deterministic function (linear, exponential, etc.).

Baseline drift correction is an indispensable part of a good chromatographic data processing procedure. The following questions have to be answered:

- What is the influence of uncorrected drift on the estimated ACF, which is used to calculate statistical quantities like the integration error variance?
- What is the integration variance after drift correction?

A complete treatment is beyond the scope of this paper, but an introduction with a simplified, but practically relevant example, will be given here. The simplest case is a linear baseline drift and the resulting model of the noisy drifting baseline is:

$$x(t) = n(t) + a + bt \tag{46}$$

where a and b are constants.
A finite measurement time from $-\frac{T}{2}$ to $\frac{T}{2}$ is chosen, which is not important for the derivation.
The estimation of the ACF is now defined as:

$$R^*_{xx}(\tau) = \frac{1}{T-\tau} \int_{-\frac{T}{2}}^{\frac{T}{2}-\tau} \Big(n(t) + a + bt\Big)\Big(n(t+\tau) + a + b(t+\tau)\Big)dt \tag{47}$$

Evaluation of this integral gives:

$$R^*_{xx}(\tau) = \underbrace{R_{xx}(\tau)}_{\text{true ACF}} + \underbrace{a^2 + \frac{b^2(T^2 - 2T\tau - 2\tau^2)}{12}}_{\text{systematic error}} + \underbrace{aP(\tau) + bQ(\tau)}_{\text{random error}} \tag{48}$$

P and Q are stochastic functions of τ, both with an expected value of zero:

$$P(\tau) = \frac{1}{T-\tau} \int_{-\frac{T}{2}}^{\frac{T}{2}-\tau} n(t)dt + \frac{1}{(T-\tau)} \int_{-\frac{T}{2}}^{\frac{T}{2}-\tau} n(t+\tau)dt \quad (49)$$

$$Q(\tau) = \frac{1}{T-\tau} \int_{-\frac{T}{2}}^{\frac{T}{2}-\tau} n(t)\ t\ dt + \frac{1}{(T-\tau)} \int_{-\frac{T}{2}}^{\frac{T}{2}-\tau} n(t)dt + \frac{1}{T-\tau} \int_{-\frac{T}{2}}^{\frac{T}{2}-\tau} n(t+\tau)t\,dt \quad (50)$$

However, assuming $T >> \tau$, their variance can be determined as the rather simple expression:

$$E[P^2(\tau)] \approx \left\{ 8 \int_0^{T-\tau} (T-\tau-t)\ R_{xx}(t)dt \right\} / (T-\tau)^2 \quad (51)$$

and

$$E[Q^2(\tau)] \approx \left(2(T-\tau)/3\right) \int_0^{\frac{T}{2}} R_{xx}(t)dt \quad (52)$$

As an example, the equations for first order noise will be given:

$$E[P^2] = 8\sigma_n^2 \tau_n/(T-\tau) \quad (53)$$

$$E[Q^2] = 2\sigma_n^2 \tau_n(T-\tau)/3 \quad (54)$$

A close look at Equation 48 leads to the following conclusion concerning the effect of uncorrected linear drift. The estimated ACF contains two systematic components, each proportional to a^2 and b^2 respectively, and two stochastic components, proportional to a and b. A final conclusion can be derived from the formulae: a considerable error in the estimation of the ACF and derived quantities can be expected if baseline drift is not corrected. This leads us to the remaining question, the determination of the integration variance after baseline drift correction.

Many correction procedures are known: linear and exponential fitting, polynomial approximation (both orthogonal and non-orthogonal), etc. In this paper the description will be restricted to the very often used linear extrapolation, again assuming a linear baseline drift. Let us consider a noisy peak with a linear drifting baseline. The usual procedure is as follows (Figure 8). Two time intervals with a time duration T_c are selected on both sides of the peak. Now each interval is fitted with a straight line. To simplify the equations the following integrals are defined:

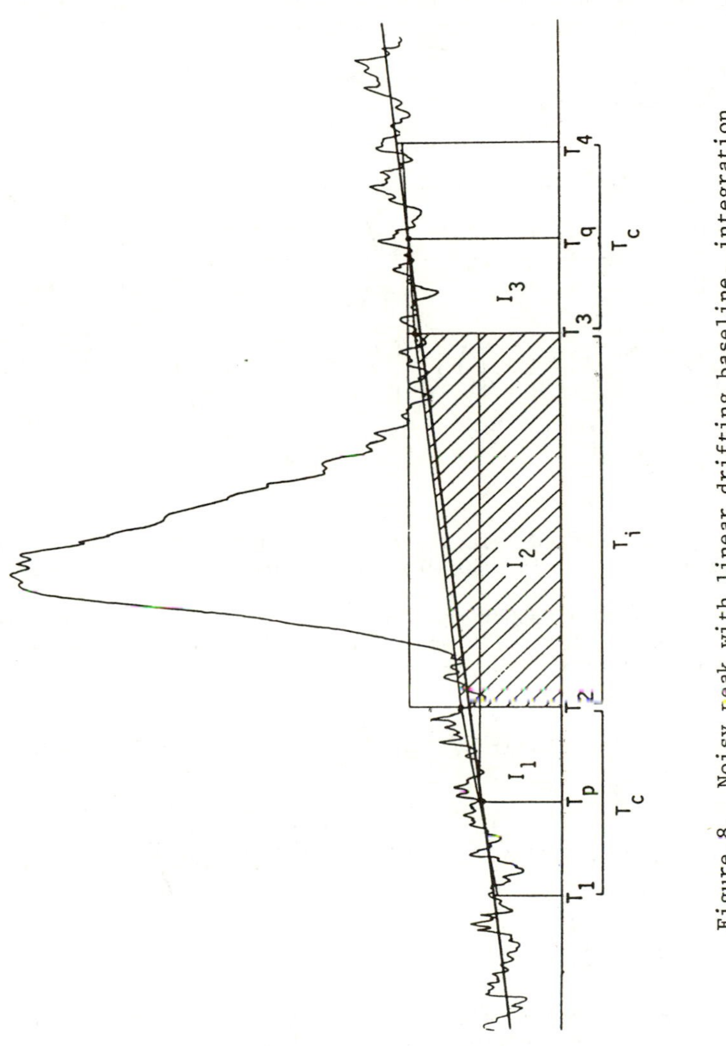

Figure 8. Noisy peak with linear drifting baseline, integration interval T_i and correction intervals T_c.

$$I_1 = \int_{T_1}^{T_2} x(t)dt \qquad I_2 = \int_{T_2}^{T_3} x(t)dt \qquad I_3 = \int_{T_3}^{T_4} x(t)dt \tag{55}$$

The length of the integration intervals are T_c, T_i and T_c, respectively. The baseline drift corrected integral is:

$$I = I_2 - \frac{T_i(I_1 + I_3)}{2T_c} \tag{56}$$

The variance of the corrected integral can be calculated:

$$E[I^2] = E\left[I_2^2 - \frac{(I_1.I_2 + I_2.I_3)T_i}{T_c} + \left(\frac{(I_1 + I_3)T_i}{2T_c}\right)^2\right] =$$

$$= E[I_2^2] + \left(E[I_1^2] + E[I_3]^2\right).\left(\frac{T_i}{2T_c}\right)^2 - \left(E[I_1.I_2] + E[I_2.I_3]\right)\frac{T_i}{T_c} +$$

$$+ E[I_1.I_3]\frac{T_i^2}{2T_c^2} \tag{57}$$

The first three expected values can be determined by already described ordinary "σ_I^2" calculations. The other three are actually cross terms.

Evaluation leads to a rather complicated formula, but nevertheless it is generally usable for all kinds of noise with known ACF. As an example the result for first order noise will be given:

$$\sigma_I^2 = \sigma_n^2 \left\{ 2T_i\tau_n - 2\tau_n^2\left(1 - \exp(-T_i/\tau_n)\right) - \left(\frac{T_i^2\tau_n^2}{2T_c^2}\right)\left\{\left(1 - \exp(-T_c/\tau_n)\right)^2 . \right.\right.$$

$$\left.\exp(-T_i/\tau_n) + 2\left(1 - \exp(-T_c/\tau_n) - T_c/\tau_n\right)\right\} - \left(2\frac{T_i\tau_n^2}{T_c}\right)\left(1 - \exp(-T_i/\tau_n)\right).$$

$$\left.\left(1 - \exp(-T_c/\tau_n)\right)\right\} \tag{58}$$

Equation 58 is graphically displayed in Figure 9, showing σ_I as a function of T_i with the correction interval as a parameter.

A remarkable result is that the curves are crossing.

A careful inspection of Equation 58 and Figure 9 leads to the following statement: If a signal with linear drifting baseline and first order baseline noise is integrated, then the optimum baseline correction interval is infinite if the integration time is greater than four times the time constant of the noise; otherwise, the optimum correction interval is zero. In the last case the use of two correction points on both sides of a peak is sufficient.

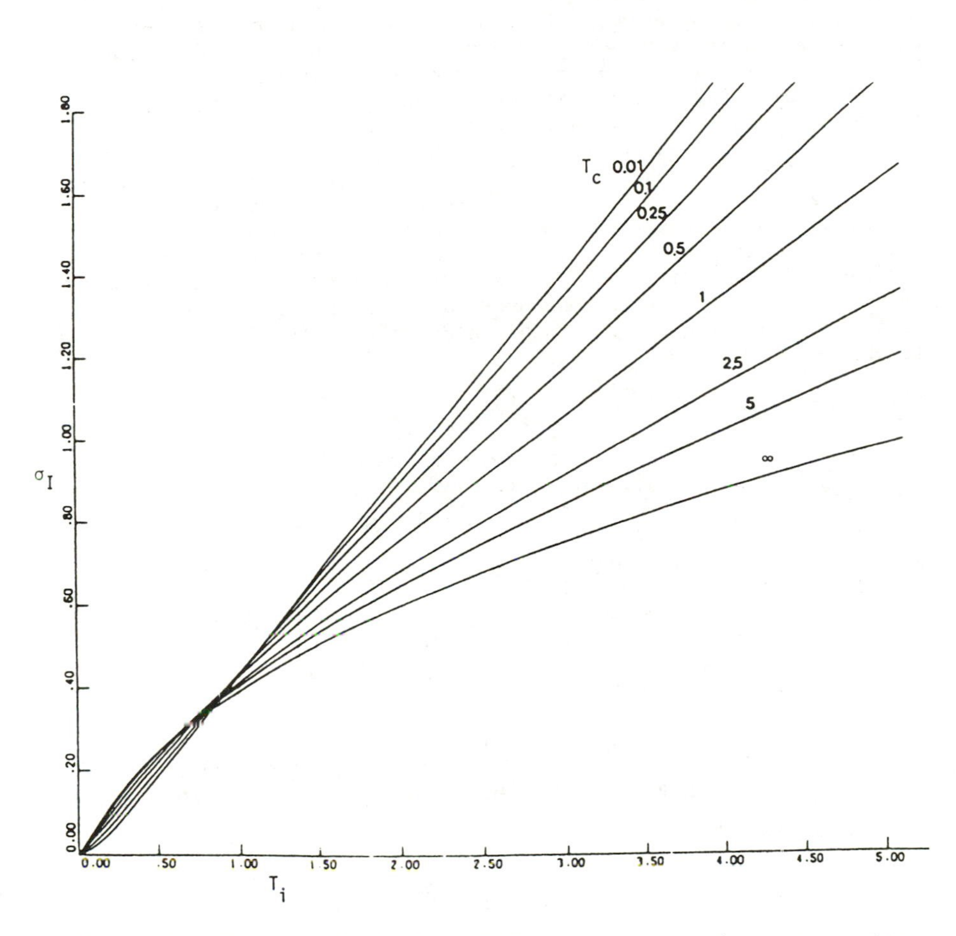

Figure 9. Standard deviation of the integrated noise after drift correction versus the integration time, with the correction interval width as a parameter.

Discussion and Conclusions

In case of a stable and stationary chromatographic system, the derived general theory is certainly usable for calculating quantitatively uncertainties and detection limits in signal integrating methods like chromatography. However, an extensive analysis of the detector noise is required and the use of a computer with a data acquisition system and special software is inevitable.

If a drifting baseline is present and possibly corrected, the formulae become rather complicated and are not directly usable in daily practice. An extension to other kinds of noise, i.e. the more realistic 1/f or flicker noise leads to even more complicated formulae. Nevertheless, it is possible to determine quantitatively detection limits in case of the application of some specific baseline drift correction procedure, if the measurement conditions are well defined and stable and good models of the noise and drift are known from an a priori analysis. Moreover, a better insight in several effects influencing the remaining uncertainty after drift correction is obtained.

Literature Cited

1. Bendat, J. S.; Piersol, A. G. Measurement and Analysis of Random Data; Wiley: New York, 1966.
2. Box, G. E. P.; Jenkins, G. M. Time Series Analysis; Holden-Day: San Francisco, 1976.
3. Papoulis, A. The Fourier Integral and its Applications; McGraw-Hill: New York, 1962.
4. Jenkins, G. M.; Watts, D. G. Spectral Analysis and its Applications; Holden-Day: San Francisco, 1969.
5. Beauchamp, K.; Yuen, C. Digital Methods for Signal Analysis; Allen & Unwin: London, 1979.
6. Smit, H. C.; Walg, H. L. Chromatographia 1975, 8, 311.
7. Smit, H. C.; Walg, H. L. Chromatographia 1976, 9, 483.
8. Duursma, R. P. J.; Smit, H. C. Anal. Chim. Acta 1981, 133, 67.
9. Laeven, J. M.; Smit, H. C. Anal. Chim. Acta 1985, 176, 77.

RECEIVED January 21, 1987

Chapter 8

Establishing Clinical Detection Limits of Laboratory Tests

Mark H. Zweig

Clinical Pathology Department, Clinical Center, National Institutes of Health, Bethesda, MD 20892

Fundamental clinical laboratory test performance can be described in terms of accuracy, or the ability to correctly classify subjects into clinically relevant subgroups. Receiver operating characteristic (ROC) curves demonstrate the limits of a given test to detect the alternative states of interest over the complete spectrum of operating conditions, providing a comprehensive and pure index of accuracy. Obtaining valid data for ROC analysis requires attention to the following important steps: (1) define carefully the specific clinical question to be addressed; (2) choose subjects who are representative of the population to which the test is ultimately to be applied; (3) perform all tests being evaluated on all subjects; (4) determine the "true" diagnosis by rigorous and complete means independent of the test(s) being studied; and (5) evaluate and compare test performance at all decision levels using ROC curves.

Swets and Pickett (1) divide test performance into a discrimination or accuracy aspect and a decision or efficacy aspect. Accuracy, on the one hand, refers to the ability of the test to classify, to correctly discriminate between alternative clinical states of the subjects under study (i.e., signals vs. noise, disease vs. non-disease, chest pain with myocardial infarction vs. chest pain without infarction, blood in stools due to malignancy vs. blood in stools from other conditions). This is accuracy or correctness relative to truth, as best as we can determine that truth. We can express accuracy as clinical sensitivity and specificity. Efficacy, on the other hand, is a measure of the actual practical value of the diagnostic information or classification - how much benefit the test provides relative to its risks and costs. Evaluating or optimizing efficacy involves decision theory and consideration of the complexities of clinical utility, rather than just accuracy.

This is part of a symposium on detection limits. In this paper I will consider limits in terms of clinical detection rather than analytical detection. By clinical detection I mean accuracy or the discriminating ability referred to in the preceding paragraph. This ability of a test, expressed as sensitivity and specificity, is nicely described and appreciated using the receiver operating characteristic (ROC) curve because it provides a pure index of accuracy, of discrimination ability. It deals with signal detection and the ability to distinguish signal from noise. The index of accuracy provided is independent of any decision criterion which might be applied or of any bias which the system might have toward one decision or another. Thus the decision aspect, which involves costs, benefits, and outcomes is separated out so as not to confound the assessment of the intrinsic ability of the test to discriminate among various states. The influence of various decision factors (prevalence, utilities) on the operation and ultimate efficacy of the test is addressed by clinical decision analysis. The formal tool of clinical decision analysis joins the estimates of the probabilities of test outcomes (true positives, false positives, etc.) provided by ROC analysis with decision factors so as to establish the decision criterion for tests and to choose the set and order of diagnostic and therapeutic steps to be taken to optimize the outcome in terms of years of life, quality of life, costs, resource utilization, etc. (2-3).

The basic job of a clinical laboratory test is to provide information about the clinical state of patients for healthcare management purposes. The goal then is to subdivide or classify seemingly similar subjects into clinically relevant management subgroups. Suppose we are talking about people who come to an emergency room with acute chest pain. Some will turn out to be having a heart attack and some won't. Laboratory tests help divide or classify those patients into subgroups - that is, lab tests help to distinguish those who probably are having a heart attack from those who aren't. The question is, what is the limit of the ability of the test to identify or detect subjects having a heart attack among those with chest pain? What are the limits of the test's powers to detect accurately the clinical state of each individual in the group? This is a signal detection theory issue.

Most diagnostic tests are imperfect and, particularly when we use a binary approach - results are either "positive" or "negative" - there are some misclassification errors, inaccuracies. Some subjects with the condition of interest will be missed or some without the condition will be mistakenly considered affected, or both will happen. The ability of a test to properly identify or classify subjects or conditions of interest can be expressed as the sensitivity and specificity of the test. For clinical purposes these are defined as follows:
SENSITIVITY (TRUE POSITIVE RATE): Fraction of all affected

subjects in whom the test result is positive; "test positivity in the presence of the disease." SPECIFICITY (TRUE NEGATIVE RATE): Fraction of all unaffected subjects in whom the test result is negative; "test negativity in the absence of the condition." These inaccuracies in terms of sensitivity and specificity can be statistically represented by the ROC curve.

This paper will discuss basic test performance in terms of accuracy, but will not deal with actual application of a test. The latter involves choosing decision levels (i.e., reference values, cut-offs, normal limits, etc.) and involves measures of utility which are beyond the scope of fundamental test performance. I will describe a set of principles or elements important for evaluating test performance and comparing tests to one another (4,5). I will particularly emphasize the power and convenience of ROC curves, an extremely effective tool for assessing and comparing tests (2,3,6-8). While the power and usefulness of ROC curves has been recognized and discussed by members of various biomedical disciplines in recent years, this tool has received little attention from the clinical laboratory community.

Signal/Noise Discrimination: Historical Perspectives

The ROC curve apparently had its origins in electronic signal detection theory. Much of this arose in the 1940's and 1950's from analysis of radar systems. During WWII, radar operators watched screens for blips which might indicate enemy aircraft for the purpose of deciding when to mobilize fighter squadrons to intercept. The problem was to distinguish between signals from hostile planes and noise from clouds, flocks of birds, etc.

They realized that in interpreting the radar signals they saw there was always a trade-off between sensitivity and specificity - as the sensitivity increased so did the rate of false positives. That is, if they lowered the threshold for which blips they interpreted as signifying enemy planes, they falsely identified clouds and migrating birds, etc., as planes more often. Specificity declined and they scrambled interceptor squadrons unnecessarily. On the other hand, raising the threshold for calling a blip "positive" (enemy bombers) meant not responding to the arrival of enemy aircraft in some instances (false negatives). They were experiencing the trade-off between sensitivity and specificity inherent in test systems.

Figure 1 shows hypothetical signals and noise in the form of peaks. Imagine this is radar information and the real planes give peaks I, II, and III. If interceptor planes are sent up when the signal exceeds criterion C, then two real signals, I and II, will be missed. However, if criterion A is used so as to catch all three real signals of enemy aircraft, a number of noise artifacts will be incorrectly classified as positives (false positives).

Signal/Noise Discrimination in the Clinical Laboratory

Figure 2 illustrates this in the form of serum myoglobin concentrations obtained 5 hours after the onset of chest pain from patients admitted to a coronary care unit with the suspicion of myocardial infarction. This test has been proposed by some as a marker for heart attacks. Some of these patients turned out to have a heart attack (solid bars) and some didn't (hatched bars). Because of the overlap between "signals" and "noise," any decision criterion we choose will result in some misclassifications. We could choose any of various decision levels, each giving a different sensitivity/specificity combination - all of the possible combinations comprising the trade-offs available with this test. This spectrum of trade-offs constitutes the detection limit of this test and is represented by the ROC curve.

We have defined the ability to identify affected individuals as sensitivity and the ability to recognize unaffected individuals as specificity and can express these abilities as percentages or decimal fractions. A perfect test would exhibit both a sensitivity and specificity of 100% or 1.0. Tests are rarely perfect. It would be rather unusual for a test to exhibit a sensitivity and a specificity of 100% at the same time. Often we hear or read that a particular test has a particular sensitivity or specificity. In reality, as noted with radar and serum myoglobin, there isn't just one sensitivity or specificity for a test, but rather a continuum of sensitivities and specificities. By varying the decision level (or "decision point," "upper-limit-of-normal," "cutoff value," "reference value," etc.), any sensitivity from 0 to 100% can be obtained. Each of these sensitivities will have a corresponding specificity. Sensitivity and specificity occur, then, in pairs. The test's accuracy is reflected in the pairs that can occur; not all pairs are possible for a particular test. A given test will have one set of sensitivity-specificity pairs in one clinical situation, but may have a different set of pairs when applied to another clinical situation where the group tested is different.

The spectrum of pairs exhibited by a test in a given clinical setting characterizes or describes the accuracy of the test. Often test users implicitly assume one sensitivity-specificity pair characterizes a test because they accept a conventional, often arbitrarily chosen, upper-limit-of-normal as the single correct decision level for that test for all circumstances. They accept the corresponding sensitivity-specificity pair as the correct one for the test. This, however, is actually only one of multiple possible operating points for the test. When the concept of varying the decision level (operating point) to generate a spectrum of sensitivity-specificity pairs is understood, then the issue becomes: How good are the pairs? Also, which pair(s) works the best for the circumstances in which the test is to be used?

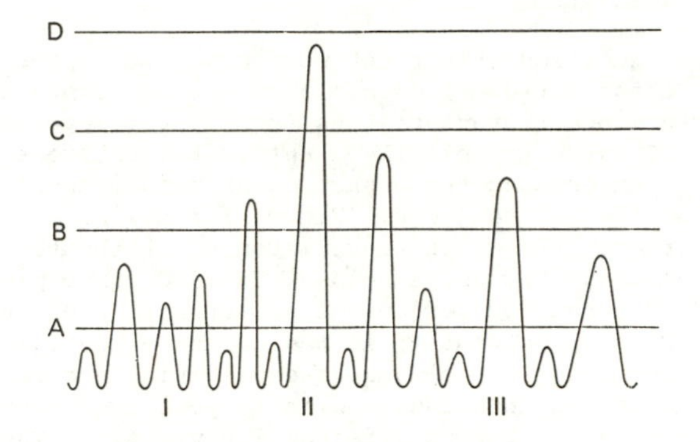

Figure 1: Diagram of a hypothetical set of peaks from a radar receiver. Peaks I, II, and III represent signals from aircraft, while all other peaks represent noise. Lines A, B, C, and D represent increasing decision level thresholds, which results in successively lower true- and false- positive rates.

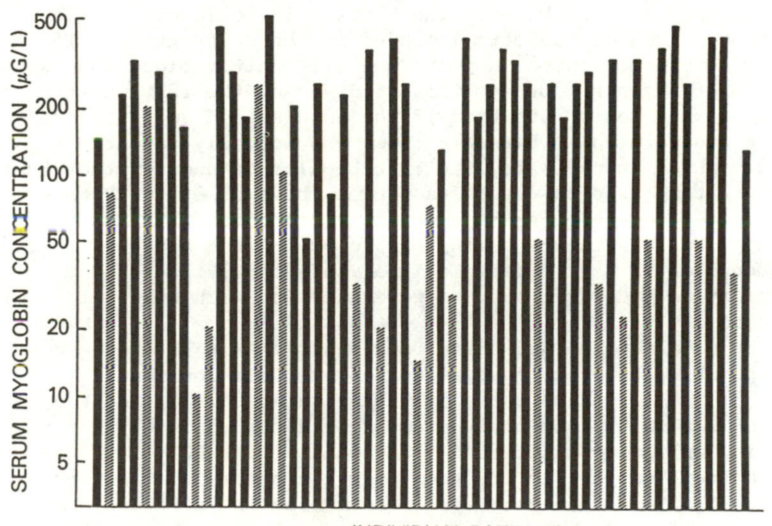

Figure 2: Serum myoglobin concentrations for 54 patients with chest pain admitted to a coronary care unit. Myoglobin was measured 5 hours after the onset of pain. Solid bars: acute myocardial infarct. Crosshatched bars: no infarct.

ROC Curves: Derivation

To answer these questions, we first need a way to represent and deal with all these different possible operating points and their resultant performance characteristics (sensitivity/specificty pairs). The ROC curve graphically displays the entire spectrum of a given test's performance for a particular sample group of affected and unaffected subjects. Figure 3 contains a hypothetical frequency distribution histogram at the top and the and the corresponding ROC curve below. The ROC curve plots the true positive (TP) rate or percentage as a function of the false positive (FP) rate or percentage as the decision level is varied. The true positive rate is the same as sensitivity and is equal to the number of affected individuals with a "positive" result divided by the total number of affected individuals. The true positive rate is also equal to 1-false negative (FN) rate. The false positive rate is the fraction of unaffected individuals who nevertheless have a "positive" test result, and is therefore related to specificity, or the ability of the test to correctly identify unaffected individuals (specificity = true negative (TN) rate = number of unaffected individuals with "negative" results/ total number of unaffected individuals = 1-false positive rate).

Both the TP and FP rates depend on the decision level chosen. Both rates also depend on the clinical setting, as reflected by the study population chosen. The FP rate is influenced by the type of nondiseased subjects included in the study group. If, for example, the nondiseased subjects are all healthy blood donors who are free of any signs or symptoms of disease, the test may appear to have a much lower rate than if the nondiseased subjects are persons who clinically resemble those who actually have the disease. Like the FP rate, the TP rate also depends on the study group. A test used to detect cancer may have a higher TP rate when applied to patients who have active or advanced disease than when applied to patients having stable or limited disease. This dependence of TP and FP rates on the study population is the reason why an ROC curve must be generated for each clinical situation.

Each point on the ROC curve represents a pair of true and false positive rates corresponding to some decision level. In Figure 3, the left hand curve of the frequency histogram (top) represents results from unaffected individuals and the right hand curve is derived from affected individuals. The ROC curve is derived from the data in the frequency histogram, so the first step is to obtain the test results from both the affected group and the unaffected group. True positive rates are calculated using the results from the affected individuals, while false positive rates are generated from the unaffected individuals' data. The ROC curve is constructed by varying the decision level from the highest test result down to zero, resulting in true and false positive rates which vary continuously. The decision level at point a in Figure 3 is higher than any observed results (see top), so at that decision level none of the results are "positive" and both true and false positive rates are zero (see bottom). As the

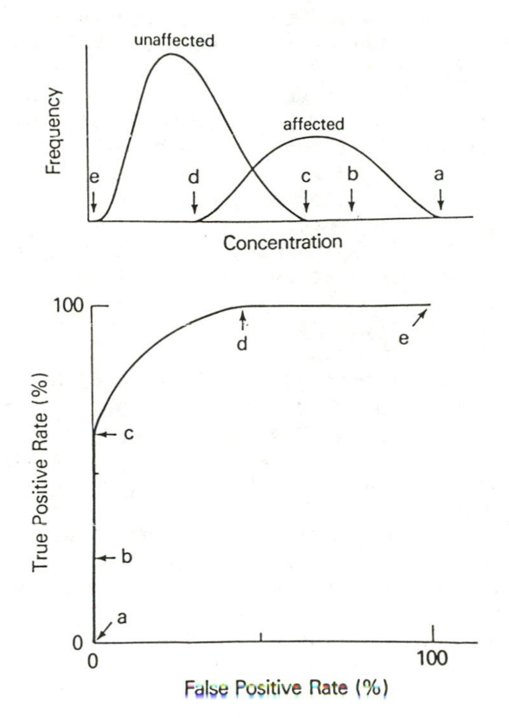

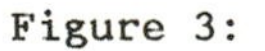

Figure 3: Top: Hypothetical frequency distribution curve. Bottom: Receiver operating characteristic (ROC) curve corresponding to data in top panel, generated by varying the decision level and then plotting the resulting pairs of true and false positive rates. Arrows at a to e mark points corresponding to decision levels in top panel. The curve from c to d describes the test's performance in the crucial overlap region.

decision level is lowered from a to b, some of the affected individuals have positive results but none of the unaffected individuals do, so the true positive rate rises while the false positive rate remains zero. Point c shows the highest true positive rate achievable (with this data) with the false positive rate still at zero. This is the edge of the overlap region (c to d). At c the ROC curve leaves the Y axis because if the decision level is lowered any further, some unaffected individuals have falsely positive results. At decision level d, all affected individuals have positive test results, so the true positive rate reaches 100%, at the expense of some percentage of false positives. This is the other edge of the crucial overlap region. The portion of the curve from c to d (where it has left the Y axis but not yet intercepted the true positive = 100% horizontal line) describes the overlap region. From decision level d to e, false positive rates increase as more and more results from unaffected individuals are incorrectly classified as positive.

ROC Curves: Interpretation

The complete ROC curve summarizes the clinical accuracy of the test by displaying the paired true and false positive rates for all possible decision levels. Good clinical performance of a test is characterized by a high true positive rate and a low false positive rate. Accordingly, as test performance improves, the ROC curve will move upward (toward higher true positive rates) and to the left (toward lower false positive rates). A perfect test would achieve a 100% true positive rate with no false positives. Thus, its ROC curve would rise vertically to the (0,100) point in the upper left corner and then move horizontally to the right along the horizontal line representing true positive rate = 100% to the (100,100) point in the upper right corner. Conversely, for a clinically useless test, which gives similar results for subjects with and without the condition, the true and false positive rates would be identical for any given decision level. Therefore, the ROC curve would be a diagonal between the lower left and upper right corners, representing the line where the true positive rate always equals the false positive rate.

Because the curve is usually above the diagonal, it starts out at the lower left with the TP rate (sensitivity) increasing faster than the false positive rate. At some point the slope begins to fall and the false positive rate starts increasing faster than the true positive rate - in other words, gains in sensitivity come at the cost of increasingly larger costs in terms of nonspecificity. This imposes a practical limit on the usable sensitivity of the test - where that limit is depends on the relative utility or benefits and the costs of true and false results and gets us beyond detection and into decision issues.

The ROC curve can also be constructed as a plot of true positive rate (sensitivity) versus true negative rate (specificity) instead of versus false positive rate

(1-specificity). This produces a mirror image of the curve shown in Figure 3, flipping the curve to the right side with the perfect point being the upper right hand corner instead of the upper left hand corner.

The ROC curve, then, provides a comprehensive picture of the test's accuracy at all possible operating points (decision levels). It does this without the need to choose a decision level or establish a normal range in advance.

Comparing Tests

Besides being valuable in evaluating a single test by demonstrating the complete spectrum of its intrinsic performance, the ROC curve is extremely useful in comparing tests to one another. Even if we are evaluating only a single new test, comparisons to existing tests are often inherent in the evaluation process. ROC curves provide an elegantly simple means for demonstrating the relative accuracy of multiple tests, comparing them at every TP rate by plotting the ROC curves for all the tests on the same graph. If the ROC curve for one test is uniformly above and to the left of the ROC curve for a second test, the first test will have a lower FP rate than the second test has for any given TP rate.

The ROC curves of Figure 4 illustrate the ambiguity involved in comparing tests at just one decision level or operating point. Consider the case in which test A has a TP rate of 98% and a FP rate of 30%, while test B has a TP rate of 70% and a FP rate of 2%. If the clinical performance of the two tests were equivalent, they would share a single ROC curve. This situation is illustrated in Figure 4, left. Test B could have achieved the same TP and FP rates as test A if a different decision level had been used. In fact either test could have achieved any of the pairs of TP and FP rates on the common ROC curve simply by changing the decision level. Thus, the two tests may in fact share a single ROC curve but initially appear to perform differently because the two decision levels used place the tests at difference points on the curve, i.e., the operating conditions were not comparable. On the other hand, the two tests may actually perform very differently, with test B clearly superior, as illustrated in Figure 4, center. Regardless of the decision level chosen for test A, it can not achieve a TP rate of 70% with a FP rate of only 2%, as did test B. In fact, when test A's TP rate is 70%, its FP rate is 10%. Similarly, the true- and false positive rates given originally would be equally consistent with the situation shown in Figure 4, right, where test A is clearly superior. These examples illustrate how the use of ROC curves avoids the ambiguity which may occur when tests are compared using only one decision level for each.

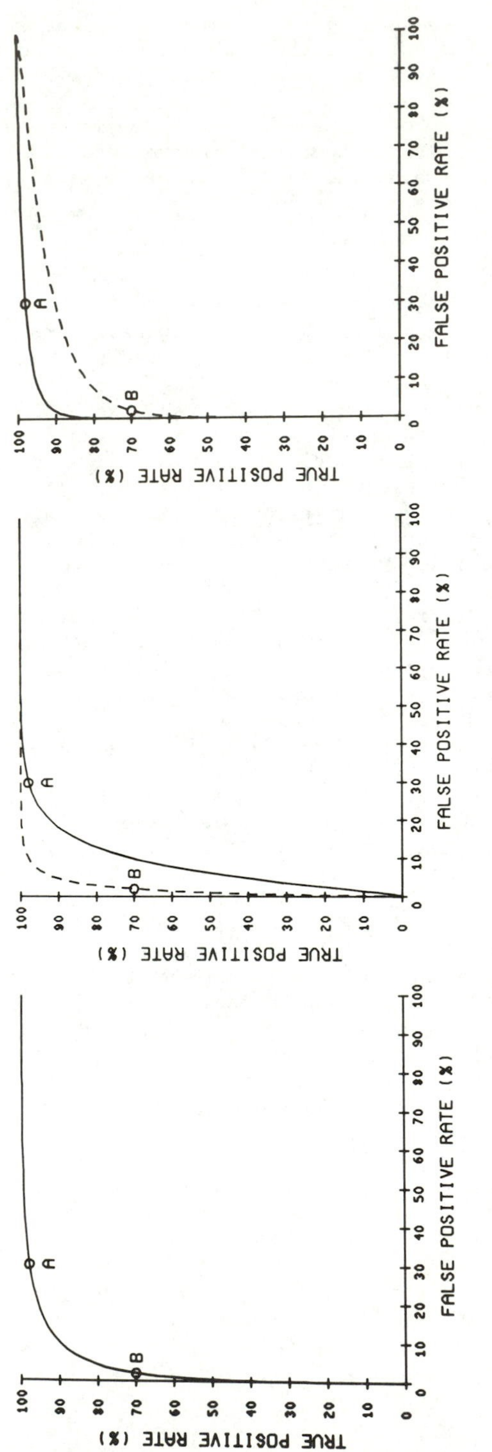

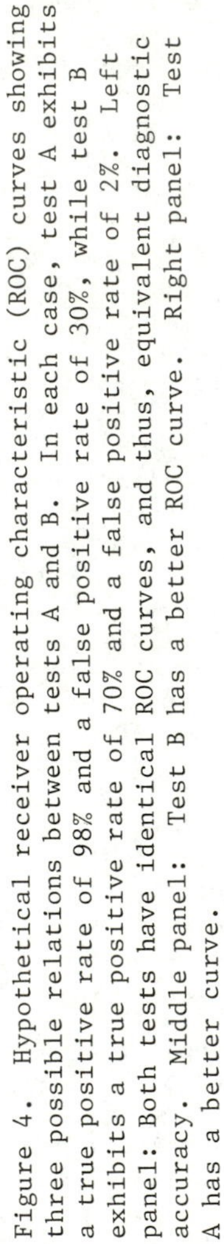

Figure 4. Hypothetical receiver operating characteristic (ROC) curves showing three possible relations between tests A and B. In each case, test A exhibits a true positive rate of 98% and a false positive rate of 30%, while test B exhibits a true positive rate of 70% and a false positive rate of 2%. Left panel: Both tests have identical ROC curves, and thus, equivalent diagnostic accuracy. Middle panel: Test B has a better ROC curve. Right panel: Test A has a better curve.

ROC Curves: Application

Figures 5 and 6 are examples of real ROC curves and illustrate how the ROC curve can represent individual test accuracy as well as compare the accuracy of multiple tests to one another. Four analytes were measured. Creatine kinase (CK) is a serum enzyme, found primarily in heart and other muscles, which has been used for some years as an early marker for necrosis. Peak serum concentrations usually occur within the first 12-24 hours after the onset of infarction. CK-MB is an isoenzyme of CK which is more specific for heart muscle than is total CK and thus has become popular in the last 10 years. CK-BB, another isoenzyme of CK found in the heart, has also been examined as a possible marker for myocardial infarction. Myoglobin, a heme containing protein found in muscle, is released into the serum with muscle injury. Serum concentrations of myoglobin appear to rise earlier than CK in patients with myocardial infarction, peaking at about 8 hours after the onset of chest pain. Figure 5 was generated by studying these four markers of myocardial injury in patients suspected of having a heart attack sampled 8 hours after the onset of chest pain. Myoglobin occupies the left-most position of the tests, and achieves the best ratio of true positives to false positives, with good absolute sensitivity (high true positive rate) and specificity (low false positive rate) simultaneously. From the ROC curve, one can make two judgements. First, myoglobin achieves the best accuracy of the four tests. Second, myoglobin probably has potential as a early marker of myocardial infarction because it's ROC curve lies quite close to the ideal location, the upper left hand corner. This indicates that it can achieve high true positive and low false positive rates at the same time. How best to use this test clinically and which decision level (i.e., where on the ROC curve to operate) to select requires clinical decision analysis with consideration of the costs of false results, the alternative tests or procedures available, the costs of the alternatives, and the utilities of the various possible outcomes (2,3). The ROC curve displays the spectrum of sensitivity/specificity pairs achievable; these pairs are the raw data needed to make the selection of decision level.

In Figure 6, the patients are sampled at 18 hours after the onset of chest pain. Myoglobin's accuracy has decreased while that of the three other tests has markedly increased to a close-to-perfect level. This reflects the fact that the serum concentration of myoglobin is not increased as much or as often at 18 hours compared to 8 hours after the onset of pain. Therefore, it is not as good at discriminating between patients having and not having an infarction. CK and its isoenzymes, on the other hand, are near peak concentrations in those patients with infarcts and lower in those without infarcts, and thus are very accurate in discriminating. In this study, the "true" diagnosis or gold standard was established by review of electrocardiographic data, clinical course, and serum lactate dehydrogenase isoenzymes, as

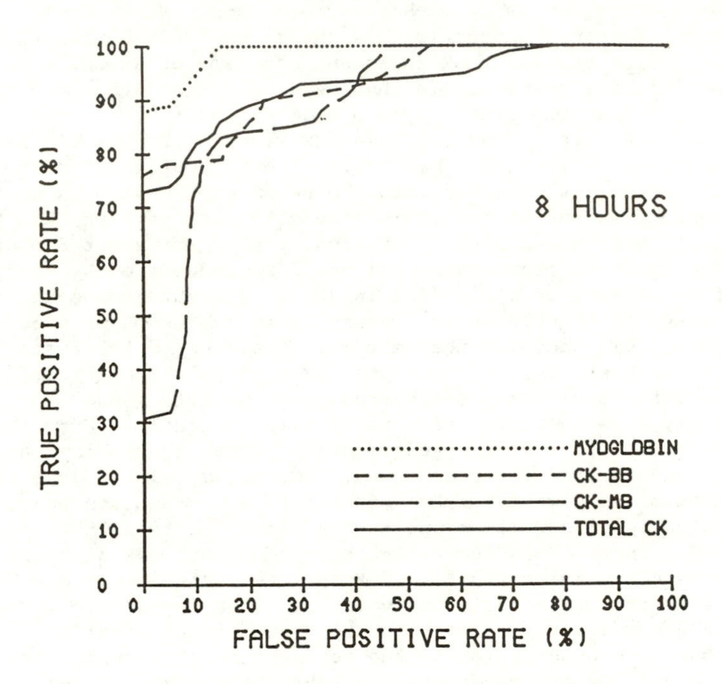

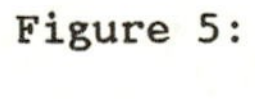
Figure 5: ROC curves of 4 serum tests 8 hours after the onset of chest pain in patients suspected of having a myocardial infarction. CK = creatine kinase; CK-BB = "brain" isoenzyme of CK; CK-MB = "myocardial" isoenzyme of CK.

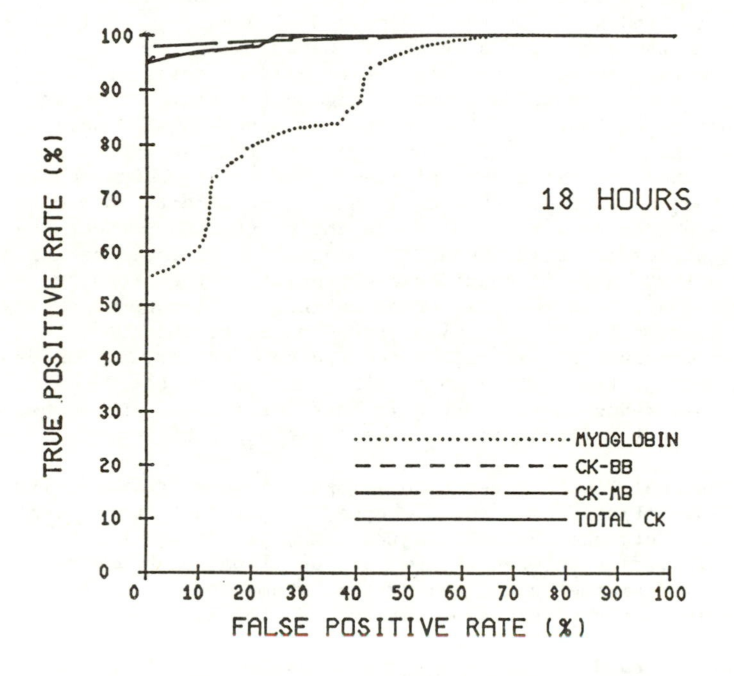

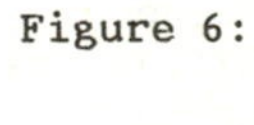

Figure 6: ROC curves of 4 serum tests 18 hours after the onset of chest pain in patients suspected of having a myocardial infarction. Abbreviations are same as for Figure 5.

well as scintigraphic findings where available. To avoid introduction of bias, the classification of patients was made without consideration of the results of any of the four tests being evaluated.

A given study provides an estimate of the ROC curve for that test and patient population. The confidence limits around the ROC curve can be calculated (8,9). Furthermore, the area under the ROC curve can be calculated for each test so as to derive a quantitative index of the test's individual accuracy and its relation to the other tests being evaluated (8,9).

ROC curves can also be used to examine the impact of analytical improvements on clinical accuracy. Figure 7 shows distributions of test results and the corresponding ROC curves, based on simulated data. For both the affected and the unaffected patients, the biological variability of the marker being measured has a standard deviation (SD) of 1 unit. The affected patients have a mean test result of 20, while the unaffected patients have a mean test result of 12. When the analytical imprecision has an SD of 4 units, there is considerable overlap in test results between the affected and unaffected patients. The corresponding ROC curve shows the poor clinical performance of the test.

If an improved analytical system reduces the imprecision of the measurement from 4 to 2 units, the overlap in test results is considerably reduced. The dramatic shift of the ROC curve upward and to the left reflects the improved clinical performance of the test.

If the analytical imprecision if again halved, reducing its standard deviation from 2 to 1, another significant improvement in clinical performance occurs. In this example, in which the biological overlap between the two groups of patients was small, the precision of the analytical system became the principal factor in determining the clinical performance of the test; substantial improvements in clinical accuracy occurred as the analytical precision improved.

In contrast, Figure 8 shows the situation in which the biological overlap is greater. In this example, the biological variation in each group has an SD of 4 units, resulting in considerable intrinsic overlap in the test results of the two groups. The figure shows this extensive overlap and the poor ROC curve for an analytical SD of 4. Decreasing the analytical imprecision (from 4 to 2 to 1) provides only a minor improvement in clinical accuracy. Thus, when the biological overlap of the two groups is large, even severalfold improvements in analytical precision may have little effect on the clinical accuracy of the test, as reflected in the ROC curve.

Principles of Test Evaluation

Once we have the basic performance data describing detection and the limits of detection as represented by the ROC curve, then we can go on to decision analysis. This involves structuring the clinical problem in the form of a decision tree, estimating utilities and costs of various outcomes, choosing decision levels

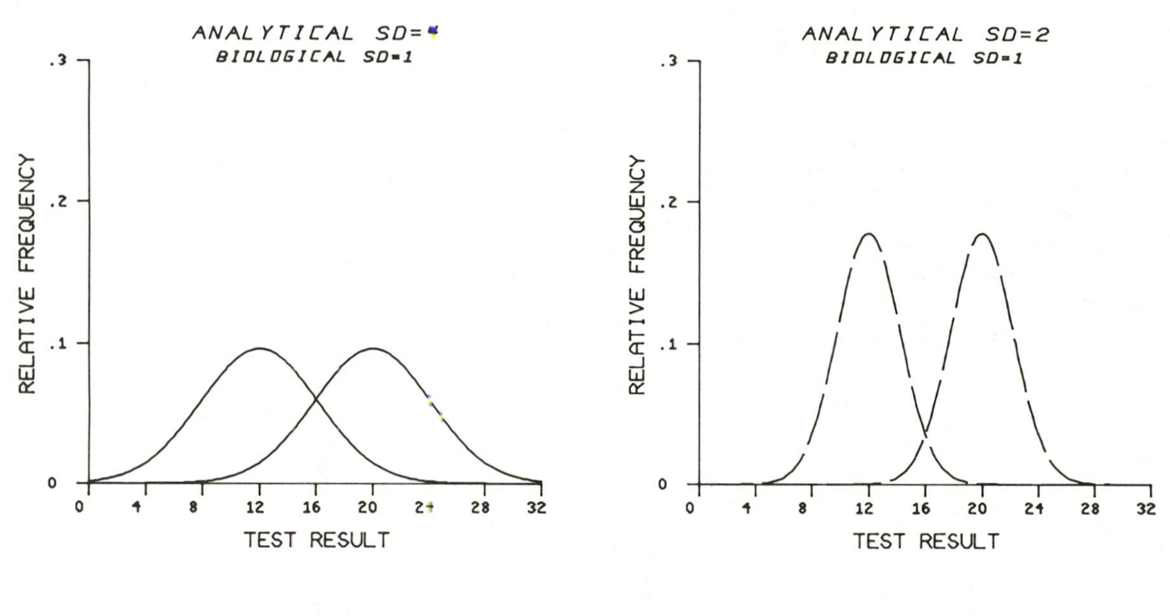

Figure 7. Frequency distribution and corresponding ROC curves for a test having a biological SD of 1 and analytical SD of 1, 2, or 4. The two peaks in the frequency distributions represent test results from diseased and nondiseased populations. Continued on next page.

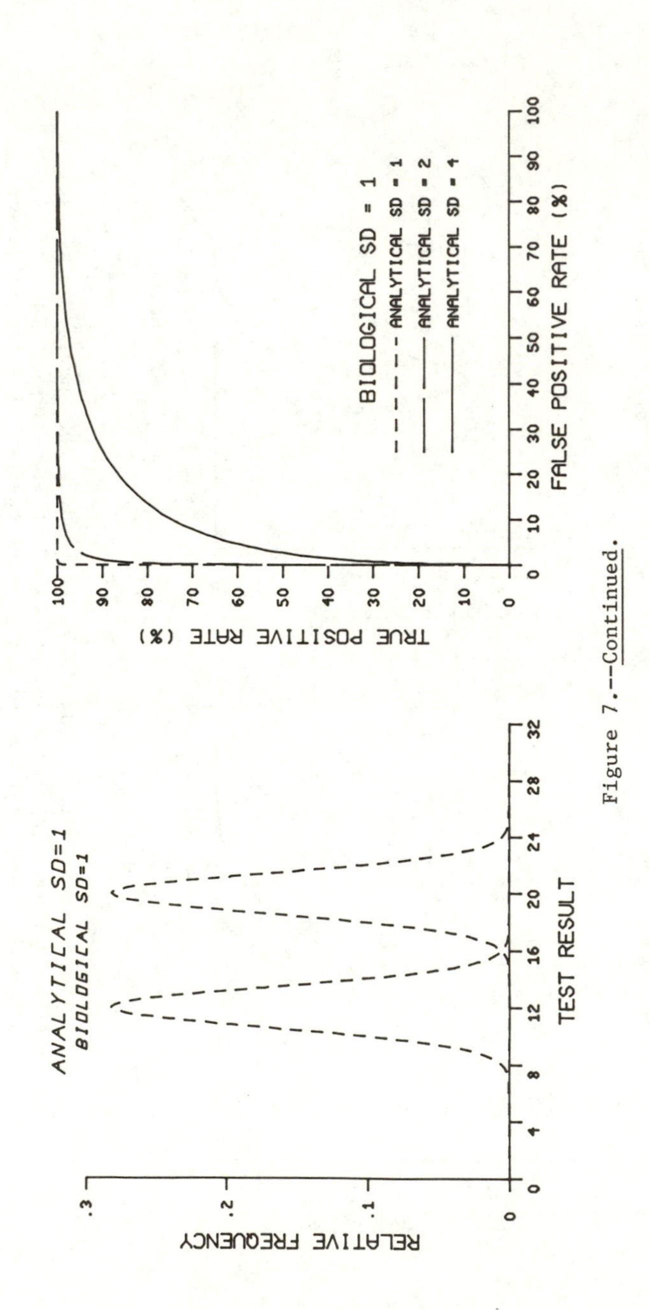

Figure 7.--Continued.

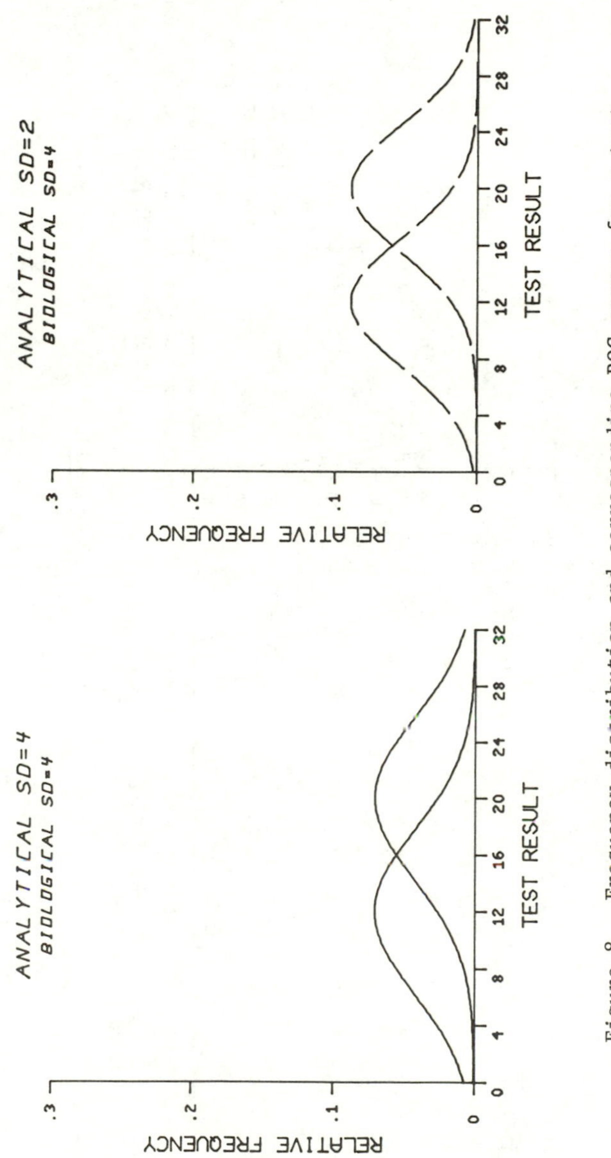

Figure 8. Frequency distribution and corresponding ROC curves for a test having a biological SD of 4. Other features same as Figure 7. Continued on next page.

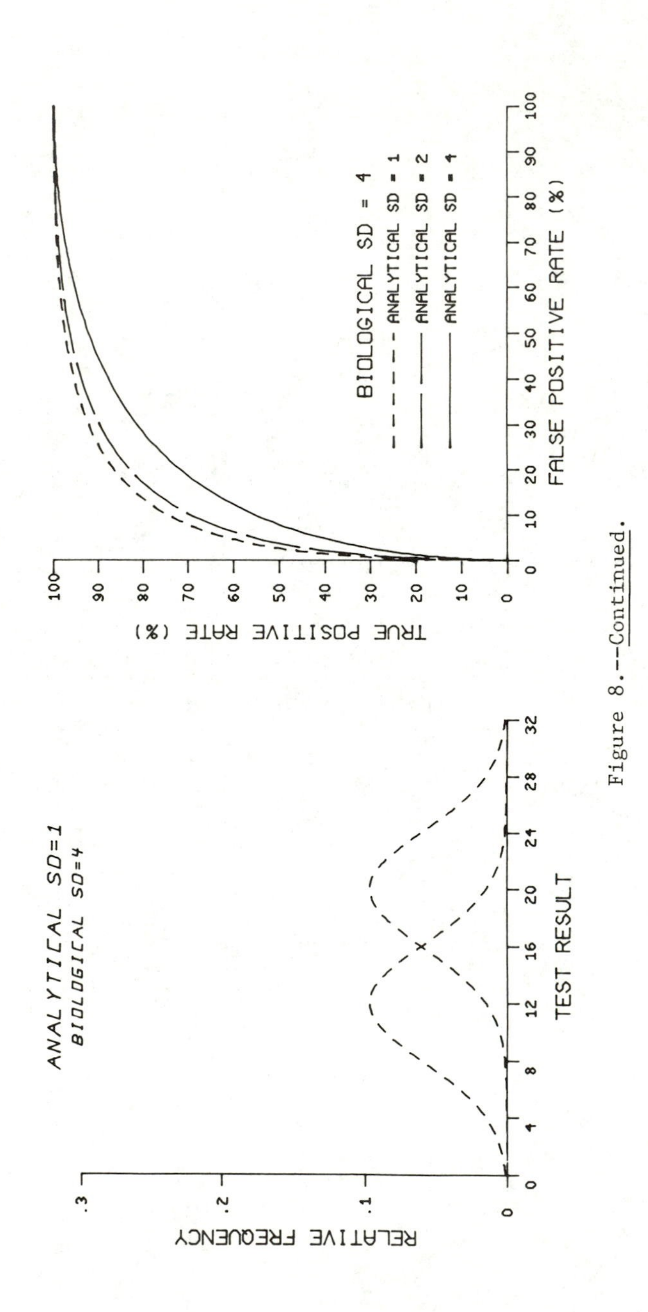

Figure 8.--Continued.

for using the test. However, obtaining valid data for the ROC curve in the first place requires attention to several common-sense principles which are suprisingly often overlooked. Table I has a list or recipe for designing a good study to evaluate the detection power of a test. The ideal study is prospective and is usually harder, longer and more expensive than the type of evaluation commonly done, but an "inexpensive" clinical evaluation may prove more costly in the long run if its erroneous conclusions lead to improper test utilization or improper patient management.

Table I. Principles of a Good Evaluation of a Laboratory Test

1.	DEFINE CLINICAL QUESTION TEST WILL BE USED FOR
2.	SELECT APPROPRIATE SUBJECTS TO STUDY
3.	CLASSIFY SUBJECTS ACCURATELY AND INDEPENDENTLY
4.	PERFORM ALL TESTS ON ALL SUBJECTS
5.	EVALUATE AND COMPARE TESTS USING ROC CURVES

The first and most important element on this list is defining specifically and carefully the clinical question or problem at which the test is to be directed. It's not enough to say "Let's look at this test for prostatic cancer or coronary artery disease and see how well it does." We need to define precisely what question of relevance to patient management is being addressed and how that test will be used in practice. Do we want to screen large numbers of people for cancer or use the test to establish the stage of cancer once we know it's there, or do we want to predict response to a particular therapy, or assess response to a particular therapy? It may provide all these functions but with varying effectiveness and requiring differing decision levels. Each of these roles must be evaluated separately because the populations are different, conditions are different, goals are different, and ROC curves may be different.

If you think about these issues, carefully and specifically defining what you are trying to establish, the rest starts falling into place.

The second element is selecting appropriate subjects. Once you have defined the question, you've pointed the way toward the proper subjects. If you want to use a tumor marker to identify colon cancer among middle aged people with bowel obstruction, occult blood loss, or unexplained anemia, then you need to look at the test performance in that group of subjects. Healthy young people aren't relevant and neither is a reference range based on them. There's no point in doing conventional normal ranges if healthy young volunteers aren't the ones for whom the test is intended.

Number three concerns establishing the true diagnosis: Once you've got a group of people with bowel signs or symptoms suggestive that cancer is possible, then you must separate them into 2 groups, those who really do have carcinoma of the colon and those who don't. This provides a gold standard for calculation of

TP rates, FP rates, etc. This diagnosis needs to be accurate as well as independent of all tests being evaluated. To the extent that either accuracy or independence is lacking, the results of the evaluation will be biased and misleading. Consider the hypothetical situation in Figure 9. The clinical question is, "Has this patient presenting at the emergency room with an acute psychiatric disorder used marijuana recently?" The routine test is sensitive enough to detect only 70% of the recent drug users; 30% of the marijuana users have falsely negative results. The routine test also suffers from various interferences, leading to false positive results in 30% of non-users. Test I represents a new test which is being evaluated. In actuality it manifests excellent sensitivity and specificity, giving positive results in all recent marijuana users and negative results in all non-users. If, however, instead of independently and accurately determining the drug-use status of each patient, the patients are simply classed as users or non-users on the basis of the routine test's results, Test I will appear to perform poorly, misclassifying 30% of the patients. In this case a perfect test appears to perform poorly simply because the clinical question was not answered accurately for each patient; i.e., the "gold standard" used for comparison was inadequate.

The opposite bias can also result from use of inadequate gold standards. Test II in Figure 9 performs even more poorly than the routine test, yielding false negative results in 40% of the marijuana users and false positive results in 40% of the non-users. If, however, the routine test's results are accepted as correct and Test II is judged on this basis, Test II will appear to misclassify only 10% of the patients -- and will have a better apparent performance than Test I!

This can occur in several ways in clinical practice. In evaluating a test for acute myocardial infarction, if the patients are classified on the basis of EKG data alone or even a combination of history, EKG findings and some cardiac enzyme results (a "routine workup"), the diagnosis may still be inaccurate and, thus, distort the apparent performance of the new test. In the case of a cancer tumor maker, if the gold standard (diagnosis or staging, etc.) is based upon clinical findings rather than surgical and/or tissue data, then the gold standard may be inaccurate and bias the apparent value of the marker. If an amniotic fluid marker for fetal lung maturity is compared to an existing imperfect marker, then even if the new marker is perfect, it will appear imperfect. The gold standard against which the new marker should be compared is the actual presence or absence of respiratory distress syndrome in those newborns delivered within a short time of measurement of the marker.

Because the validity of a clinical evaluation's conclusions is critically dependent on the accurate determination of the answer to the clinical question for each subject, routine clinical diagnoses are likely to be inadequate for test evaluation studies. Definitive determination of a patient's true clinical subgroup may require such procedures as biopsy, surgical exploration, autopsy examination, angiography, or long term follow-up of response to therapy and clinical outcome.

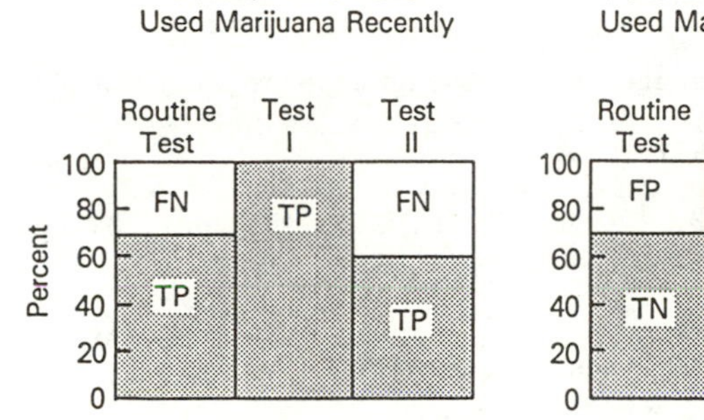

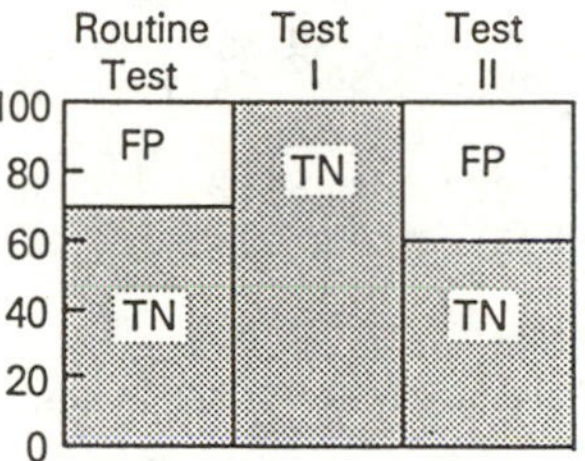

Figure 9: Hypothetical performances of three tests for marijuana use in two subgroups of patients, one which has used marijuana recently and one which has not. Assumes that the routine test gives correct results in 70% of subjects.

The next item is performing all tests being evaluated on all the subjects being used. This may sound reasonable but is very often overlooked. If the specimens or subjects aren't identical for all tests being examined, observed differences in test performance could simply be reflections of differences in the subjects rather than true differences in performance.

The last element is evaluating and comparing tests using ROC curves, extensively discussed above. The ROC analysis is a powerful tool which provides a pure index of accuracy, of discrimination capability, clearly describing the limits of clinical detection possible for a given test in a given clinical setting. Adherence to the recipe in Table I, including ROC analysis, should maximize the likelihood of obtaining a valid assessment of laboratory test accuracy.

Literature Cited

1. Swets, J.A., and Pickett, R.M. Evaluation of Diagnostic Systems. Methods from Signal Detection Theory; Academic Press: New York, 1982; Chapter 1.

2. McNeil, B.J.; Keeler, E.; Adelstein, S.J. N. Engl. J. Med. 1975, 293, 211-215.

3. Weinstein, C.; Feinberg, H.V. Clinical Decision Analysis; W.B. Saunders Co.: Philadelphia, 1980.

4. Zweig, M.H.; Robertson, E.A. Clin. Chem. 1982, 28, 1272-1276.

5. Robertson, E.A.; Zweig, M.H.; Van Steirteghem, A.C. Amer. J. Clin. Pathol. 1983, 79, 78-86.

6. Metz, C.E. Semin. Nucl. Med. 1978; 8, 283-298.

7. Turner, D.A. J. Nucl. Med. 1978, 19, 213-220.

8. Beck, J.R.; Shultz, E.K. Arch. Pathol. Lab. Med. 1986, 110, 13-20.

9. McNeil, B.J.; Hanley, J.A. Med. Dec. Making. 1984, 4, 137-150.

RECEIVED December 24, 1986

Chapter 9

Perspectives on Detection Limits

for Nuclear Measurements in Selected National and International Programs

Lloyd A. Currie[1] and Robert M. Parr[2]

[1]Center for Analytical Chemistry, National Bureau of Standards, Gaithersburg, MD 20899
[2]International Atomic Energy Agency, A-1400 Vienna, Austria

Issues involving the definition and practical significance of Detection Limits are discussed in the light of US and international programs in which the concept plays a central role. The US program relates to the "Lower Limit of Detection" (LLD) which forms a part of the Technical Specifications for nuclear power reactors, as required by the US Nuclear Regulatory Commission for measurements of effluent and environmental radioactivity. The programs of the International Atomic Energy Agency, which are oriented toward coordinated research and technical cooperation, similarly require common understanding and use of the "Limit of Detection" (LOD) as a practical and meaningful performance characteristic for measurements of trace elements in bioenvironmental matrices. Efforts to meet the needs of these two programs to formulate realistic and practicable detection limits will be reviewed, with special focus on problems-in-common such as the treatment of the blank, decision criteria, algorithm and assumption dependence, and the reporting of subliminal results.

Detection limits have practical significance in a number of important societal contexts, where measurement processes must possess adequate detection capability to meet specific diagnostic, regulatory, or research needs. Two such contexts, where specific requests were made to formulate meaningful and reliable approaches to detection limits, involved: a) the requirements of the U. S. Nuclear Regulatory Commission [NRC] for the detection of effluent and environmental radioactivity, and b) the requirements of the International Atomic Energy Agency [IAEA] for the detection of trace elements in a wide range of bioenvironmental matrices. The NRC project related directly to the formal regulatory requirement, as set forth in the technical specifications for operating nuclear power reactors, that the radioactivity measurement processes have detection limits meeting specified standards for the purpose of protecting the public from excessive releases. Typical requirements

0097-6156/88/0361-0171$06.75/0

for the "Lower Limit of Detection" [LLD] are given in Table I (1). The IAEA effort was focussed on measurements of essential and toxic trace elements in bioenvironmental samples globally, as found in their laboratory intercomparison and coordinated research programs. Detection capabilities in this case relate directly to the ability to detect deficiencies (of essential elements) or excesses (of toxic elements) in foods or biological samples from different geographical regions. Figure 1, which is drawn from the IAEA's current list of available reference materials, indicates some typical matrices and elements of interest (2,3). Concentrations varied widely with element and matrix; by way of illustration, certified values on a dry weight basis in sample H-8 (horse kidney) ranged from 0.91 mg/kg for Hg to 12.6 g/kg for Cl.

The two projects shared some features in common. Reliable and comparable measures of detection capabilities were required on a continuing basis among many laboratories, and they had to be developed in a manner that would be suitable for practical application to a broad range of analytes (trace elements, radionuclides) and sample matrices. Detection limits had to be absolute in the sense that actual concentration or radioactivity levels may be linked to specific nutritional or health effects. At the outset of each study it was observed that varying definitions and evaluations of detection limits were in use, with the result that stated detection limits were not only non-comparable but in some cases in error by orders of magnitude. Such discrepancies are already most serious from a scientific or metrological standpoint, but they invite dangerous misinterpretation and confusion when viewed by the lay public or interpreted by special interest groups.

Approach to NRC and IAEA Studies

Initial States. An outline of the approach taken for each study, together with an indication of the initial detection limit definitions in use by the two organizations is given in Table II. Apart from the difference in terminology -- Lower Limit of Detection [LLD], and Limit of Detection [LOD] -- the formulations differ in coefficients and in underlying principles. (For the purposes of this chapter we shall represent the detection limit by the single symbol L_D.) The initial NRC expression explicitly recognizes the two hypothesis testing errors, (5%) false positives and false negatives, though with a trivial rounding error [4.66 rather than 4.65]. The initial IAEA formulation, which was taken from Keith (4), explicitly takes into account only the false positive error, but at a greatly reduced (1-sided) significance level (as little as 0.13% in contrast to 5%). This is a problem: ignoring false negatives does not make them go away! In fact, nearly all such formulations invite false negatives at a rate of 50%.

The initial NRC and IAEA formulations differed also with respect to the denominator. That is, Reference 4 treats detection in terms of "signals", the denominator which converts from signal units to concentration being only implied. In contrast, the denominator for the LLD shows explicitly factors for detector E(fficiency), sample V(olume), chemical Y(ield), and radioactive D(ecay). The estimated standard deviation of the blank (s_B) is the

Table I. Detection Capabilities for Environmental Sample Analysis[a]
Lower Limit of Detection (LLD) $[\alpha = 0.05 = \beta]$

Analysis	Water (pCi/L)	Airborne Particulate or Gas (pCi/m^3)	Fish (pCi/kg,wet)	Milk (pCi/L)	Food Product (pCi/kg,wet)
gross beta	4	0.01			
H-3	2000*				
Mn-54	15		130		
Fe-59	30		260		
Co-58,60	15		130		
Zn-65	30		260		
Zr-Nb-95	15				
I-131	1	0.07		1	60
Cs-134	15	0.05	130	15	60
Cs-137	18	0.06	150	18	80
Ba-La-140	15			15	

[a] Adapted from Reference 1.
*If no drinking water pathway exists, a value of 3000 pCi/L may be used.
___ = "simple" counting.

SAMPLE CODE	MATRIX	ELEMENTS OR NUCLIDES REFERENCED	CONCENTRATION OR ACTIVITY LEVEL	SAMPLE SIZE (1 UNIT)	CLASS OF SAMPLE*	REMARKS
BIOLOGICAL MATERIALS (terrestrial)						
A-11	Milk powder	Trace elements: Ca, Cl,Cu,Fe,Hg,K,Mg,Mn, Na,P,Rb,Zn	Environmental levels	25 g	RM	Non-certified information values for: Ag, Al, As, Au, B,Ba, Br, Cd, Cr, Cs, F, I, Li, Pb, Pt, Sb, Sc, Si, Sn, Sr, Ti, U, V
A-12	Animal bone	Ra-226 Sr-90	5mBq/g (0.14 pCi/g) 55mBq/g (1.48 pCi/g)	80 g	CRM	
A-13	Freeze dried animal blood	Trace elements:Br,Ca, Cu,Fe,K,Na,Rb,S,Se,Zn	Normal levels	25 g	CRM	Non-certified information values for: Mg,Ni,P,Pb
A-14	Milk powder	Sr-90, Cs-137 Na, K, Ca	Environmental levels	250 g	CRM	Non-certified information value for Sr (total)
V-8	Rye flour	Trace elements: Br, Ca,Cl,Cu,Fe,K,Mg,Mn, P,Rb,Zn	Environmental levels	50 g	RM	Non-certified information values for: Al,Au,Ba,Cd,Co,Cs, Mo,Na,S,Sb
V-9	Cotton cellulose	Trace elements: Ba, Ca,Cl,Cr,Cu,Hg,Mg,Mn, Mo,Na,Ni,Pb,Sr	Environmental levels	25 g	CRM	Non-certified information values for: Al,Br,Cd,Fe,Ga, Hf,Li,S,Sc,Se,Sm,Sn,Th,U,V

V-10	Hay (powder)	Trace elements: Ag, As, Ba, Br, Ca, Cd, Ce, Co Cr, Cu, F, Fe, Ga, Hg, I, K Mg, Mo, Na, Ni, P, Pb, Si Sn, U, V, Zn	Environmental levels	50 g	CRM	Non-certified information values for: Al, Cs, Eu, Hf, La, Mn, Sb, Se, Sm, Th
H-4	Animal muscle	Trace elements: Br, Ca, Cl, Cs, Cu, Fe, Hg, K, Mg, Mn, Na, Rb, Se, Zn	Normal levels	20 g (2 vials)	CRM	
H-5	Animal bone (including mineral <u>and</u> organic components)	Trace elements: Ba, Br, Ca, Cl, Fe, K, Mg, Na, P, Pb, Sr, Zn	Normal levels	30 g (2 vials)	RM	
H-8	Horse kidney	Trace elements: Br, Ca, Cd, Cl, Cu, Fe, Hg, K, Mg, Mn, Mo, Na, P, Rb, Se, Zn	Normal levels	30 g	RM	Cd concentration similar to human levels Non-certified information values for: Co, Cs, S, Sr
H-9	Mixed human diet	Multielement	See remarks	30 g	RM	Representative of diet consumed in Finland. Available from March 1986

* RM for Reference Material, CRM for Certified Reference Material

Figure 1. Biological Reference and Certified Reference Materials available from the International Atomic Energy Agency. (Reproduced with permission from Ref. 2. Copyright 1986, 1987 International Atomic Energy Agency.)

Table II. Detection Limits: Practical Needs

Nuclear Regulatory Commission [NRC];
International Atomic Energy Agency [IAEA]

APPLICATION:

NRC -- Detection limits specified in the Technical Specifications for nuclear power reactors for the detection of effluent and environmental radioactivity.

IAEA -- Detection limits needed for trace elements [essential, toxic] in bioenvironmental matrices. [Coordinated Research Programs]

OBJECTIVE:

Method-independent definition and formulation; real samples. ["cookbook" manual]

INITIAL STATE[a]:

NRC -- LLD = $4.6\underline{6}\ s_B/(E \cdot V \cdot Y \cdot D)$

IAEA -- LOD = $3\ s_B$

APPROACH:

NRC -- Literature research; site visits to assess problems and practices. [NRC regional offices; power reactor; contracting labs]

IAEA -- Multidisciplinary consultants' meeting [IAEA HQ, Dec. 1985]

[a] s_B represents the standard deviation of the background or blank. Other symbols indicate: Efficiency, Volume, Yield, and Decay factor, resp.

scaling parameter in both expressions, though little guidance is given for its determination. Also the NRC definition is slightly ambiguous in stating that s_B is "the standard deviation of the background...or...of a blank sample as appropriate" (1). Neither approach explicitly treats possible non-normality, degrees of freedom, uncertainties in the denominator [overall calibration factor], systematic error, or the application to multicomponent systems which exhibit interference and matrix effects and often require more sophisticated computational procedures.

Effects of the relatively simplistic formulations are seen, for example, in Table I and Figures 2 and 3. In the table, we see that only two of the radionuclides listed are commonly measured by "simple counting" where a signal minus background operation obtains. Far more commonly LLD's must be determined for multicomponent gamma ray spectrometry, which may involve relatively complicated computations and attendant assumptions regarding interfering components, baseline shapes, etc. Even in the best of circumstances, with isolated gamma ray peaks, estimation and detection algorithms are not always consistent and erroneous detection limits follow. One practical illustration of the pitfalls attending the application of the "simple counting" expression to relatively simple gamma ray spectrometry has been described by Reichel (5). In this investigation, commercial spectrum analysis software was employed to calculate the detection limits of Hg-203 (279 keV) and Cr-51 (320 keV) on a Compton continuum from interfering Co-60. The recommended detection limit option which "has a probability of 95% of [a peak] being detected," produced detection limits of 1557 and 1511 counts for Hg-203 and Cr-51, respectively. Yet when peaks for these two nuclides were added to the continuum at levels of about 1580 and 1910 counts, respectively, from a mixed source, neither was detected by the software, though both were clearly visible to even an untrained eye! See Figure 2. The third peak at 310 keV, deriving from the Co-60 Compton interactions, was also undetected. The peaks remained undetected until their levels were approximately twice the presumed detection limits, at which point a good result was obtained for Hg-203, but a poor one [too small] for Cr-51. This latter result no doubt followed from an erroneous baseline assumption, even though the program option for "double peak evaluation" was chosen.

Figure 3 illustrates the problem faced by the IAEA in the broader context of their trace element laboratory intercomparison program. These data show the reported results of 16 laboratories for measurements of arsenic in the horse kidney intercomparison sample (H-8), based on various versions of atomic absorption spectrometry, optical emission spectrometry, neutron activation analysis, and induced X-ray emission analysis. The objective of the horse kidney intercomparison was to assess (and refine) analytical methods for the determination of essential and toxic trace elements in this surrogate for human kidney (7). Kidney, as the main target organ which accumulates toxic elements, was of special interest with respect to cadmium. Horse kidney, which contains similar levels of cadmium to the human kidney cortex, was selected for the development and maintenance of methods having a demonstrated level of quality to assure reliable biological monitoring of this element. Participants were invited to analyze some 24 additional trace elements, however,

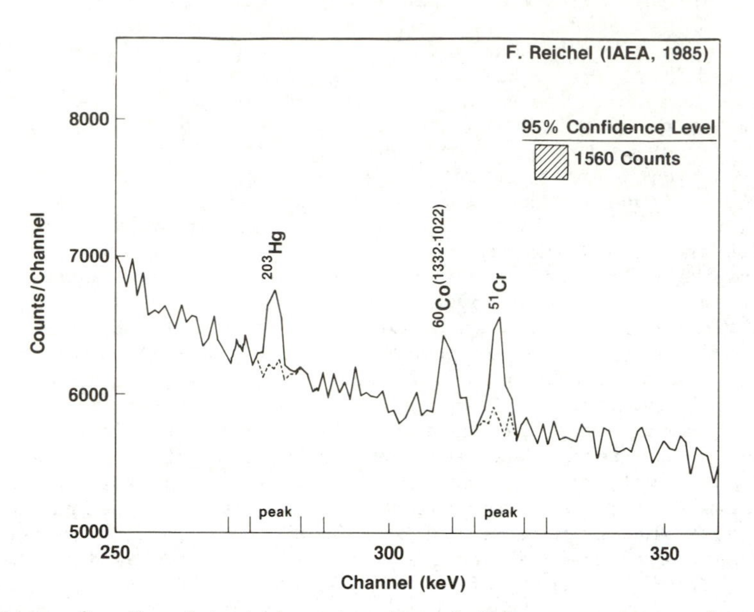

Figure 2. Non-detected, yet quite visible gamma ray peaks of Hg-203 and Cr-51 from IAEA practical examination of software performance (5). The continuum, shown dashed beneath the Hg-203 and Cr-51 peaks, is due to Co-60.

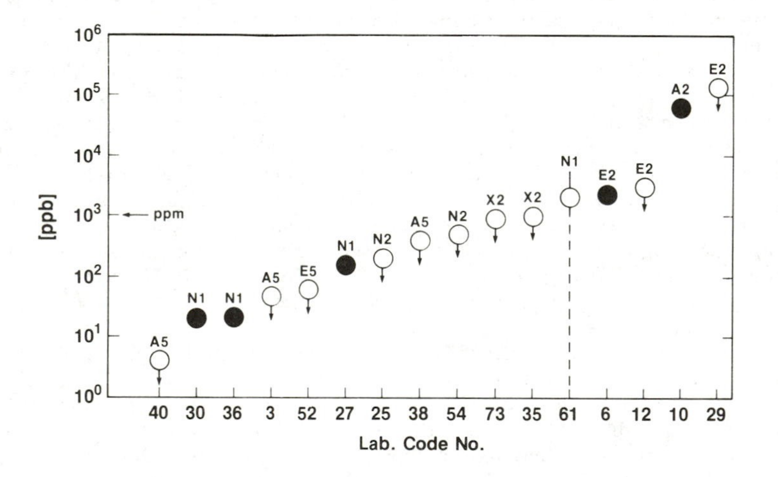

Figure 3. Interlaboratory results for As, from the IAEA intercomparison of cadmium and other elements in horse kidney (H-8) (7). Filled circles represent quantitative results (uncertainties not exceeding circle diameter); open circles correspond to reported detection limits.

most of which are shown in Figure 1. Results for cadmium, at a level of about 190 mg/kg were quite satisfactory, but the much lower concentrations of arsenic presented difficulties. The 5 filled circles in figure 3 represent "quantitative" results -- ie, those whose reported uncertainties do not exceed the diameter of the circle. These are clearly inconsistent, by more than three orders of magnitude. Open circles represent "detection limits" for non-detected [ND] results; their spread exceeds four orders of magnitude. The crowning inconsistency is the fact that arsenic has been "measured" by some laboratories at levels considerably in excess of the detection limits of others reporting ND -- clearly a logical impossibility.

To derive actual levels and establish quality measurement for trace elements in foodstuffs, materials such as milk powder (A-11) and representative regional dietary blends (eg, USDIET-1) have been provided (8,9). Consistency of interlaboratory results among "independent" analytical procedures makes possible the certification of such materials or the establishment of dietary intakes. The nature of the initial measurement problem is illustrated by the results of the milk powder (A-11) intercomparison, where the means, relative ranges (max/min, dimensionless) and medians of 17 atomic absorption spectroscopy [AAS] and 7 radiochemical neutron activation analysis [RNAA] laboratory results for manganese were as follows.

	mean (mg/kg)	max/min	median (mg/kg)	(95% CI)
[AAS]	3.99	279	0.67	(0.45 - 1.27)
[RNAA]	0.295	4.8	0.26	(0.12 - 0.58)

It is interesting to note that the ratio of means (AAS/RNAA) is 13.5, whereas the ratio of medians is but 2.6. Reliability is increased through the use of the median as a robust statistic, especially when the number of replicates n (here, laboratories) is relatively large. (When n>8, the 95% CI of the median does not directly depend upon the extremes, so some additional, automatic outlier protection is afforded.) It is noteworthy, however, that a large number of replicates does not guarantee quality! The relatively few RNAA measurements yield a more reliable result than would a very extensive set of AAS measurements (representative of this particular intercomparison). The basic issue is one of contamination and interference, as implied for example by the skewness represented in the ratio of mean to median. This "blank" issue is brought out at this point in our discussion because it is probably the single most important limitation to reliable measurement and reliable detection for low-level, multicomponent measurements in complex matrices. As such, it has been a major consideration in addressing the question of "real sample" detection limits for both radiological and trace element measurements in bioenvironmental samples.

Experts and Concerned Groups. The foregoing observations -- involving diverse formulations and interpretations for radiological detection limits, and diverse and discrepant interlaboratory results for low level trace element measurements -- led the NRC and the IAEA, respectively, to question the validity or at least the universality of detection limit formulations as then applied. In order to address the question in an efficient manner the two organizations employed problem solving approaches of demonstrated effectiveness: viz., (a) site visits and field interviews with each of the affected sectors [NRC], and (b) an intensive workshop involving experts representing each of the respective analytical disciplines [IAEA, (10)]. The importance of these approaches deserves emphasis, for it transcends the particular scientific problem under consideration. In the case of the NRC, for example, the varying needs and perceptions of the utilities and their operators, the contract (radiological) laboratories, the cross-check (control) laboratory, the inspectors, and NRC headquarters personnel could not otherwise have been discerned. As a result it was possible to identify and to some extent sort out measurement from policy issues, and attempt to clarify measurement and detection limit concepts and terminology which were subject to ambiguous interpretation. A summary of the findings of the NRC site visits is given in Table III.

Table III. NRC Findings

* LLD Manual Needed -- for diverse backgrounds

* Wide ranging nomenclature and formulations
  [LLD, MDA, ...]

* Socio-political-economic issues
  [biased reporting (public perceptions), LLD requirements for minor components when high interference levels,...]

* Detection decisions: false negatives ranged from 5% to 50%

* Blank: ambiguity in initial draft document, blank variations often excluded from LLD

* Non-counting errors generally ignored, especially sampling

* Simple counting formulation only [signal - blank]; inapplicable to many cases of nuclear spectrometry

* Gamma spectrometry: multiple detection decisions; occasionally hidden, changing algorithms; erroneous parameters

* Appropriate [low-level] and double blind QA samples needed

The calling together of complementary or multidisciplinary teams of experts is a tradition in a number of international organizations, such as the IAEA. Since the IAEA objective with respect to detection limits was explicity "method independent", such an approach was mandatory. That is, expert knowledge concerning the nature and sources of error, such as the blank, matrix effects, and detector characteristics for the methods of interest (NAA, AAS,

ICP,...) was essential so that the approach to detection limits would at least be broad enough to encompass the characteristics of these particular methods. A theoretical formulation was not enough; a common, practical approach to detection, suitable for these several methods of trace analysis was called for. Table IV gives the composition and objectives of the IAEA task force.

Outcomes

The basic concepts of detection decisions and detection limits, based upon hypothesis testing as outlined earlier in this volume, necessarily served as the foundation for the NRC and IAEA programs. For the probabilistic part of the problem, false positives and false negatives were taken equal to 0.05 in each case. This means that for normal random errors, the detection decision is made once the observed signal exceeds 1.645 times the standard error of the null signal [standard error of the blank for a well known blank, or $\sqrt{2}$ larger for paired observations]. When the number of degrees of freedom is small, of course, the z-statistic is replaced by Student's-t. For simple measurements the detection limit is just twice this critical level. In addition to this relatively straightforward normal, "statistical" treatment of detection, however, it was necessary to pay attention to the special characteristics of the real measurement processes involving the detection of low levels of mixed radionuclides, and trace concentrations of multiple elements in complex matrix materials. This matter, which was an explicit component of both projects, cannot be overemphasized. If one calculates statistical detection limits for interference-free measurements in pure solutions, estimated detection limits will be too low, often by orders of magnitude. Similarly, detection limits which are predicted for an alteration or optimization of a measurement process will be unrealistically optimistic if one applies just a simple statistical formula to calculate the change in detection limit, eg, with increased replication or increased counting time. Unless tests are made with known or common materials comparable in composition to the samples of interest, these erroneous estimates will not be exposed.

NRC Special Topics. The estimation of practical, real sample detection limits requires attention to all of the sources of error that must be faced in deriving confidence, or more correctly, uncertainty intervals (11). In the NRC study, special attention was given to the following: a) bounds for uncompensated systematic error in the background and calibration factor, b) non-Poisson random error (that which exceeds "counting statistics"), c) deconvolution and model error connected with alpha or gamma-ray spectrum analysis of multicomponent mixtures, and d) special limitations from the non-normality of the Poisson distribution for extreme low-level counting, such as occurs with monitoring of actinides or other alpha emitters. Each of these factors required extension of the "simple-counting" expression originally found in the draft NRC Technical Specifications [see table II]. A full discussion, supplemented with appropriate formulas, references, and examples may be found in Ref. 12; and one of the more broadly applicable results will be treated here. In table V we give an extension of the

Table IV. Consultants' Meeting on Limit of Detection

Dates and Place

2-4 December 1985; IAEA Headquarters, Vienna

Participants

Austria:	W. Wegscheider
Belgium:	J.P. Op de Beeck R. Van Grieken
F.R.G.	M. Stoeppler
U.S.A.	L.A. Currie (Chair)
IAEA	R.M. Parr (Scientific Secretary) F. Reichel R. Schelenz H. Vera-Ruiz

METHODS

Topics

- Activation Analysis and gamma-ray spectrometry (J.P. Op de Beeck; F. Reichel)
- XRF and PIXE (R. Van Grieken)
- AAS (M. Stoeppler)
- ICP-AES (W. Wegscheider)
- Voltammetry and other methods (M. Stoeppler)

GOALS

General discussion

- Method-independent definition of the limit of detection (LoD)
- Practical determination and use of the LoD
- Special problems associated with individual analytical methods
- Decision making and reporting of data below or near the LoD

Preparation of meeting report

Table V. Design and Optimization
LLD (L_D) Variations with background (B) and Counting Time (t)[a]

	$t \ll \tau$	$t \gg \tau$
$B \lesssim 1$	$L_D \propto t^{-1}$	L_D = const
$1 \lesssim B \lesssim 1000$	$L_D \propto t^{-1/2}$	$L_D \propto t^{1/2}$
$B \gtrsim 1000$	L_D = const′	$L_D \propto t$

[a]Units of B are counts. R_B and t represent background rate and sample counting time, resp., and η is a pure number determined by relative sample and background counting times. τ represents the mean life ($t_{1/2}$/Ln2). Special limitations derive from the relative systematic uncertainty in the calibration factor [ϕ_A], deviation of the Poisson distribution from normality [P ≠ N], and the systematic uncertainty in the blank [Δ_B].

$$L_D = \left(\frac{\overset{[\phi_A]}{f}}{2.22\ YEV}\right)\left(\frac{\overset{[P \neq N]}{2.71} + 3.29\sqrt{R_B \eta t} + \overset{[\Delta_B]}{0.10\ R_B t}}{\tau\ [1-e^{-t/\tau}]}\right)$$

simplified expression, which now takes into account bounds for possible systematic background and calibration error, as well as moderate deviation of Poisson statistics from normality. The symbols have the following meaning: L_D is the detection limit in units of "concentration" -- ie, specific activity (pCi/L), following Reference 1; Y, E, V, t, and τ represent radiochemical yield, counting efficiency, sample volume, counting time, and mean (decay) life, respectively; $R_B t$ = B represents the number of background counts; η is a pure number, ranging from 1 to 2, which takes into account the effect of sample vs background counting times or (for simple spectrometry) counting channels. Moderate deviations of Poisson statistics from normality are covered by the term 2.71; and the terms ϕ_A [= f-1] and Δ_B provide for relative and absolute systematic error bounds in the overall calibration factor [=YEV] and background, respectively.

The standard deviation of the null signal in this expression is given in terms of counting statistics; if Poisson statistics are not likely to account for most of the random counting error, then it would be prudent to deduce σ_B from a moderate number of replicates -- ie, replace the second term in the numerator of the second factor by $2t's_B\sqrt{\eta}$, where t′ is Student′s-t and s_B is the estimated standard deviation for the blank (counts). Bounds for systematic error should be based on sound experience or analysis of the measurement process; default values that reflect much low-level radionuclide measurement experience are set at 1% [baseline], 5% [blank], and 10% [calibration], respectively. Poisson deviations from normality are adequately accounted for by this expression down to B = 5 background counts; below this, tabulated values for the Poisson distribution should be used [(12), pp 84ff].

The importance of the systematic error bounds may be appreciated as the counting time, sample size or number of replicates is increased. Without the added term for blank systematic error (last term in the numerator of the second factor), the detection limit would decrease indefinitely, for a long-lived nuclide, simply by extending the counting time. At the cost of longer counting, a false sense of security would result. By way of example, the combined effects of a finite half-life and realistic systematic error bounds have been shown to restrict the improvement in detection limit for a specific I-131 measurement to a reduction of about 25% even though the counting time increased by a factor of ca 100 (from 200 min to the near optimal 2 weeks) [12, p. 137]. Table V illustrates the impact of systematic error, non-normal random error [Poisson] and decay constant [mean life, τ] on counting-time induced variations of radionuclide detection limits. It is evident that variations can range widely, from $\propto t^{-1}$ to $\propto t^{+1}$, depending on mean life and background level.

Before leaving the topic of systematic error bounds, two points should be made. First, as is perhaps obvious, the probabilistic meaning of false positives and false negatives is necessarily altered. These "errors" or risks are now inequalities ["no greater than..."], and their validity rests greatly on that of the systematic error bounds, just as in the case of uncertainty intervals for high level signals. Second, estimation of non-Poisson random error and systematic error empirically, by comparison and replication is not an easy task. One can show that at least 15 and 47 replicates, respectively, are necessary just to detect systematic and excess random error components equivalent to the (Poisson) standard deviation [(12), p 25f; (13)].

Most of the effluent and environmental radioactivity measurements are made using gamma-ray spectrometry. This is a far more cost effective approach than radiochemical analysis; the instrumental measurement can be readily automated, and detection decisions can be made more or less simultaneously for many radionuclides. The validity of those decisions, and of the corresponding detection limits, however, requires either that the peaks be isolated and lie on a linear baseline, or that a detection limit model be employed which is more complex than that used for "simple" counting. Baseline or interference model uncertainties should be included, and an iterative solution is required to estimate the detection limit when spectrum deconvolution is involved. Details are beyond the scope of this chapter, but a relatively simple limiting estimate can be derived by treating the estimated standard error for a low level radionuclide peak of interest as though it were the null standard error, σ_o [12, p. 81].

Multicomponent gamma-ray spectrometry is subject to several additional detection limit pitfalls which will simply be noted here. If the algorithm changes in passing from peak detection to peak estimation, an invalid detection limit will be given, unless algorithm switching is properly taken into consideration. Incorrect false positive and/or false negative errors will result from nonlinear peak search routines, erroneous peak/baseline models, and subtle but oft-hidden deviations from the basic hypothesis testing principles. Since algorithms are not always available for examination, and since even the raw spectral data may be non-retrievable, these automatic spectral evaluation techniques can sometimes generate misleading detection limit estimates. (See the foregoing discussion of Fig. 2.) The impact of multiple detection decisions on false positives and negatives is another issue that must be faced in large bandwidth (many potential component) systems (14). For example, if one were to search a large spectral region for, say, 100 peaks which are truly absent, the probability of a false positive would be practically unity (99.4%) unless the critical level L_C were appropriately adjusted.

IAEA Special Topics. Unlike the NRC, the IAEA was concerned primarily with the actual detection limits obtained by program participants, as practical performance characteristics of their methods as applied to complex samples. In this case a predefined regulatory limit (L_R) did not exist as a driving force for method optimization or redesign. Also, sample matrices and analytes and trace analysis methods were somewhat broader than those considered by the NRC. That, plus the requirement for a "method independent" approach, meant that the IAEA had to give somewhat more scrutiny to the foibles of the several analytical methods, especially with respect to blanks, losses, interference and matrix effects, and detector characteristics. Methods experts, referred to above, provided critical information on explicit problems, such as uncertainties in baseline or peak overlap models in NAA and XRF methods of analysis. Such added sources of error mean, for example, that detection limits derived strictly from spectrum fitting variance will be too small. Numerous additional method-specific sources of error, ranging from dissolution and recovery to matrix-induced radiation scattering were developed by the experts, and will be summarized in the subsequent report (10).

The blank and contamination problem was highlighted in the IAEA deliberations. It seems likely that this is the most significant cause of the orders of magnitude discrepancies in detection limits reported in IAEA intercomparisons. To deal effectively with this problem, several recommendations were made. First, future intercomparison participants will be encouraged to measure and report results for a minimum number of blanks, perhaps nine. Second, detection limits will be based on paired comparisons of sample measurements with equivalent blank measurements, and the average of at least three or four (but preferably nine or more) such paired comparisons will be used (with Student's-t) for detection decisions. The philosophy underlying this approach is that: a) paired comparisons force symmetry in the estimated net signals, yielding a better chance to approach normality through averaging [central limit theorem], plus the possibility of using the Gauss inequality (in contrast to the wider Tschebyscheff inequality) for distribution free limit estimation; and b) with nine or more observations, the median and its 95% confidence interval can be used for robust estimation with little influence from possible outliers. Finally, the detection limit can be estimated as approximately twice the critical level or $2t\sigma_o$. When σ is estimated as 's' via replication, and one wishes a conservative upper limit for L_D, this can be formed through the combined use of the Gauss inequality and the χ^2 distribution.

Let us illustrate. Assume that a set of ten paired observations of the total signal (T = S + B) and the blank (B) for the analysis of trace levels of Zn in animal bone (in units of mg/kg) were as follows

B: 164.9 144.7 135.1 139.1 167.6 246.3 228.8 111.9 150.3 153.0
T: 203.9 277.7 288.7 202.9 164.2 227.4 241.3 262.2 238.4 250.3

The estimated means and standard errors are

Mean blank = 164.2	SE = 13.3	(mg Zn/kg bone)
Mean $\hat{S}$ (T-B) = 71.5	SE = 20.0	(mg Zn/kg bone)

For 10 degrees of freedom Student's-t at the (1-sided) 5% significance level is 1.81, so the critical level is (1.81)(20.0) = 36.2, and one would conclude that the sample Zn was significantly greater than the blank -- ie, Zn was detected. (Note that s_o was taken as SE of $\hat{S}$ above, a valid procedure, so long as S is small, ie, below the detection limit. Alternatively, it could have been taken as $\sqrt{2}\cdot SE_B$ = 18.8, with 9 degrees of freedom.) The complementary question is: "What is the smallest concentration of Zn that could be detected with 95% probability, given the above critical level?" The answer, namely the estimated detection limit for the measurement process under consideration, is approximately twice the critical level or 72.4 mg/kg. Note that 71.5 is an experimental outcome (which we tested for statistical significance), whereas 72.4 is a measure of the inherent detection capability of the measurement process. An upper limit for L_D may be computed using the upper limit for σ/s. Using χ^2 with 10 degrees of freedom, we find a value of 1.59 for this ratio (1-sided, 5% significance level), hence an upper limit for L_D of 115 mg Zn/kg bone. The true detection limit is likely (95% chance) smaller than this value, but a more precise estimate of σ would be required to better determine it. A still more conservative value, incorporating the Gauss inequality where $[t + (1.5\sqrt{2\beta})^{-1}]\sigma_o$ replaces $2t\sigma_o$, would raise this upper limit for the detection limit by an additional 8%, to 124 mg/kg. The <u>major constraint</u> on precise knowledge of the detection limit is thus knowledge of σ_D; for normally distributed data, with s based on replication, the uncertainty range (at the 90% confidence level) for σ/s falls below a factor of 2 once the number of replicates exceeds a dozen.

Two points merit emphasis in the above exercise: a) The statistical confidence interval for the outcome is based on $\hat{S}$ and its SE (using a 2-sided Student's-t); SE but not $\hat{S}$ is used also for the estimation of L_D. b) The confidence interval, and L_C and L_D (and its upper limit) are correct for normally distributed random errors. Paired T, B comparisons and a moderate number of replicates tend to make these assumptions reasonably good; this is an important precaution, given the widely varying blank distributions of such difficult measurements. Perhaps the <u>most important consequence</u> of the paired comparison induced symmetry, is that the expected value for the null signal $[\hat{B} - \hat{B}']$ will be zero -- ie, unbiased. Systematic error bounds, some deeper implications of paired comparisons, and distribution free techniques are treated in the extended discussion of the above example (10). Consideration of reasonable systematic error bounds for the blank and overall calibration factor led to a final detection limit [upper limit for] of about 140 mg/kg. Use of the median as a more assumption-free estimator yielded a 95% confidence interval extending from -3 (ergo, 0) to 150 mg/kg. (The numerical values for this simulated example were inspired by actual IAEA intercomparison data for Zn in animal bone [H-5], as reported in Table I of Ref. 6.)

Generating authentic blanks for measurement is another matter. The central importance of the blank for reliable analyte detection, and the complexity of the blank in multicomponent trace element analysis of bioenvironmental matrices are such that the IAEA gave this matter special attention. A three faceted approach was devised, comprising the "ideal blank," the "simulation or surrogate blank," and finally "propagation of the blank." Full discussion, including method specific information, may be found in Ref. 10, but a brief exposition follows. The ideal blank is defined as that which reflects the sample and the measurement process in every respect save one: the absence of the analyte of interest. For relatively simple matrices and relatively low levels of interference, the ideal blank may be approached. Failing this, one must devise surrogate blanks which closely simulate the real sample. This process, as well as the alternative blank propagation technique, requires true expertise in the relevant analytical science as opposed to statistics. Remembering that the "blank effect" -- ie, true analyte blank together with unresolved interferants -- may be associated with the sample, itself, dissolution procedures, reagents, and instrumental (and even software) artifacts, we note that a good surrogate blank requires the introduction of analyte or interferant at the same stages of the measurement process and in the same amounts as occur with the actual sample. Spectral interferences, matrix effects, recoveries, sensitivities, and reagent quantities and sample sizes should be sufficiently similar. The allowable degree of departure, again is a sample-method specific scientific issue which must be determined by the respective experts, perhaps with the aid of experimental ruggedness tests.

An interesting illustration of the subtlety of surrogate blank pitfalls is found in the analysis of trace elements in milk by neutron activation analysis. Knowledge of trace constituents of milk, especially human milk, is of considerable importance in IAEA cooperative research programs involving global trends in human health and nutrition (3,8). Lacking sufficient information regarding 1) baseline interference in NAA using gamma ray spectrometry, and 2) trace element composition of presumably similar biological materials, one might select cows' milk as a potential surrogate. Such a choice would be misleading for a number of activated gamma emitters, however, because cows' milk contains phosphorus at concentrations which exceed those in human milk by about a factor of 10, and the bremsstrahlung from the high energy beta emitting neutron activation product P-32 would cause a much increased instrumental baseline contribution which would not be representative of the true, human milk blank (15). For other methods, such as AAS or XRF, this problem would not arise.

Space does not permit a detailed discussion of the propagation approach to the blank and its variability, but in essence it consists of considering each sequential step of the measurement process, together with the introduction and propagation of analyte blank and interferants through each step, taking into account the respective recoveries. In the final, instrumental measurement step perturbations of the response (shape and sensitivity) for analyte, interferants, and matrix absorption and scattering must all be considered. Propagation of components of the blank thus represents an excellent independent approach to devising a surrogate blank. Its success, however, depends upon the estimation of appropriate recovery and sensitivity factors for the three trouble makers: adventitious analyte, interferants, and matrix effects. Excellent experimental techniques which may help in the task include multiple standard additions, isotope dilution, and multiple *sample* additions (16).

Reporting of Low-Level Data, and Quality Assurance

Reporting and quality issues transcend the specific NRC and IAEA programs, and they are treated in some detail elsewhere in this volume. Low-level data are far more subject to information loss and bias than data corresponding to large signals. This follows because of a tendency to report observations which are not statistically significant as zero, or "trace", or as various types of upper limits. Averaging or combining or even comparing sets of results so reported is either impossible or not possible without bias. Worse still, the same results could be reported according to different prescriptions, leading at the very least to misconceptions by the lay public. The recommendation given both the NRC and the IAEA was that observed values and their uncertainties be reported, together with appropriate detection limit information, even when the detection decision is negative.

Quality control or cross check samples at low levels are essential for assuring quality measurements in these same concentration regions. Internal, known controls in similar matrices and having similar interferences should be the first step toward attaining control. Certified, natural matrix materials are the best, for one can then test not just consistency or interlaboratory comparability, but also accuracy. Use of external, blind control samples, perhaps including laboratory intercomparisons, may be the only way to reliably assess performance under routine conditions (17). Similarly, test data sets can be most important for algorithmic quality assurance of low level multicomponent spectrometry (18,19). It was felt that the objectives of both the NRC and IAEA programs could be better met by taking careful measures to assure truly blind control samples that would be faithful surrogates with respect to analyte concentration and matrix composition for the samples and analytes of concern. Distribution of blind *blanks* could further assure valid claims concerning detection limits.

Conclusion

The NRC and IAEA programs share the common practical goal of seeking to establish accurate values for the detection limit performance characteristic for their respective measurement processes for trace radionuclides and trace elements, respectively. The driving forces for establishing meaningful detection limits for the actual measurement process were, in the first case for radiological monitoring to prevent the release of radioactivity levels of environmental concern, and in the second, to increase our knowledge of concentrations and geographic variations of essential and toxic trace elements in biological and environmental samples.

Detection limit formulations which appeared similar, but which were fundamentally different, were initially found. In both cases the expressions were also somewhat limited in applicability, in that they did not explicitly treat systematic errors, or those associated with many of the "real-life" problems of low-level measurement such as overlapping peaks, matrix effects, blank introduction at various stages of measurement, instrumental artifacts, or non-normal random errors. A synopsis of the issues, together with an indication of the crucial role of scientific expertise, is given in Table VI. The benefits of sound measurement design, of course, extend beyond the derivation of meaningful detection limits. In Figure 4, to be contrasted with Figure 3, we illustrate a "success story." Here, the improvement in measurement quality has made possible the discrimination of significant geographical variations in the trace

Table VI. Outcome and Desiderata

NRC -- Method-independent σ_o formulation; attention to non-counting random and systematic error. Special treatment for few counts

IAEA -- Paired observations; report an adequate number of "ideal" blanks; use of "t", central limit theorem, and σ/s (limit)

1. Sound, conceptual basis [not *ad hoc*]

2. Method-independent terminology, formulation

3. Responsible reporting [low-level data]

4. Use of the full power of statistics [hypothesis testing, robust estimation, error propagation...]

5. GOOD SCIENCE

   Knowledge of the Measurement Process in place of empiricism ["local" matrix calibration and large replications, are costly]

   Expert knowledge re: simulation blank, propagation of components of the blank.

   Special attention to error bounds [including model-error], validation, contamination, interference, and matrix effects

   Analytical Quality Assurance via low-level Certified Reference Materials, and Simulation Test Data [algorithmic QA].

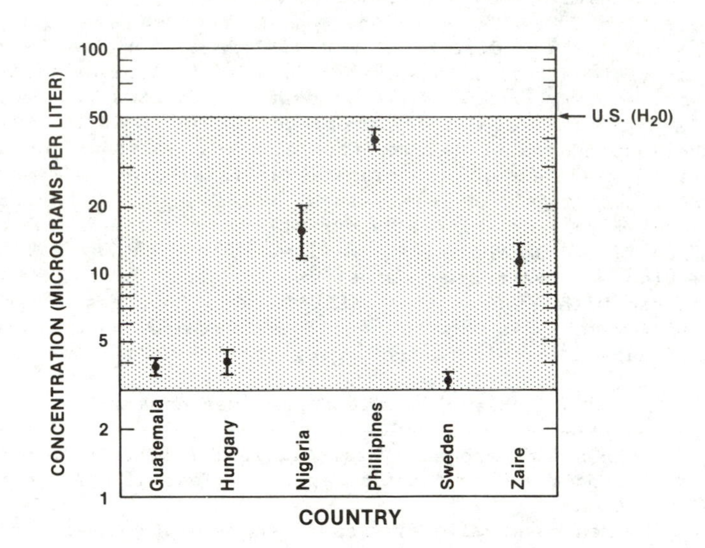

Figure 4. Manganese concentrations in human milk from different countries, medians ± SD (median) (20). Shaded region superposed represents the "inner range" of the A-11 milk powder intercomparison (8), for 27 results excluding one major outlier and the two remaining extreme values, translated by median matching. That is, the median of 21 "accepted" A-11 results, in units of μg/g of milk powder, was translated to match the median of the six regional liquid human milk results, in units of μg/L, in order to compare relative ranges. Differences in the numerical values of the two medians are due in large part to the water content of the liquid milk. The range of the 21 "accepted" A-11 results was smaller, equivalent to a factor of 6, but this still would have been inadequate for detection of the geographic variations shown. The arrow at 50 μg/L corresponds to the suggested limit for Mn in U.S. drinking water (21).

element composition of human milk (20). Knowledge of this sort, as stated earlier, constitutes one of the primary goals of the IAEA coordinated research programs. It can lead to better understanding of nutritional deficiences through correlation with states of health, and more importantly, can lead the way toward improved global health through appropriate dietary supplements during the crucial first years of life. The shaded region in Figure 4 shows the connection with the earlier detection and measurement capabilites for the same element (Mn) in the IAEA A-11 milk powder intercomparison (8). This region constitutes the inner range (excluding a major outlier plus the two remaining extremes) representing 27 laboratories, scaled by median matching. That is, the median of 21 "accepted" results for Mn in the A-11 Milk Powder, 0.32 μg/g, was aligned with the median of the geographic results for human milk, ca 7 μg/L, in order to compare the relative ranges. Clearly, the important geographic variations would not have been detected at the A-11 intercomparison level of precision. Shown also in Figure 4, for comparison, is the "Suggested limit that should not be exceeded" for Mn in U.S. drinking water (21).

Looking beyond the question of detection limits for a given method-sample combination, it becomes interesting and important to consider detection in a larger sense, ie, detection of a phenomenon, a state of health, an environmental process, etc., as manifest by chemical or radionuclide composition. This immediately requires us to face questions of scientific understanding and modeling of the phenomenon under consideration. Given that, one can then address the issue of the appropriate sampling design and measurement to guarantee adequate detection of the phenomenon. It is at this level that detection limit concepts, appropriately linked with scientific understanding and insight, can have their greatest impact. To cite one example, the bioassay of Zn, a trace element of comparable importance to Fe for vertebrates (22), can be quite misleading in the absence of "biological insight." Iyengar (23), for example, has illustrated how sampling variations can generate quite erroneous conclusions [false positive and negative levels] regarding this element depending on such blood sampling conditions as fasting, intake of normal food, low Zn food, high Zn food, stress, and pregnancy. The link between chemical analysis and the detection or interpretation of biological states is further confounded by complex element interactions. For example, due to bioavailability variations, a given concentration of "Cu in animal feedstuff can be deficient, adequate or toxic depending upon the Mo and S content of the feedstuff" (22).

Our conclusion must be that a sound conceptual basis for detection limits in chemical and radiological analysis, linked with a sound scientific understanding of the respective measurement process -- especially the nature of the blank, is a prerequisite to addressing important societal questions involving biological and environmental systems. If we are to gain full benefit from such measurements, however, we must work in a multidisciplinary atmosphere, so that we can design our measurement processes to exhibit detection (and quantification) limits which are adequate to speak to these larger issues.

Acknowledgment

Colleagues who shared information concerning extant detection limit practices, and NRC and IAEA program needs, are gratefully acknowledged. Included in this particular acknowledgment are other participants in the IAEA Consultants' Meeting (see Table IV), and hosts for the NRC site visits (see p. 9, in Ref. 12). Special thanks go also to W. W. Meinke and V. Iyengar for many stimulating discussions and exchanges of information, and to F. Reichel for permission to present some of her findings in Fig. 2.

Literature Cited

1. "Standard Radiological Effluent Technical Specifications for Pressurized Water Reactors"; U. S. Nuclear Regulatory Commission, NUREG-0472, Rev. 3, September 1982.
2. "Analytical Quality Control Service Programme; Intercomparison Runs, Certified Reference Materials, Reference Materials"; International Atomic Energy Agency, 1986-87.
3. Parr, R. M. "IAEA Biological Reference Materials"; In "Biological Reference Materials: Availability, Uses, and Need for Validation of Nutrient Measurement"; Wolf, W. R., Ed.; Wiley: 1985; Chap. 3.
4. Keith, L. H.; Crummett, W.; Deegan Jr, J.; Libby, R. A.; Taylor, J. K.; Wentler, G. "Principles of Environmental Analysis," Analyt. Chem. 1983, 55, 2210.
5. Reichel, F. "Practical Estimation of the Limit of Detection in Gamma Spectrometry Using a Commercially Available Program"; Chemistry Unit, IAEA Laboratories, 1986. See also K. Heydorn in Ref. 6, pp. 179ff.
6. "Quality Assurance in Biomedical Neutron Activation Analysis"; IAEA, IAEA-TECDOC-323, 1984.
7. "Intercomparison of Cadmium and Other Elements in IAEA Horse Kidney (H-8)"; IAEA, Progress Report No. 2, 1985.
8. Dybczynski R.; Veglia A.; Suschny, O. "Report on the Intercomparison Run for the Determination of Inorganic Constituents of Milk Powder"; IAEA, IAEA/RL/68, 1980. See also Ref. 3.
9. Iyengar, G. V.; Tanner, J. T.; Wolf, W. R.; Zeisler, R. "Preparation of a Mixed Human Diet Material for the Determination of Nutrient Elements, Selected Toxic Elements and Organic Nutrients: a Preliminary Report"; submitted to The Science of the Total Environment, 1986.
10. Parr, R. M. convenor, "IAEA Consultants, Meeting on Limit of Detection," Vienna, December 1985.
11. Natrella, M. G. 'The Relation between Confidence Intervals and Tests of Significance,' in Ku, H., Ed.; "NBS Spec Publ 300," 1969.
12. Currie, L. A. "Lower Limit of Detection: Definition and Elaboration of a Proposed Position for Radiological Effluent and Environmental Measurements"; U. S. Nuclear Regulatory Commission, NUREG/CR-4007, 1984.
13. Currie, L. A.; DeVoe, J. R. 'Systematic Error in Chemical Analysis,' in Devoe, J. R., Ed.; "Validation of the Measurement Process"; American Chemical Society, Washington, 1977; Chap 3.

14. Currie, L. A. "Quality of Analytical Results, with Special Reference to Trace Analysis and Sociochemical Problems"; Pure & Appl. Chem. 1982, 54, 715. (See especially the discussion surrounding Fig. 19.)
15. Iyengar, G. V.; Kasperek, K.; Feinendegen, L. E.; Wang, Y. X.; Weese, H. "Determination of Co, Cu, Fe, Hg, Mn, Sb, Se and Zn in Milk Samples"; The Science of the Total Environment, 1982, 24, 267.
16. Parr, R. M. "Technical Considerations for Sampling and Sample Preparation of Biomedical Samples for Trace Element Analysis"; J. Res. NBS 1984 and reference 48 therein: Damsgaard; Heydorn, K, RISØ Rep Risø-M-1633, 1973.
17. "Quality Assurance in Biomedical Neutron Activation Analysis"; Analyt. Chim. Acta. 1984, 165, 1, IAEA, Report of an Advisory Group.
18. Parr, R. M.; Houtermans, H.; Schaerf, K. 'The IAEA Intercomparison of Methods for Processing Ge(Li) Gamma-Ray Spectra'; in "Computers in Activation Analysis and Gamma-Ray Spectroscopy"; U. S. Dept. of Energy, Sympos. Ser. 49, 1979, 544.
19. Currie, L. A. "The Limitations of Models and Measurements as Revealed through Chemometric Intercomparison"; J. Res. NBS 1985, 90, 409.
20. Iyengar, G. V. "Concentrations of 15 Trace Elements in Some Selected Adult Human Tissues and Body Fluids of Clinical Interest from Several Countries"; Juelich Nuclear Research Center, Rep. No. 1974, Feb 1985.
21. "Standard Methods for the Examination of Water and Wastewater"; 13th Ed, Amer. Public Health Assoc., Washington, 1973, p. 33..
22. Iyengar, G. V. "Human Health and Trace Element Research"; Sci. Total Environ. 1981, 19, 105.
23. Iyengar, G. V. "Normal Values for the Elemental Composition of Human Tissues and Body Fluids: a New Look at an Old Problem," In "Trace Substances in Environmental Health XIX"; Hemphill, D. D., Ed.; Univ of Missouri, Columbia, 1985, 277.

RECEIVED September 28, 1987

Chapter 10

Effects of Analytical Calibration Models on Detection Limit Estimates

K. G. Owens[1], C. F. Bauer, and C. L. Grant

Department of Chemistry, University of New Hampshire, Durham, NH 03824

Detection limit estimates derived from confidence bands around analytical calibration curves are highly dependent on the experimental design and on the statistical data treatment. Procedures are described for testing the linearity of data and whether the intercept differs significantly from zero. Insensitivity of the correlation coefficient for the evaluation of goodness of fit of calibration models is emphasized. Unweighted linear models with an intercept often yield overly conservative detection limits. Frequently, an unweighted zero-intercept model is justified on both theoretical and statistical grounds. This model yields confidence bands and detection limits consistent with experiment. When the variance of signal measurements increases with concentration, more realistic confidence bands and detection limits are produced by weighting the data.

Despite numerous papers dealing with the specification of analytical method detection limits (see, for example 1-6), much disagreement remains about the choice of both experimental and computational procedures. In part, these disagreements appear to be related to the technical objectives of the experimenter. For example, a detection limit (DL) might be estimated from some multiple of the standard deviation of blank solution signals measured during a short time interval using a painstakingly optimized instrument. Such a DL estimate clearly provides useful information but it would be unrealistic to expect to maintain an equivalent DL during an extended analysis program with real samples. Unfortunately, these differences are frequently ignored, thereby encouraging unproductive controversy. It is important to remember that a DL is not an intrinsic property but rather, it is

[1]Current address: Chemistry Department, Indiana University, Bloomington, IN 47405

0097-6156/88/0361-0194$06.00/0

the interactive product of many variables including the method, the instrumentation, the nature of the samples, and the experimenter.

This discussion will be restricted to estimation of DL's appropriate for routine application of methods in which a calibration function relates a signal to concentration for a series of standards. No attempt will be made to evaluate the merits of various terms such as limit of quantitation, method detection limit, lower limit of reliable assay measurement, and others. For a discussion of those issues, the reader is referred to the articles cited here and to other papers from this symposium.

Consider a typical procedure such as the spectrophotometric determination of an analyte in groundwater samples. Quite likely, a single calibration curve will be used to cover a concentration range that extends from below the regulatory limit (hopefully) to some elevated concentration far removed from the limit. In this situation, a DL can be based on confidence limits (CL) around the calibration curve (7-14). The DL estimate produced in this fashion can then reflect the combined uncertainties in sample analysis and calibration.

Clearly the design of the calibration procedure and the statistical analysis of the data are both important considerations. Questions which require attention are conveniently divided into two groups; those pertaining to the experimental design and those pertaining to the statistical analysis. Design questions include:

a) What concentration range should be covered?
b) How many standards and blanks should be used?
c) How should standards be distributed over the range of interest?
d) How many replicate measurements should be made on standards and blanks, in what order, and over what time frame?
e) How should standards be prepared?

Statistical analysis questions may include:

a) Are the signal measurements (y_i) normally distributed at each concentration level used?
b) Are the variances of the signals (S^2_{yi}) homogeneous, i.e., independent of concentration?
c) Is a linear model justified or is curvature indicated?
d) If linear, is the intercept (b_o) significantly different from zero, i.e., is a zero-intercept model suitable?

Calculation of CL's should proceed only after these questions have been answered. Too frequently, a linear model of the form $\hat{y} = b_o + b_1 x$ is fitted using least squares procedures and CL's are calculated without proper attention to these issues. If the calibration model is inappropriate, it almost certainly follows that DL's based on CL's around that model are inaccurate. Therefore, a major portion of this paper will deal with the selection of a calibration model. At the end of the paper, we return to a consideration of the effects of that model on DL estimates.

Experimental Design Questions.

Concentration Range. Choice of the concentration range to be covered is generally dictated by the nature of the samples to be analyzed, precision and accuracy requirements, and the inherent limitations of the procedure. For example, an environmental pollutant may be present in samples at concentrations near the DL on the fringes of an area but at substantially elevated concentrations within the most severely impacted area. Low concentrations must be accurately estimated in order to define the geographic distribution in accordance with regulatory guidelines. However, it is also necessary to determine high concentrations accurately to plan cleanup strategy. In these circumstances, it may be best to produce two calibration curves; one for the low concentration range from which the DL can be estimated, and one for the full concentration range. If only the latter calibration is performed, it is likely that DL estimates will be unsatisfactory.

Number of Standards. The number of standards required depends in large measure on the concentration range and the nature of the anticipated functional relationship. A zero intercept linear model for a limited concentration range may be adequately defined with three standards although most investigators prefer four or five. In contrast, a thorough evaluation of a curvilinear model requires at least five standards and more may be desirable when spanning a wide concentration range.

Distribution of Standards. The location of the calibration points on the concentration axis exerts far more influence on DL estimates than has been generally recognized. Common practice is to space standards equidistant across the entire range of interest. However, for a specified number of standards, lower estimates of DL's are obtained without compromising the reliability of the high concentration range when an unsymmetrical distribution favoring low concentrations is used. These effects will be illustrated in a later section where CL computations are discussed.

Replication of Standards. An obvious benefit of replication is improved reliabiliy of the results. Other benefits are the ease of testing the goodness of fit of the calibration model and the opportunity to intersperse measurements on standards randomly across an entire lot of samples to which that calibration curve will apply. By this arrangement the standard deviation estimated from the calibration data will usually correspond closely to the value estimated from replicate measurements on samples. Otherwise, reproducibity of samples may be poorer than for standards. Of course, this assumes no systematic drift of signal response during the course of the measurements. If drift is suspected, it can be checked by making several measurements on standards over an extended time period. Any procedure should be demonstrated to be performing normally and in control before starting on a lot of samples.

The question of how many replicate measurements to make must include consideration of the magnitude of variability, the time required, the cost of each measurement, and the reliability required in the final result. For many situations we find duplicates or triplicates are quite adequate.

Preparation of Standards. To meet the requirement of independence of errors, each replicate and each standard should be prepared separately and with great care (15). Subsequent regression analysis will assume no error in the concentration (or at least, that the error is small in comparison to the error in signal measurement) and that the errors are independent. Thus, solution standards should ideally be prepared from more than one stock solution using a variety of pipets and volumetric flasks.

Choosing a Calibration Model.

An assumed regression model may be either linear or nonlinear. Choosing the correct model is crucial to obtaining accurate results. Although several recent papers (16-18) extoll the virtues of nonlinear calibration curves as a means of improving accuracy or to extend the range of concentrations covered, this discussion will consider only linear models. Evaluation of nonlinear models is an extension of the linear case with similar conceptual framework.

Unweighted least squares curve fitting is based on the assumptions that (a) measurement errors follow a Gaussian distribution and that (b) variances are independent of concentration, i.e., they are homogeneous. In typical calibrations, insufficient data are collected to rigorously test either assumption. Fortunately, modest violations do not cause serious errors but Garden et al. (19) warned that incorrect use of an unweighted least squares analysis could cause gross errors in the estimation of trace concentrations. In the initial portion of this discussion we will consider examples where both assumptions appear valid. Later we will examine the effects of nonuniform variance.

Linear Model With Intercept. There are two distinct linear first-order regression models that are generally encountered in analytical calibration. The non-zero intercept model is the most familiar, and it is given by Equation 1.

$$\hat{y} = b_o + b_1 x \quad (1)$$

The estimates of intercept (b_o) and slope (b_1) are calculated so as to minimize the sum of squares (SS) of the deviations of the observed signals (y_i) from the predicted value ($\hat{y}$) at any concentration (x) without constraints. For some determinations, however, theory predicts that the response of the instrument should be linear with concentration and should also be zero when there is no analyte present. Thus, if the instrument has been calibrated correctly, the calculated line should pass through the origin by definition. The proper regression model would then be the zero intercept model shown as Equation 2.

$$\hat{y} = b_{1_o} x \quad (2)$$

The estimate of the true slope, b_{1o}, is calculated so as to minimize the SS of deviations from the line with the restriction that the line must pass through the origin. Each of these models will be considered in the following paragraphs.

To facilitate this discussion, we have fabricated some typical spectrophotometric calibration data employing duplicate absorbance measurements at five concentrations. A reagent blank was used to set zero absorbance. Two sets of data are shown in Table I. The absorbance values at each of the 4 lowest concentrations are identical for each set. The difference occurs in the absorbance values for the highest standard where a negative deviation from Beers Law is represented by a reduced absorbance for case II. The regression equations and correlation coefficients were calculated according to standard equations available in any text on regression analysis.

Table I. Data and Statistical Analysis for Spectrophotometric Calibration

Concentrations of Standards (x)	Measured Absorbances (y) Case I	Measured Absorbances (y) Case II
0.500	0.054, 0.050	0.054, 0.050
1.00	0.103, 0.109	0.103, 0.109
2.00	0.202, 0.192	0.202, 0.192
5.00	0.494, 0.514	0.494, 0.514
10.00	0.975, 1.005	0.915, 0.945
Least Squares Model With Intercept	$\hat{y}=0.00431+0.0988x$	$\hat{y}=0.0149+0.0927x$
Correlation Coefficients	r = 0.9996	r = 0.9988
Models Through Origin	$\hat{y} = 0.0994x$	$\hat{y} = 0.0948x$

Goodness of Fit. The fitted model with intercept for Case I is seen to have a correlation coefficient of 0.9996 which would often be interpreted to mean that the equation fits the data very well. However, we shall see from the Case II data set that the correlation coefficient is not a sensitive method of evaluating curve fit. Hunter (20) notes that in statistical theory, correlation is a measure of the relationship between two random (dependent) variables. In a calibration problem, however, it is assumed that there is a definite functional relationship between the dependent and independent variables. Correlation, in its strict statistical sense, does not exist. Van Arendonk et al. (21) point out that the correlation coefficient is an insensitive tool for use in evaluating the quality of the fitted equation, and its use in such a manner may lead to erroneous conclusions.

We believe that it is far more instructive to perform a regression analysis in which the sources of variation are fractionated into the sums of squares (SS) attributable to regression and the SS for residuals. When replicate measurements have been made, the residual SS can be further fractionated into a systematic error component and a random error component. The systematic error component is designated the SS due to lack of fit (LOF) because it arises from the inadequacy of the fitted regression model to describe the experimental points. Table II-a gives the equation for calculating the SS of residuals with N-2 degrees of freedom (d.f.), since two regression coefficients were fitted. Many statistical analysis programs routinely provide the SS of residuals. The SS for random error (SS error) is independent of the regression model employed, i.e., it depends solely on the distribution of replicates around the mean response for each standard. When duplicate measurements have been acquired for each standard, the SS error is calculated as shown in Table II-a where d = difference in signal for each set of duplicates.The total d.f. in this error estimate would be equal to the number of duplicate sets since each would contribute 1 d.f. The SS for LOF is obtained by difference between the residual SS and the random error SS. Similarly, the d.f. associated with LOF is obtained by difference.

These calculations are illustrated in Table II for linear models with intercepts fitted to the data sets of Table I. Inspection of Table II reveals that the F-ratio for LOF for the Case I results is not significant as expected, i.e., the model is an adequate description of the data. For the Case II results, however, the LOF is significant at the 0.10 significance level despite finding a correlation coefficient of 0.9988! With the high probability that the linear model does not properly fit the data for Case II, it seems unreasonable to use such a calibration curve without trying to resolve the problem. Note that the nature of this test is such that the LOF will not show significance if large random errors are present. In fact, when random error is large, it is difficult to detect systematic variations that might result in LOF. In this example, however, random error is the same for each case. The LOF is caused by a negative deviation of absorbance for the highest concentration standard. The problem can be resolved by reducing the concentration of the highest standard to the upper limit of the linear range or possibly by fitting a nonlinear model. The important point is that the correlation coefficient provides no real insight concerning the extent or nature of residuals whereas the LOF test does.

It is important to note that an observation (or set of observations) on a standard may be rejected as an outlier only if it is not at the extreme ends of the calibration curve. If the lowest (or highest) standard appears to be an outlier, it can not be determined from the data collected whether the concentration of the standard is in error or if the response of the instrument is beginning to deviate from linearity. The "outlying" observation would have to be retained unless additional measurements made on a standard of lower (higher) concentration indicates that the deviation from the calculated line is not due to non-linearity of the response function.

When replicate measurements are not available, a thorough analysis is required of the residuals: the individual differences between the experimental points and the calculated regression line. Patterns in residual plots provide insight concerning the validity of the fitted equation and possible causes when the fit is

Table II-a. Formulation of Regression Analysis Table Using The Calibration Data of Table I and A Linear Model With Intercept

Source of Variation	Sum of squares (SS)	Degrees of freedom (df)	Mean square (MS)	F-ratio (F)
Residual	$\left[\Sigma y^2 - \frac{(\Sigma y)^2}{N}\right] - b_1^2\left[\Sigma x^2 - \frac{(\Sigma x)^2}{N}\right]$	8	$\frac{\text{resid. SS}}{8}$	-
Error	$\frac{\Sigma d^2}{2}$	5	$\frac{\text{SS error}}{5}$	-
Lack of Fit (LOF)	Residual SS - Error SS	3	$\frac{\text{SS LOF}}{3}$	$\frac{\text{MS LOF}}{\text{MS error}}$

Table II-b. Regression Analyses With LOF Tests For Table I Calibration Data Using Intercept Model

Source of Variation	SS	df	MS	F-ratio*
Case I				
Residual	0.000872	8	0.000109	
Error	0.000726	5	0.000145	
LOF	0.000146	3	0.0000487	0.34
Case II				
Residual	0.002518	8	0.000315	
Error	0.000726	5	0.000145	
LOF	0.001792	3	0.000597	4.12

*The F-ratios required for 3 and 5 df at various significance levels are 3.62 for 0.10, 5.41 for 0.05.

poor. This subject is comprehensively discussed in Draper and Smith (15).

Zero Intercept Model. For spectrophotometric determinations, theory predicts that response of the instrument should be linear with concentration and that the response should be zero when there is no analyte present. The zero intercept regression model (Equation 2) provides parameter estimates which meet this restriction. The expression used to calculate the slope of the line through the origin is:

$$b_{1_o} = \frac{\Sigma xy}{\Sigma x^2} \tag{3}$$

Fitted models through the origin are shown in Table I for the two sets of data previously discussed.

Before the equation of the line calculated using the zero intercept model is employed to evaluate unknowns, it must be tested to determine if the model is adequate to describe the experimental data. Regression analysis tables are constructed prior to testing the statistical validity of the assumption that the intercept of the line is zero. The format for calculation of the regression analysis tables is shown in Table III-a and the analyses of the Table I data are shown in Table III-b.

Inspection of these tables shows that the LOF test results are very similar to those for the models with intercepts. Comparison of Tables II-b and III-b reveals that the SS residuals are somewhat larger for the zero intercept models than for the models with an intercept. This difference can be used to test the hypothesis that the intercept is zero. First, it must be demonstrated that the LOF is not significant since it would not make good sense to test the zero intercept hypothesis for linear models shown not to fit the data. Furthermore, the SS error and SS(LOF) should not be combined as SS residuals when LOF is significant. These requirements are met by the Case I results. To test the hypothesis that the intercept does not differ significantly from zero, calculate:

$$F = \frac{(\text{SS residual for zero intercept model}) - (\text{SS residual of model with intercept})}{\text{MS residual of model with intercept}} \tag{4}$$

For the Case I data, $F = \frac{0.000960-0.000872}{0.000109} = 0.81$

The d.f. in the numerator will always be 1 because (N-1-(N-2)=1 and, therefore the difference in these SS are divided by 1 to get the MS. The d.f. in the denominator are N-2 or 8 in this example. At the 0.05 significance level, the required F value with 1 and 8 d.f. is 5.32. Clearly, we can not reject the hypothesis that the intercept is zero and consequently we conclude that this model is consistent with the data.

It can be very advantageous to achieve a calibration that has a zero intercept if it can be demonstrated that this condition can be sustained on a long term basis. We find that some systems that are carefully zeroed on blanks will meet this requirement. Under

Table III-a. Formulation of Regression Analysis Table Using The Calibration Data in Table I and a Zero Intercept Model

	Sum of squares (SS)	Degrees of freedom (df)	Mean square (MS)	F-ratio (F)
Residual	$\Sigma y^2 - \frac{(\Sigma xy)^2}{\Sigma x^2}$	9	$\frac{\text{resid. SS}}{9}$	-
Error	$\frac{\Sigma d^2}{2}$	5	$\frac{\text{SS error}}{5}$	-
Lack of Fit	Residual SS-Error SS	4	$\frac{\text{SS LOF}}{4}$	$\frac{\text{MS LOF}}{\text{MS error}}$

Table III-b. Regression Analyses With LOF Tests For Table I Calibration Data Using Zero Intercept Model

Source of Variation	SS	df	MS	F-ratio*
Case I				
Residual	0.000960	9	0.000107	
Error	0.000726	5	0.000145	
LOF	0.000234	4	0.0000585	0.40
Case II				
Residual	0.003577	9	0.000397	
Error	0.000726	5	0.000145	
LOF	0.002851	4	0.000713	4.92

*The F-ratios required for 4 and 5 d.f at various significance levels are 3.52 for 0.10, 5.19 for 0.05.

such conditions, calibration for each lot of samples to be analyzed requires only that replicates of the highest standard be run. The mean of duplicate or triplicate measurements of instrument response is checked to see if it falls within the confidence intervals established for the original calibration curve. When the mean is within the intervals (as it will be most of the time), the system is considered to be in control and the original calibration curve is employed. If the mean is outside the intervals, further instrument calibration (adjustments) must be made to return the response to a state of control before analysis of samples is attempted.

Although theory predicts that the calculated line should pass through the origin by definition, sometimes the experimental data indicate that the zero intercept model is not adequate, i.e., the intercept of the line is statistically different from zero. When this happens, valuable diagnostic information about the analytical method is available. For a spectrophotometric system such as the one described here, a positive intercept may reveal the presence of uncorrected background interferences or that the data have not been adjusted for the blank. Undetected nonlinearity can also cause the intercept to deviate from zero. Another possibility is if the concentrations of calibration standards are high by a constant amount, it will shift the calibration curve to the left and produce a positive intercept. If the intercept of the calculated equation is negative, it may indicate that the concentrations are actually lower than what is calculated. Thus knowledge of the method being used, coupled with a thorough statistical analysis of the data, could indicate chemical problems with the method that caused it to deviate from theory.

Heterogeneity of Variances. Earlier in this discussion, we promised to return to the question of non-uniform variance of signal measurements over the concentration range used. Often it is not clear by inspection whether variances are heterogeneous. One way to test this assumption is by Bartlett's Chi-square test. Because the calculations for this test are quite extensive we use a simpler test based on the comparison of ranges (22). The test involves calculating the range between the highest and lowest responses reported for each standard and calculating the ratio $R_{max}/(R_1 + R_2 + --R_k)$ where R_1, R_2---R_k are the ranges for each of k standards and R_{max} is the largest range. This test can be applied with as few as two replicates at each concentration, but it is more reliable and more sensitive when a greater amount of replication is available. When the calculated ratio exceeds the tabular value for a 0.05 significance level, we reject the hypothesis that the variances are homogeneous.

A variety of approaches have been recommended to deal with the problem of heterogeneity of variances. We favor the use of weighting because "the accuracy of the calibration curve is almost invariably increased when weighting factors are incorporated, taking into account the experimentally determined variances at each measurement point" (17). Weighting can be approached using different degrees of sophistication, but all methods are based on obtaining estimates for the variance across the measurment range.

For instance, Garden et al. (19) recommend that a functional relationship be fitted to plots of variance and standard deviation versus the independent variable (concentration). A weighting procedure we often find useful when a large number of replicates are available is an empirical one in which the weighting factor is simply the reciprocal of the variance (13). This scheme gives lower emphasis to more highly variable observations. The change in slope of the fitted model is usually quite minimal compared to an unweighted estimate, but there is often a large reduction in the standard deviation and consequently in the DL.

Equations used for the calculation of regression coefficients and for statistical analyses are the same as for unweighted data except that each datum is modified through multiplication by the appropriate weighting factor. (It could be argued that a weighting factor of 1.0 is used throughout when performing unweighted calculations). These calculations are easily performed using a variety of software programs. For details and equations, refer to Draper and Smith (15) and Oppenheimer et al (13).

Another recommended approach to deal with this problem is the use of transformations, especially the log transform (12). In our hands, this procedure has been less satisfactory than weighting but it represents an alternative approach deserving of consideration.

Confidence Limits and Detection Limits.

The widths of confidence intervals around calibration curves depend not only on the variability in the data, but also on the regression model chosen (14). For the non-zero intercept model, the best absolute precision occurs at $\bar{x}$, $\bar{y}$ which is the centroid of the regression line (Figure 1a). This circumstance pertains largely because $\bar{x}$, $\bar{y}$ is the axis of rotation for the uncertainty in the fitted slope of the calibration curve. In contrast, the best absolute precision for the zero intercept model occurs when the concentration (x) is zero (Figure 1b).

One of the commonly used methods for estimation of a DL with a non-zero intercept model is that of Hubaux and Vos (8). Equations for calculating the CL's and the DL are given in the original paper. In Figure 1a, the point where the upper band intersects the Y axis is designated Y_{DL}. Below this signal, an individual measurement cannot be distinguished from zero based on the calibration data collected. According to Hubaux and Vos, the lowest concentration that can be distinguished from zero is given by the intersection of the horizontal constructed at Y_{DL} with the lower confidence band, designated X_{DL} in Figure 1a.

Several problems arise when this procedure is followed. Because the centroid of the calibration data is far removed from zero, the CL's diverge greatly near zero due to the long extrapolation associated with uncertainty in the slope. Furthermore, if the variance tends to increase with concentration, the pooled standard deviation estimate is inflated relative to the actual standard deviation near the DL. As a consequence, the estimated DL is frequently above the lowest standard. Bailey et al. (23) reported recoveries up to 97% for spiked concentrations of

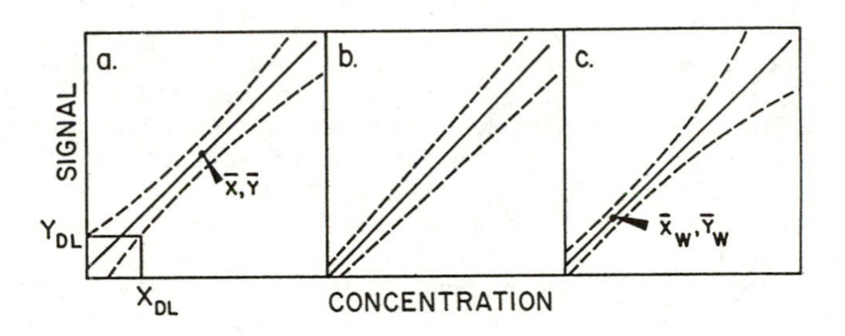

Figure 1. Confidence Limits For (a) Unweighted Non-Zero Intecept, (b) Unweighted Zero Intercept, and (c) Weighted Non-Zero Intercept Regression Models. Y_{DL} is the minimum detectable signal and X_{DL} is the concentration detection limit.

dye that were below the DL calculated by this procedure! Clearly this method of estimating the DL is overly conservative.

Hubaux and Vos recognized that when the variance increased with concentration, an inflated estimate of the DL would result. Their suggestion to cope with this problem was to recommend that most of the replication be conducted with the low concentration standards, a procedure that is somewhat analogous to weighting. In extreme cases, the high standards are not replicated at all. Such an approach will normally reduce the size of the standard deviation and it also moves the x, y point much closer to zero, thereby reducing the length of extrapolation. Predictably, DL estimates drop when this is done but the quality of calibration at higher concentrations also suffers.

In our experience, the unweighted zero-intercept model is often justified. As shown in Figure 1b, CL's for this model reflect our intuitive expectations and the DL estimates are usually consistent with our experimental evidence. Thus, we recommend careful study of the suitability of this model as described earlier.

When a non-zero intercept model is required and when variances are not homogeneous, then weighted procedures are favored. Equations for the calculation of CL's using weighting are given by Oppenheimer et al. (13). The general shape of these bands is shown in Figure 1c. Because the weighted centroid X_w, Y_w is much closer to the origin than for unweighted data, and because the weighted standard deviation estimate reflects more strongly the repoducibility at low concentrations, the DL limit estimates are substantially lower than for unweighted data. In our experience, estimates derived in this manner more accurately reflect the real capability of an analytical method.

Summary.

Careful experimental design is shown to be an essential prelude to analytical calibration. Such experiments lend themselves to thorough statistical evaluation and to improved confidence in the experimental estimates derived therefrom. Large variations in D.L. estimates can be explained by differences in calibration models and the associated assumptions and computations.

Acknowledgment.

Thanks are due to Mr. Thomas F. Jenkins, US Army Cold Regions Research and Engineering Laboratory, Hanover, NH for numerous helpful discussions of the concepts in this paper.

Literature Cited.

1. Currie, L.A. Anal. Chem. 1968, 40, 586-93.
2. Kaiser,H. Two Papers on the Limit of Detection of a Complete Analytical Procedure; Hafner: New York, 1969.
3. Boumans, P.W.J.M. Spectrochim. Acta 1978, 33B, 625-34.
4. Glaser, J.A.; Foerst, D.L.; McKee, G.D.; Quave, S.A.; Budde, W.L., Environ. Sci. Technol. 1981, 15, 1426-35.

5. Winefordner, J.D.; Long, G.L. Anal. Chem. 1983, 55, 712A-24A.
6. Kirchmer, C.J. Environ. Sci Technol. 1983, 17, 174A-81A.
7. Linnig, F.J.; Mandel, J. Anal. Chem. 1964, 36, 25A-32A.
8. Hubaux, A.; Vos, G. Anal. Chem. 1970, 42, 849-55.
9. Schwartz, L.M. Anal. Chem. 1977, 49, 2062-68.
10. Agterdenbos, J. Anal. Chim. Acta. 1979, 108, 315-23.
11. Agterdenbos, J; Maessen, F.J.M.J.; Balke, J. Anal. Chim. Acta. 1981, 132, 127-37.
12. Kurtz, D.A. Anal. Chim. Acta. 1983, 150, 105-14.
13. Oppenheimer, L.; Capizzi, T.P.; Weppelman, R.M.; Mehta, H. Anal. Chem., 1983, 55, 638-43.
14. Schwartz, L.M. Anal. Chem., 1986. 58, 246-50.
15. Draper, N.R.; Smith, H. Applied Regression Analysis, 2nd Ed.; John Wiley & Sons, Inc., New York, 1981
16. Schwartz, L.M. Anal. Chem. 1983, 55, 1424-26.
17. Jonckheere, J.A.; De Leenheer, A.P.; Steyaert, H.L. Anal. Chem. 1983, 55, 153-55.
18. Barnett, W.B. Spectrochim. Acta 1984, 39B, 829-39.
19. Garden, J.S.; Mitchell, D.S.; Mills, W.M. Anal. Chem. 1980, 52, 2310-15.
20. Hunter, J.S. J. Assoc. Off. Anal. Chem. 1981, 64 574-83.
21. Van Arendonk, M.D.; Skogerboe, R.K.; Grant, C.L. Anal. Chem. 1981, 53, 2349-50.
22. Pearson, E.S.; Hartley, H.O., Eds. Biometrika Tables for Statisticians, Cambridge Univ. Press, 1966, Table 31b.
23. Bailey, C.J.; Cox, E.A.; Springer, J.A. J. Assoc. Off. Anal. Chem. 1978, 61, 1404-14.

RECEIVED December 24, 1986

DETECTION LIMITS IN PRACTICE: CHARACTERIZATION AND APPLICATION OF ULTRASENSITIVE METHODS

Chapter 11

Critical Assessment of Detection Limits for Ion Chromatography

William F. Koch and Walter S. Liggett

National Bureau of Standards, Gaithersburg, MD 20899

The statistical basis for ion chromatography detection limits is investigated through the analysis of chromatograms by time series methods. Time series methods reveal two important chromatogram noise components, a cyclic variation caused by the pump and some large low-frequency variations with obscure origins. The component due to the pump can be removed from the chromatogram. The causes of the low frequency component should be investigated because these causes may not satisfy the prerequisites of statistical inference. Detection limit assessment depends on the choice of a peak detection algorithm. This algorithm must include a method for separating the low frequency component from the peak of interest and the a method for locating the peak in time. Algorithms that search for the peak in time cannot be assessed in the same way as algorithms that involve no search. This difference is discussed.

The detection limit indicates the performance of an instrument at low analyte concentrations. This indication may be used as a guide to instrument optimization, as a gauge of the suitability of an instrument for a particular application, or as a criterion for the interpretation of low concentration measurements. This paper concentrates on the latter use of detection limits and expands the discussion to include all aspects of statistical inference on low concentration measurements. In this case, the use of the detection limit is confined to the measurements in question and to the study at hand. The use of detection limits for instrument optimization and for suitability judgments requires a broader perspective that covers the various conditions under which the instrument might be used.

The question of what detection limit is achieved in the course of a set of measurements is both comfortingly specific and very demanding. The question is specific in that only the operating conditions used for the set of measurements need to be considered. Further, good practice suggests that these operating conditions be limited as much as possible. Thus, the amount of data that is needed is relatively modest. The appropriate data could be collected as part of the laboratory quality assurance program. The question is demanding in that conclusions with believable probabilities of error are often needed. This means that the measurements used to derive properties of the error must have the same error properties as the measurements about which conclusions are to be drawn. In other words, the unknown samples and the quality control samples must both result from a measurement process that is under control. In many cases, this condition is not easily achieved because of subtle differences between the conditions under which the quality control measurements are made and the conditions under which the real measurements are made. Specific causes of such differences, sample-to-sample carryover and mechanical transients, are discussed below.

This paper considers ion chromatography, which is a form of liquid chromatography based on ion exchange separation of analytes followed by conductimetric detection and quantitation (1). In particular, this paper is based on data from a Dionex Model 2020i. (See the disclaimer.) This instrument was set up for the measurement of nitrate and sulfate at concentrations below 1 mg/L. Except in the cases noted below, the instrument was configured as follows: The anion separator column was number AS4A (Dionex); the eluent was an admixture of 0.75 mmol/L $NaHCO_3$ and 2.0 mmol/L Na_2CO_3; the flow rate was 2.0 mL/min; and the sample loop volume was 20 µL. The background conductance of the eluent was chemically suppressed with a hollow fiber chemical suppressor (Dionex) with a 0.0125 mol/L H_2SO_4 regenerant flowing at 2.8 mL/min. Under these conditions, the background conductance was approximately 16 µS/cm. Most of this paper is relevant to other ion chromatography configurations. In some ways, this paper is relevant to all instruments that detect a peak on a noisy baseline. However, ion chromatography differs substantially from gamma spectroscopy and other measurement techniques based on radioactive decay because the randomness of radioactive decay has no analog in ion chromatography.

An issue of considerable importance in the discussion of detection limits is the choice of software for peak identification and integration. In practice, the choice of a data analysis algorithm can have as large an effect on the detection limit as the choice of instrument configuration. The data analysis in this paper has been done with general purpose statistical software rather than with one of the proprietary packages available for ion chromatography. General purpose software has the advantage of allowing flexibility in the data analysis. Also, general purpose software is based on algorithms that are known precisely. Unfortunately, proprietary packages often do not come with an exact specification of the algorithms employed. This is troublesome in work on detection limits because, as illustrated below, detection limits can be very sensitive to the choice of algorithm. Of course, the general purpose software used in this paper is not as convenient for routine laboratory use, as fully developed, or as well tested as the proprietary packages available.

The purpose of this paper is to demonstrate a method for exploring low-concentration performance and to illustrate the data characteristics that such a method typically reveals. Adaptations of the method can be used by any laboratory for the exploratory analysis of chromatograms. There are, however, chromatograms that require some sophistication in data analysis. Exploratory methods such as those discussed in this paper can be misleading. This paper does not discuss all the ways that the method might be misleading nor all the ways that the method might be improved. In the next section, some of the needed data analysis steps for the analysis of a single chromatogram are presented. In the third section, three properties of chromatograms, the pump cycle, the underlying white noise, and a low frequency component of unknown origin are discussed. In the fourth section, the question of how assessment of the detection limit is influenced by this low frequency component and by the choice of peak identification and integration algorithm is considered.

Initial Processing of the Chromatogram

The detection limit for a particular analyte is determined by the peak in the chromatogram due to the analyte and by other variations that obscure this peak. A chromatogram often has additional components that must be removed before the variations that determine the detection limit can be analyzed. First, a chromatogram often has a water dip, some large peaks due to other analytes, and other gross variations. Second, a chromatogram often exhibits a slowly varying baseline. Third, a chromatogram from a system with a pump might exhibit a cyclic variation due to the pump. In the chromatograms we consider, these components can easily be distinguished from the peak of interest. The water dip and large peaks due to other analytes do not coincide with the peak of interest; the baseline is much smoother than the peak of interest; and the pump cycle repeats regularly in time whereas the peak of interest does not. For these reasons, these components can be removed from the chromatogram with negligible effect on the detection limit. In other words, these components can be removed in such a way that the resulting adjusted chromatogram can be analyzed as if these components were never present.

To illustrate the removal of these components, we consider a chromatogram that is the result of a sample consisting of 0.005 mg/L nitrate and 0.025 mg/L sulfate. The first 116 seconds of this chromatogram consists of a short interval before sample injection, the water dip, and a large peak immediately following the water dip due perhaps to a solvent inadvertently mixed with the sample. Since the variations in the first 116 seconds are so large, we have excluded them from Figure 1 and from our analysis of this chromatogram. Otherwise, Figure 1 shows the chromatogram as it was produced by the instrument. As in other figures in this paper, we retain the origin for the time scale that was set by the instrument. The most obvious feature is the sulfate peak. What evidence there is of nitrate precedes the sulfate peak by about 120 seconds.

To specify our estimates of the baseline and the pump cycle, we introduce some notation. Let $y^{(0)}(t)$ denote the chromatogram as delivered by the instrument. The conductance units are nS/cm. Let $[t_0, t_0 + T - 1]$ denote the interval selected for analysis. For this chromatogram, we have $t_0 = 117$ and $T = 799$. Shifting the time origin to $t_0 - 1$, we obtain

$$y^{(1)}(t) = y^{(0)}(t + t_0 - 1), \quad t = 1, \ldots, T$$

In our estimation of the baseline and pump cycle, we must exclude the sulfate peak. To specify the computations, we define a weight function $w(t)$. This weight function includes the cosine-bell tapering of the ends of the intervals needed to reduce the bias in spectral estimation (2). Let the interval to be excluded because it has the sulfate peak be denoted by $[t_a, t_b]$. We let

$$
\begin{aligned}
w(t) &= 0 && \text{if } t_a \le t \le t_b \\
&= (1 - \cos(\pi t/21))/2 && \text{if } 1 \le t \le 20 \\
&= (1 - \cos(\pi(t-t_a)/21))/2 && \text{if } 1 \le t_a - t \le 20 \\
&= (1 - \cos(\pi(t-t_b)/21))/2 && \text{if } 1 \le t - t_b \le 20 \\
&= (1 - \cos(\pi(t-T-1)/21))/2 && \text{if } 1 \le T-t+1 \le 20 \\
&= 1 && \text{otherwise}
\end{aligned}
$$

If we had decided to exclude other segments of the chromatogram, we would have set $w(t)$ equal to zero for these segments, and we would have set $w(t)$ equal to the cosine-bell taper in the adjacent 20 point segments. The choice of a 20 point taper seems reasonable, but another choice might be better.

To remove the baseline, we first fit the baseline with a cubic spline (3). To specify the spline, we choose $n - 2$ interior knots, which we denote by $t_2, t_3, \ldots, t_{n-1}$. Let $1 = t_1 \le t_2 \le \ldots \le t_n = T$. In the example being considered, we chose two interior knots, $t_2 = 173$ and $t_3 = 484$. These knots divide the interval $[1, T]$ into approximately equal segments. A cubic spline can be characterized as a function with a continuous second derivative that is a cubic polynomial in each segment (t_j, t_{j+1}). We fit the cubic spline by linear least squares using all the points in the chromatogram for which $w(t) \neq 0$. Let the fitted baseline be denoted by $b(t)$. The chromatogram with baseline removed is given by

$$y^{(2)}(t) = y^{(1)}(t) - b(t)$$

Because the knots are far apart, the fitted baseline is so smooth that removing the fitted baseline has no effect on our ability to distinguish relatively narrow peaks from other similar variations.

If Figure 1 were plotted again with an expanded time axis, the periodic variation due to the pump would be visible. To remove the pump cycle, we first estimate its fundamental frequency from the spectrum of the chromatogram. The Fourier transform of the chromatogram that we use in our estimation is actually the Fourier transform of $w(t)$ times $y^{(2)}(t)$,

$$z(f) = \sum_{t=1}^{T} w(t)y^{(2)}(t)\exp(-i2\pi f(t-1))$$

Note that $z(f)$ equals the complex conjugate of $z(1-f)$ and that $z(f)$ equals $z(f-n)$, where n is an integer. Thus, we can compute $z(f)$ for $0 \le f \le .5$ and obtain its values for all f. Using the fast Fourier transform algorithm, we computed $z(f)$ for $f = j/N$, where $N = 6720$ and $j = 0, \ldots, 3360$. We chose a value of N much larger than T so that we could estimate the fundamental of the pump cycle with sufficient accuracy. The fundamental frequency of the pump cycle, which we denote by f_0, can be estimated from the peaks of the spectral estimate $|z(f)|^2$. There is reason to be concerned about the properties of this spectral estimate because of the gap in the series where the sulfate peak was. Spectral estimation for series with missing data have been discussed (4). For our example, we obtained $f_0 = 1129/6720$. This frequency corresponds to a period of approximately 6 seconds.

Our correction for the pump cycle is based on the premise that the pump cycle is perfectly stable over the entire chromatogram, although, as shown below, this is not always true. Based on the spectrum of the chromatogram, we have concluded that the fundamental and the first four harmonics contain virtually all of the pump cycle. For these reasons, we use as an estimate of the pump cycle

$$p(t) = 2\mathrm{Re}\left[\sum_{k=1}^{5} z(kf_0)\exp(i2\pi kf_0(t-1)) \right] / \left[\sum_{t=1}^{T} w(t) \right]$$

If $2kf_0$ is an integer, then this formula must be adjusted by substituting $z(kf_0)/2$ for $z(kf_0)$. We obtain as our adjusted chromatogram

$$y^{(3)}(t) = y^{(2)}(t) - p(t)$$

If the pump cycle is indeed stable, then this correction for the pump cycle does not distort peaks of interest because the estimate p(t) is obtained from a long record, the entire chromatogram.

The adjusted chromatogram for our example is shown in Figure 2. The sulfate peak stands out. Some evidence of the nitrate peak can be seen about 120 seconds before the sulfate peak. Comparison of Figures 1 and 2 shows that removal of the pump cycle does reduce what in Figure 1 appears to be random noise.

Models of Chromatogram Noise

The adjusted chromatogram, which is illustrated in Figure 2, provides a starting point for detailed modeling of the variations in the chromatogram that interfere most with the detection of analytes at low concentrations. In this modeling, we concentrate on three components, the pump cycle, vestiges of which may remain because of pump cycle instability, the underlying white noise, and some low frequency variations that might be mistaken for peaks of interest.

A direct way to investigate the stability of the pump cycle is to estimate the pump cycle at various points along the chromatogram. One approach to this estimation is inverse Fourier transformation of the Fourier coefficients z(f) that lie close to the pump cycle fundamental and its harmonics. This approach is a version of complex demodulation (5). Let M - 1 be the number of harmonics to be included; and let

$$z_M(f) = \begin{cases} z(f) & \text{if } |fN - \mathrm{mod}(kf_0N,N)| \le 24 \text{ or } |fN - \mathrm{mod}(N-kf_0N,N)| \le 24 \\ & \text{for } k = 1 \text{ or } 2 \text{ or } \ldots \text{ or } M \\ 0 & \text{otherwise} \end{cases}$$

An estimate of the pump cycle that allows some variation over the chromatogram is given by

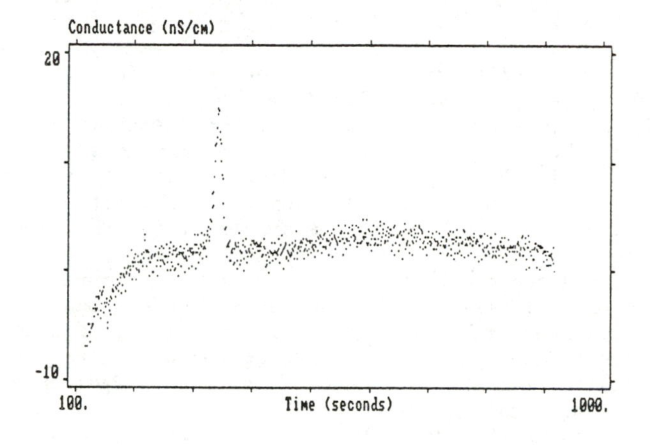

Figure 1. Chromatogram produced by the instrument.

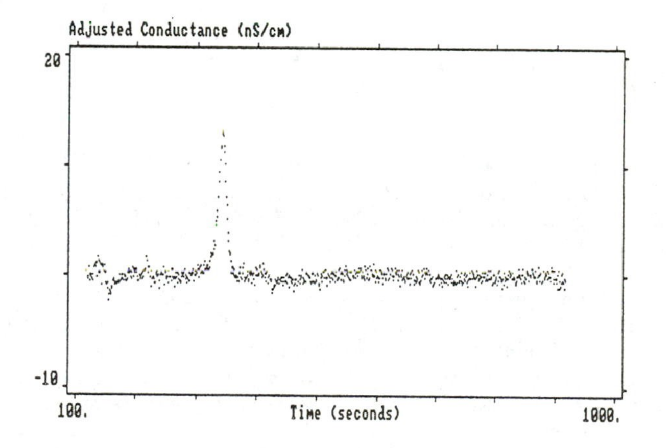

Figure 2. Chromatogram initially adjusted for baseline and pump cycle.

$$\tilde{p}(t) = (1/N) \sum_{j=0}^{N-1} z_5(j/N)\exp(i2\pi j(t-1)/N)$$

Figure 3 shows the pump cycle for a chromatogram different from the one considered in the previous section. The fundamental of the pump cycle in this case is 1128/6720. The shape of the pump cycle is shown every 50 periods, which is approximately every 300 seconds. In the figure, successive traces are offset by 1 nS/cm. The shapes shown are typical of those produced by the two piston pump that we used. This chromatogram shows considerable instability. The pump cycle in the chromatogram considered in the previous section is much more stable as is the pump cycle in another chromatogram that we consider in detail below. If the pump cycle is unstable, then an estimate such as $\tilde{p}(t)$ may be more appropriate for the removal of the pump cycle than the estimate used in the previous section.

The dependence of the pump cycle on the configuration of the instrument is of interest. As noted above, the chromatograph is generally operated with a flow rate of 2.0 mL/min. Reduction of the flow rate is accomplished by lengthening the period between strokes. The reduction to 1.5 mL/min lengthens the period from approximately 6 seconds to approximately 8 seconds.

Variations in the strength of the pump cycle are also of interest. Because the pump cycle is potentially unstable, we consider the power in frequency bands about the fundamental and the first two harmonics instead of the amplitude of the pump cycle as in the previous section. Our index of pump cycle power is given by

$$\left[\sum_{j=1}^{N-1} |z_3(j/N)|^2 \right] / \left[K_3 \sum_{t=1}^{T} w^2(t) \right]$$

where K_3 is the number of values of j in the sum for which $z_3(j/N) \neq 0$. The pump cycle power can be quite variable. The chromatogram discussed in the previous section has a pump cycle power of 6.6. Two other chromatograms obtained under the same conditions showed pump cycle powers of 5.8 and 12.3.

A series of chromatograms were obtained to investigate the dependence of the pump cycle power on the eluent. These chromatograms were obtained with the guard column AG4A (Dionex) in place. The first and last chromatograms, which were run with the standard eluent, gave pump cycle powers of 5.9 and 2.6. The second chromatogram, which was run with a weaker eluent 0.75 mmol/L Na_2CO_3, gave a pump cycle power of 4.0. This eluent gives a background conductance of 9.6 µS/cm instead of the standard 16 µS/cm. The third chromatogram, which was run with deionized water, gave a pump cycle power of 0.1. The background conductance of deionized water is approximately 2.6 µS/cm. This third chromatogram provides support for the hypothesis that the pump cycle power is lower for weaker eluents. Additional research is planned to investigate these aspects more fully.

A random noise component with flat spectrum is evident in the adjusted chromatograms. This component seems to be caused by the electronics of the instrument because its standard deviation does not vary with chromatographic conditions. The standard deviation of this underlying white noise is approximately 0.4 nS/cm. This value can be compared to the size of the pump cycles shown in Figure 3.

The noise shown in Figure 2 does not have a flat spectrum, however. The power spectrum is higher at low frequencies. To display the low frequency part of the adjusted chromatogram, we smooth the adjusted chromatogram using a filter with impulse response

$$h_1(t) = (1 + \cos(\pi t/8))/16 \qquad \text{if } -7 \le t \le 7$$
$$= 0 \qquad \text{otherwise}$$

This cosine bell is 8 seconds wide at half height and thus seems to match the peak width expected of a small nitrate peak. Figure 4 shows the smoothed adjusted chromatogram given by

$$\sum_{j=-7}^{7} h_1(j)w(t-j)y^{(3)}(t-j)$$

Note that we have included the weight function $w(t)$ to suppress the sulfate peak. Figure 4 has two equal height peaks on the left side. The right most of these is the nitrate peak. This peak can be compared to the other peaks in this smoothed chromatogram. Clearly, there are several peaks that might be mistaken for a peak due to an analyte of interest.

Figure 5 shows another chromatogram that we have adjusted and smoothed in the same way as Figure 4. This chromatogram is the first of the series we obtained to investigate the dependence of the pump cycle on the eluent. No sample was injected in the generation of this chromatogram. Nevertheless, this chromatogram also shows several peaks that might be mistaken for analyte.

The cause of the low frequency noise shown in Figures 4 and 5 is an important question. In most applications of ion chromatography, two possibilities can be suggested. One is sample-to-sample carryover due either to contamination in the sample injection loop or to a slowly eluting organic left on the separator column by a previous sample. The other is mechanical transients. The existence of the pump cycle suggests that variations in flow past the conductivity detector cause variations in the chromatogram. Mechanical transients can also cause variations in the flow. Other sources of noise are also possible, and will be investigated in future research.

Can this low frequency component be treated as though it were generated by a random mechanism so that a statistical statement can be made about it? Sample-to-sample carryover can be treated statistically only under very special conditions on the order in which the samples are analyzed. Mechanical transients can be treated statistically only if their source is random in some sense. The best solution to this low frequency component is to reduce its size until the question of its randomness is no longer important. This might be done by reducing its magnitude or alternatively by injecting each sample twice or more. In any case, any laboratory that does extensive low concentration ion chromatography analysis should

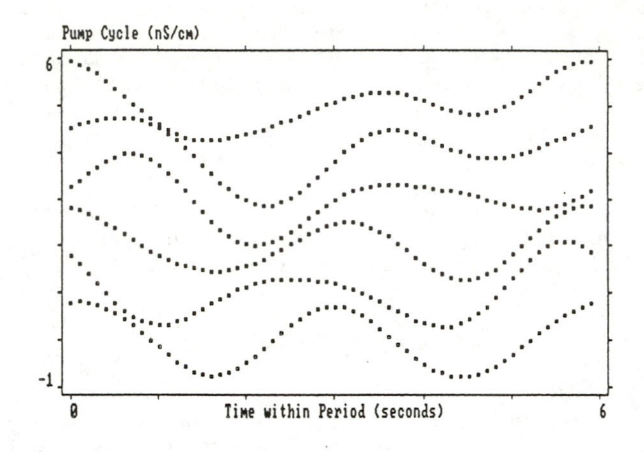

Figure 3. Shape of the pump cycle every 50 periods.

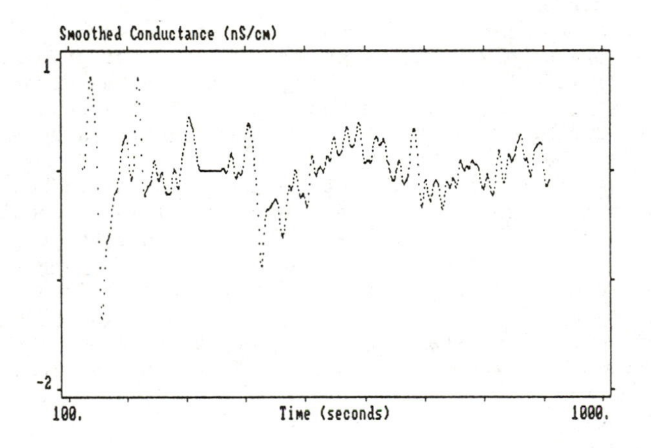

Figure 4. Chromatogram initially adjusted and smoothed, example with nitrate peak.

experiment with this component to see what type of statistical inferences are valid.

Detection Limits

Under the assumption that the low frequency component is a stationary Gaussian random process, let us proceed with the specification of a detection algorithm. The detection limit indicates the performance of the detection algorithm. Detection of peaks of interest in the adjusted chromatogram involves removal of the low frequency component to the extent possible and perhaps the search in time for the desired peak.

Figures 4 and 5 show variations with time that are much more gradual than the peaks of interest. These variations can be removed in various ways. Consider first the use of a filter with the following impulse response

$$h_2(t) = h_1(t) - (1 + \cos(\pi t/16))/32 \quad \text{if } -15 \le t \le 15$$
$$= 0 \quad \text{otherwise.}$$

This filter is the difference between the smoother used to obtain Figures 4 and 5 and a broader cosine-bell smoother. Applying this filter to the adjusted chromatogram discussed in the second section, we obtain the filtered chromatogram shown in Figure 6. Note that since we have suppressed the sulfate peak as we did in Figure 4, the nitrate peak is now the largest peak. The peaks in Figure 6 can be compared to a threshold appropriate to the underlying white noise component. The filtering changes the standard deviation of this component from 0.4 to 0.064. Thus, the appropriate threshold for a significance level of 0.05 is $1.645 \cdot 0.064 = 0.11$. Figure 6 suggests that this threshold is too low and that the filtering has not removed the low frequency component entirely. Nevertheless, this threshold is nearly right, and the underlying white noise seems to be the major contributor to the filtered chromatogram.

Another way to remove the low frequency component is to fit a spline that is not as smooth as the spline fit in the second section. We fit a quadratic spline (one that is a quadratic polynomial between knots and that has a continuous first derivative). We choose knots every 50 seconds and fit the spline by least squares to the adjusted chromatogram. Consider the adjusted chromatogram that was smoothed and shown in Figure 5. After subtracting the fitted quadratic spline from this adjusted chromatogram, thereby further adjusting it, and then smoothing the result with the filter $h_1(t)$, we obtain the smoothed, adjusted chromatogram shown in Figure 7. The threshold for this figure based on the underlying white noise is 0.20. Once again, the threshold seems to be too low but seems to be close enough to suggest that the underlying white noise is the dominant contributor to the smoothed, adjusted chromatogram shown.

Most software for ion chromatography includes the option to search in time for the peak of interest. Consider the following simple algorithm for finding the nearest peak:

1. Initialize left maximum and right maximum with the value at the time where the peak is expected.
2. Search left for a higher maximum until a value is encountered

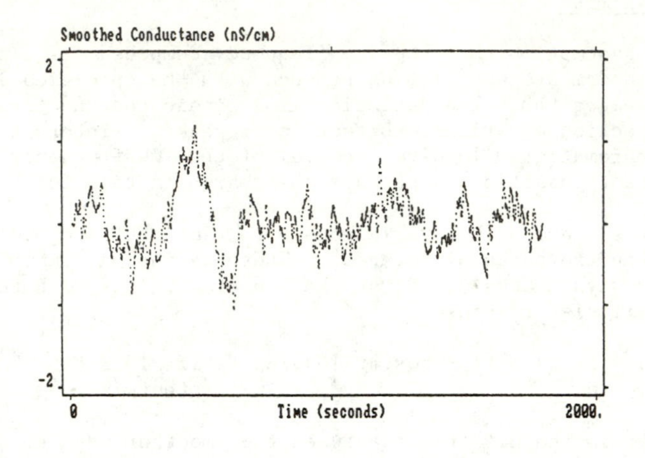

Figure 5. Chromatogram initially adjusted and smoothed, example without nitrate peak.

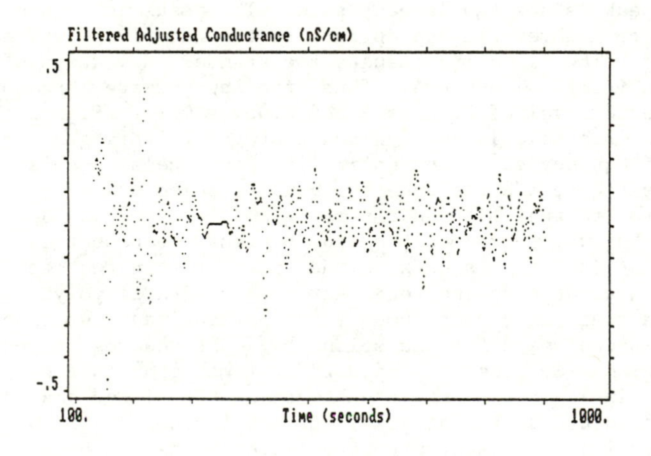

Figure 6. Chromatogram initially adjusted and filtered, example with nitrate peak.

that is less than the current left maximum by a specified amount.

3. Search right for a higher maximum until a value is encountered that is less than the current right maximum by a specified amount.
4. Choose the larger of the left and right maxima.

We have applied this algorithm to the smoothed, adjusted chromatogram shown in Figure 7 using as starting points a sequence of times that are 30 seconds apart. The specified amount used to end the search was 0.07. We can compare the distribution of the results of this to the distribution of all the points shown in Figure 7. We do this by plotting the quantiles of the distribution of nearest peaks against the corresponding quantiles of the distribution of all the points (6). The result is shown in Figure 8. If the two distributions were the same, then the points would lie along the 45 degree line through the origin. The distribution of the peaks is offset with respect to the distribution of all the points, a fact which is not surprising. Interestingly, the slope of the points is close to 1 suggesting that the variances of the two distributions are nearly the same.

Figure 8 provides one illustration of the difference between a detection algorithm that searches and one that does not. With an algorithm that does not search, any point in Figure 7 might be the noise contribution to the observed analyte peak. With an algorithm that searches, the distribution of the noise contribution depends on the concentration of the analyte. With no analyte, the distribution of the noise contribution is the distribution of noise peaks. However, the presence of analyte influences the search. With sufficient analyte, the analyte determines the result of the search. For this reason, in the case of an algorithm that searches, the distribution of the noise contribution to the observed analyte peak can be thought of as lying between the two distributions compared in Figure 8. This fact complicates the statistical inference for measurements near the detection limit. It also suggests that the determination of detection limits from measurements on the blank may not be a valid procedure when the software employed searches for the peak.

If no search for the nitrate peak is needed, if the low frequency component can be treated as stationary and Gaussian, and if the contribution to the chromatogram from the analyte does not vary, then the results in this paper provide an assessment of the nitrate detection limit. Figure 6 shows that 0.005 mg/L is close to the detection limit. If we add to the standard deviation of the underlying white noise an amount to compensate for the low frequency component and multiply by the usual factor 3.29, we obtain a comparable number. The amount to be added can be judged from Figures 6 and 7. Of the three conditions on the validity of this assessment, the first two are discussed above. The third, the condition on the contribution from the analyte, is also important. Variation in the contribution from the analyte might be caused by the separator column, for example. Several chromatograms with replicate sample injections with the same concentration of nitrate are needed to assess this variation. This will form the basis of the second phase of this research in the evaluation of detection limits.

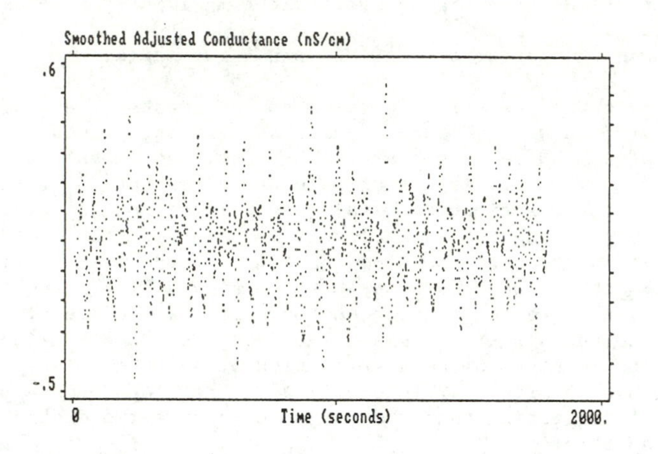

Figure 7. Chromatogram initially adjusted, finely adjusted for baseline, and smoothed, example without nitrate peak.

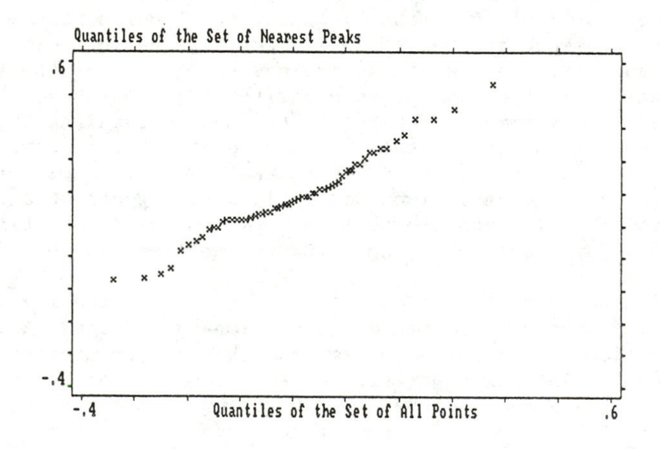

Figure 8. Quantile-quantile plot of nearest peaks and all points.

Disclaimer

Certain commercial equipment, instruments, or materials are identified in this paper in order to specify adequately the experimental procedure. Such identification does not imply recommendation or endorsement by the National Bureau of Standards, nor does it imply that the materials or equipment identified are necessarily the best available for the purpose.

Literature Cited

1. Wetzel, R. A.; Pohl, C. A.; Riviello, J. M.; MacDonald, J. C. In Inorganic Chromatographic Analysis; MacDonald, J. C., Ed.; John Wiley and Sons: New York, 1985; p 355.
2. Tukey, J. W. In Spectral Analysis of Time Series; Harris, B., Ed.; John Wiley and Sons: New York, 1967; p 25.
3. De Boor, C. A Practical Guide to Splines; Springer-Verlag: New York, 1978; Chapter 14.
4. Marquardt, D. W.; Acuff, S. K. In Applied Time Series Analysis; Anderson, O. D.; Perryman, M. R., Eds.; North Holland Publishing Co.: Amsterdam, 1982; p 199.
5. Bloomfield, P. Fourier Analysis of Time Series: An Introduction; John Wiley and Sons: New York, 1976; Chapter 6.
6. Chambers, J. M.; Cleveland, W. S.; Kleiner, B.; Tukey, P. A. Graphical Methods for Data Analysis; Wadsworth International Group: Belmont, California, 1983; Chapter 3.

RECEIVED December 24, 1986

Chapter 12

Risk Assessment as a Tool

for Verifiable Detection and Quantification of Fusarium Trichothecenes in Human Blood at Low Parts-Per-Billion Concentrations

D. J. Reutter, S. F. Hallowell, and E. W. Sarver

Research Directorate, Development and Engineering Center, U.S. Army Chemical Research, Aberdeen Proving Ground, MD 21010–5423

Analytical chemists are asked often to determine the presence of a chemical in a specific matrix at some stated concentration. Examples of such analytical procedures include the determination of controlled substances and toxins in the biological fluids of victims of alleged poisonings. Reports of the analytical findings in these cases frequently become the object of legal decisions or international policies. The definitions of detection limits in these instance can be significantly different from those commonly used in research and industrial laboratories. Stringent quality assurance and quality control throughout the entire analytical process is required to establish and maintain statistically definable limits of detection and quantification and the certainty of identification. The quality assurance plan developed and employed by this Center for the determination and quantification of Fusarium mycotoxins in human blood is reported. Under this plan, the analytical method is divided into four steps; (1) sample and standard handling, (2) sample preparation and clean-up, (3) derivatization and (4) analysis. Each step is evaluated to ascertain where errors occur, and specific quality control procedures are introduced in each step to detect, isolate and correct errors during the analysis. Limits of detection, verification and quantification are individually determined and validated.

Recently, few topics in analytical chemistry have occupied the scientific community more than the ability of chemical laboratories to reliably determine at the low parts-per-billion level the presence of Fusarium trichothecenes in environmental and toxicological samples. This paper provides a systematic approach for developing and implementing a quality assurance and quality control program for a complex analytical method in which human error and system failure can occur. The application of this approach to the problem of determining the presence of nine naturally

Published 1988 American Chemical Society

occurring mycotoxins and their metabolites in human blood at low part-per-billion (ppb) concentrations is given. The key to a successful program is the development of a risk analysis for the analytical procedure which explicitly recognizes the potential for human error. A necessary step in developing an analytical strategy is the re-statement by the analyst of program requirements necessary to delineate the intended use of data and the significance of reporting false positive, false negative, and imprecise results. It is only after these requirements are defined, and a reliable analytical method is developed, that detection limits which are appropriate for the analysis can be calculated. Then, specific quality control measures that eliminate or quantify errors must be developed and implemented. The implementation of such a plan for the analysis of a set of samples prepared by an independent laboratory and analyzed blindly by this laboratory follows.

Trichothecene mycotoxins are secondary metabolites of various fungal species. Structures of some trichothecene mycotoxins of interest to the US ARMY are given in Figure 1. Several methods have been reported for the analysis of these toxins (1-11, 15). Of these, mass spectrometry techniques are both sensitive and definitive when applied to toxicologic and environmental samples. With current technology, the most sensitive and qualitatively definitive analytical technique for the determination of these toxins is derivatization with an electron deficient moiety followed by analysis with negative ion chemical ionization gas chromatography-mass spectrometry (NICI-GC/MS).

The analytical procedure that is used by this laboratory for the analysis of simple Fusarium mycotoxins will be reported separately. However, the analytical scheme is outlined in Figure 2. The method is very arduous due to several sample clean-up steps which necessitates transfer of the sample between containers. The trichothecenes and their derivatives have a tendency to adhere to glass and can be quantitatively transferred only with numerous methanol washes. While the analytical method is both sufficiently sensitive and definitive for the program requirements, the sheer amount of human manipulation required for the completion of this analysis makes it somewhat unreliable if implemented without a responsible quality assurance and quality control program.

In an emerging and complex analytical methodology, three factors must always be considered. First, as always, the detection criteria must be established a priori., along with the criteria for the risk the laboratory will accept in reporting the results. This includes determination of the probable rate of reporting false positives and false negatives for the analysis. Second, a full risk assessment must be conducted to explore on paper, even before entering the laboratory, where catastrophic failures in the analytical methodology might occur. Third, in view of the risk assessment, a quality assurance/quality control program which will prevent catastrophic failures and measure the performance of the analysis must be implemented.

DETECTION, VERIFICATION AND QUANTIFICATION CRITERIA

The development of detection criteria always results from program requirements initiated by some general management guideline. In

T-2 TYPE

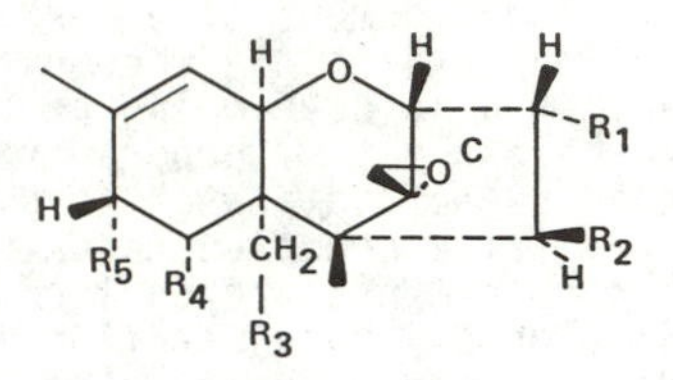

Trichothecene	R_1	R_2	R_3	R_4	R_5
Basic Trichothecene	H	H	H	H	H
Trichodermol (roridin C)	H	OH	H	H	H
Trichodermin	H	OAc[a]	H	H	H
* Verrucarol	H	OH	OH	H	H
* Scirpentriol	OH	OH	OH	H	H
Monoacetoxyscirpenol (MAS)	OH	OH	OAc	H	H
* Diacetoxyscirpenol (anguidine)(DAS)	OH	OAc	OAc	H	H
7-Hydroxy DAS	OH	OAc	OAc	OH	H
Calonectrin	OAc	H	OAc	H	H
15-Diacetylcalonectrin	OAc	H	OH	H	H
Dihydroxy trichothecene	H	OH	H	H	OH
* T-2 tetraol	OH	OH	OH	H	OH
Neosolaniol (solaniol)	OH	OAc	OAc	H	OH
Monoacetylneosolaniol	OH	OAc	OAc	H	OAc
7,8-Dihydroxy DAS	OH	OAc	OAc	OH	OH
* HT-2 toxin	OH	OH	OAc	H	b
* T-2 toxin	OH	OAc	OAc	H	b
Acetyl T-2 toxin	OAc	OAc	OAc	H	b

* - Toxins used in Study

a - $O-\underset{\underset{O}{\|}}{C}-CH_3$: Acetate

b - $O-\underset{\underset{O}{\|}}{C}-CH_2-CH(CH_3)_2$: Isovalerate

NIVALENOL TYPE

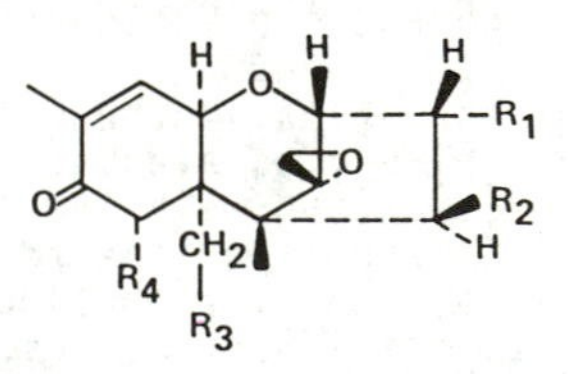

Trichothecene	R_1	R_2	R_3	R_4
Nivalenol	OH	OH	OH	OH
* Monoacetylnivalenol (fusarenon-X)	OH	OAc	OH	OH
Diacetylnivalenol (DAN)	OH	OAc	OAc	OH
* Deoxynivalenol (DON) (vomitoxin)	OH	H	OH	OH
Monoacetyl DON	OAc	H	OH	OH
Diacetyl DON	OAc	H	OAc	OH
Trichothecin	H	a	H	H
Trichothecolone	H	H	OH	H

Figure 1. Structure of Some Simple Trichothecene Mycotoxins

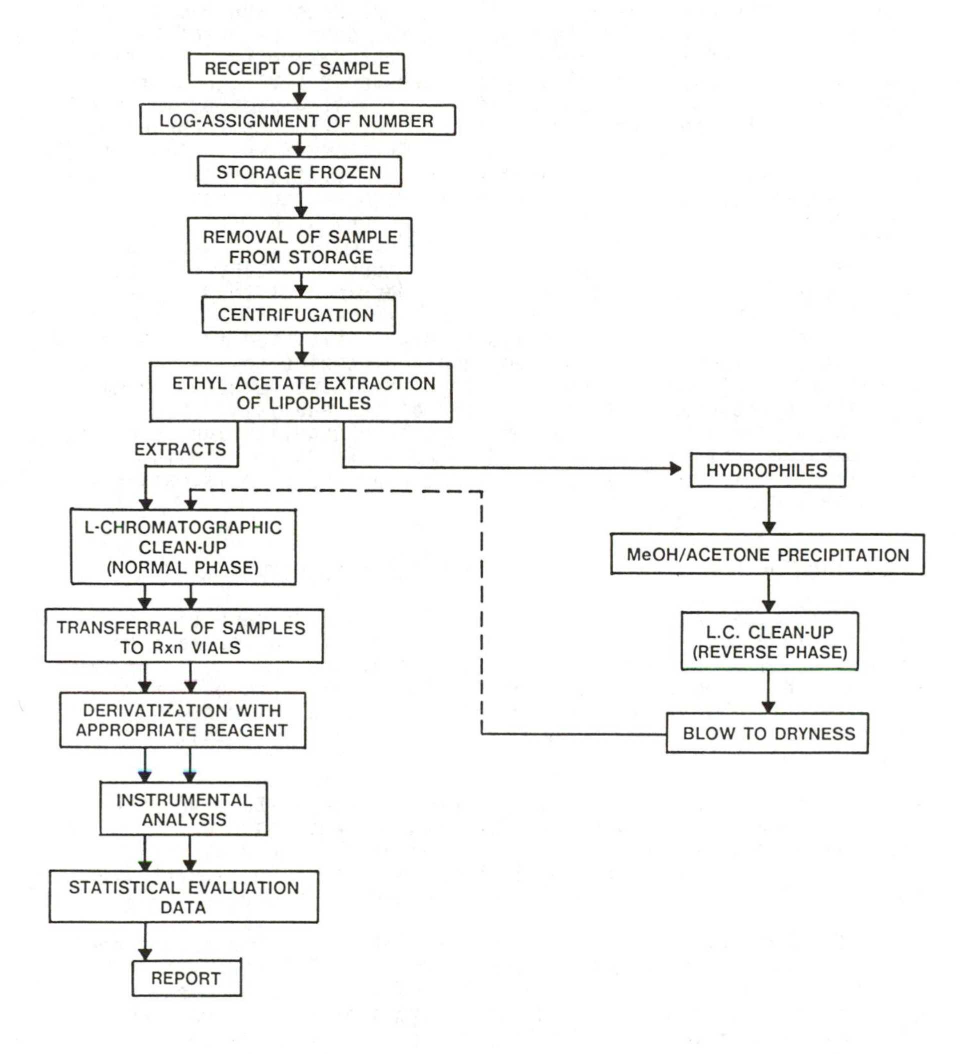

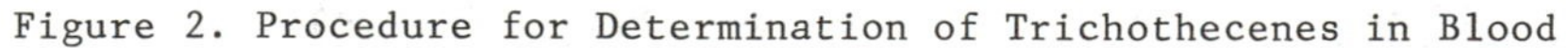
Figure 2. Procedure for Determination of Trichothecenes in Blood

this case, toxicological studies have established that concentrations as low as 10 ppb can be significant for some trichothecene mycotoxins when present in the blood or tissues of humans or other mammals(12-15). This is the detection limit criterion which served as the benchmark for development of the analytical methodology; however, it does not define the detection limit for the analysis as it is practiced and reported out of the laboratory. One of the most important tasks of a program manager is to translate the general program requirements into technical criteria. The program requirements for this study were simply to develop an analytical methodology which was sufficiently sensitive to detect and validate the presence of trichothecenes at toxicologically significant levels with a very high degree of confidence. This was translated into technical requirements as follows. 1) The quality of the data is more important than sample throughput. 2) False positives are more detrimental than false negatives however both modes of failure are very serious. 3) Minimum detectable limits and minimum limits of verification must be in the same range or lower than reported levels of toxicological significance. 4) Quality control must be run at a level which will demonstrate the quantitative precision of the analysis. 5) Qualitative certainty is more important than quantitative accuracy. The specific criteria which were developed to define minimum detection limit, verification limit and range of quantification are given in Table I.

EVALUATION OF SELECTIVITY FOR AN ANALYTICAL PROCEDURE

Due to political and international implications of reporting a false positive result, this type of error were considered the most serious error possible in this analysis. Therefore, the analytical procedure was examined to determine the potential for this type of error to occur. The best method of doing this is by experimentally analyzing a large number of samples which include a significant number in which the analyte is known to be below the detection limit of the procedure. In practice, this is seldom done both because of the cost involved in performing such a study and the probable nonavailability of standard samples which are well characterized at the concentration level at which a new, very sensitive analytical technique is being applied. An alternative to this is to examine the analytical procedure theoretically.

Using Kaiser's (16) definition for calculating the "informing power" of an analytical procedure, Fetterolf and Yost(17) have made calculations of the qualitative specificity of several mass spectrometric techniques and weighed them against the requirements of some analytical tasks. While their treatment makes several assumptions which may not be born out in practical analysis, it is useful for gauging a particular methodology to an analytical task. Fetterolf and Yost (17) calculated that capillary GC/MS at 1 amu resolution using full mass scans over a mass range of 1000 amu gave sufficient informing power by itself to identify with certainty any chemical compound in a chemical mixture.
In our methodology, the mass scans were, by necessity, limited to at most five ions in order to meet the milestone for sensitivity. This greatly reduces the informing power of the GC/MS portion of the analytical procedure; however, this is more than compensated for by the sample preparation steps in the

procedure which include extraction, liquid chromatography and derivatization. This is not wasted information, but is necessary to insure the qualitative reliability and precision of the analysis thus fulfilling the program requirements.

Johnson and Yost (18) have demonstrated the relationship between the intensity of the instrumental response to an analyte and the signal-to-noise ratio as a function of the number of steps in an analysis. With each additional step in the procedure, the magnitude of instrumental response decreases due to unavoidable loss of the analyte with each manipulation as the signal-to-noise ratio increases. The increase in the signal-to-noise ratio is due primarily to the elimination of interfering compounds which contribute to the "chemical noise". The obvious conclusion to draw is that when trying to achieve very low limits of detection the analytical methods should involve as many steps as possible before the absolute signal level is reduced to an undetectable level. This approach ignores the errors introduced by multiple manipulations of a sample which could potentially destroy the integrity of the analysis. Figure 3 shows, that while instrumental responses are generally lowered by additional steps, the overall sensitivity improved as the reliability of the analysis generally deteriorates. Thus, while analytical methodologies dealing with many steps may push detection limits to heretofore unachievable levels, they do so at the risk of producing unreliable data. The effect of human error on the analysis is particularly difficult to handle because such errors are not amenable to mathematical modeling, however human error is the most likely contributor to the reporting of false positive or false negative values, which is the most serious error made in an analysis.

RISK ASSESSMENT

One responsible means of addressing the effect of human error on an analytical methodology is to perform a risk assessment on the analytical procedure. A risk/cost/benefit analysis is based on a host of factors, including reasoning, guess work and past performance. The purpose of a risk assessment is to make sure that calamities happen first on paper, not in reality. Although almost all analytical chemists perform an informal risk assessment of some form or another, in a case where the analytical process must be used to make decisions of gravity, it is appropriate to bring as much analytical formality as possible to bear. There is a well developed science devoted to risk assessment; it is extensively used in the development of sophisticated hardware items or systems of linked hardware items. Examples include the risk assessment studies which are performed during the design phase of a nuclear power plant or an aircraft to determine the possible modes of failure. Some organizations require a risk assessment study for any development item resulting from any major program for research, development and acquisition of materiel items or systems (19). Although engineers must perform risk assessments, analytical chemists are not required to perform nor are they knowledgeable of such requirements, even though a failure of an analytical methodology can have implications and long range effects at least as serious as a similar failure of a hardware item.

To associate a risk with a complex analysis it is necessary to break

TABLE I
MASS SPECTROMETRY
DETECTION CRITERIA

• LIMIT OF DETECTION:	SINGLE ION MONITORING (SIM) PEAK • RETENTION TIME ± 5 SECONDS • SIGNAL/NOISE RATIO ≥ 5
• LIMIT OF VERIFICATION:	• RETENTION TIME ± 5 SECONDS. • MINIMUM OF 3 IONS (PREFERABLY 5-7) WITH S/N ≥ 2 • ION INTENSITY RATIOS WITHIN 20% (RELATIVE TO STANDARD.)
• LIMITS OF QUANTITATION (RANGE):	• ESTABLISHED BY HUBAUX-VOS* • UPPER AND LOWER LIMITS BOTH DEFINED

* Procedure described in Anal Chem 42 (1970) 849-885

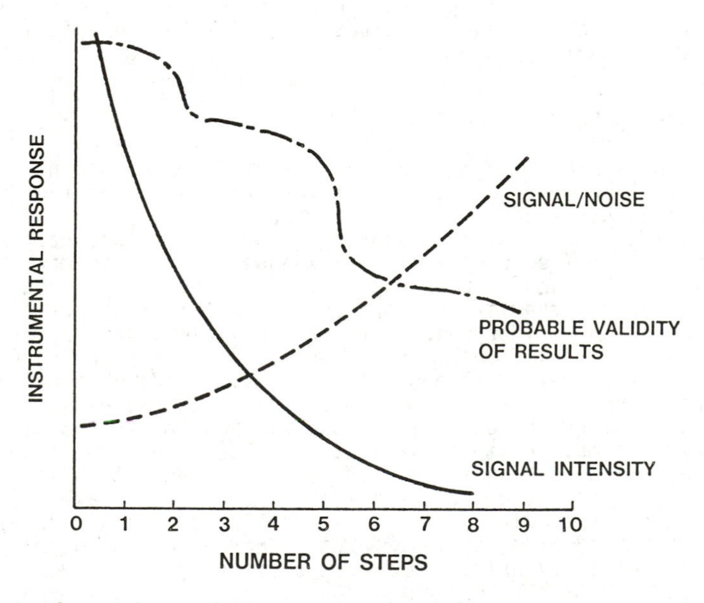

Figure 3. Quality of Data vs. Number of Steps in Procedure

down the procedure into individual actions, or potential failure causing events. Risk is usually defined as:

$$\text{Risk} = \text{Probability} \times \text{Severity} \quad (20)$$

The magnitude of risk from some event depends on the product of how often the analyst thinks an event will occur and how seriously the event impacts on the overall process. Therefore, it is incumbent on the scientist to develop a quantitative sense of where the risks in an analysis exist, and how serious they are. The best systems analyst cannot perform this function; only the person who the is most knowledgeable about the analytical procedure can function as the risk assessor. This person is normally the research chemist who developed the methodology and not the analyst who may run the procedure routinely. He or she is most familiar with the emerging methodology and has a basis (whether it be historical, intuitive or reasoned) to assign a factor of risk to the individual components of the analysis. Typical mechanisms for risk assessment studies include either the use of a "Fault Tree", which uses lists of major failures and associated minor failures which might cause them, or a "Failure Modes and Effects Analysis Model" (21) which uses lists of the ways a system can fail and the results of each failure. For this study, the "Failure Modes and Effects Analysis Model" was chosen.

This risk assessment was conducted on the analytical scheme outlined in Figure 2. Each step in the method was examined to identify potential errors that could occur if performed by a "competent analyst". (It should be noted that it is not useful or informative to do this type of risk assessment if one assumes the analytical procedure is going to be carried out by an incompetent or overly negligent individual. The first step in any quality assurance program in an analytical laboratory should be to identify such individuals and exclude them from participation in any critical programs.) Wherever a possible error was identified, the probable consequence of that error was also identified. Each consequence was rated and placed into one of three categories, again using the management guidelines as the basis for the severity of the rating. The most severe rating was critical, which included any error that could result in a total failure of the analysis or the reporting of a false positive value. The second rating was subcritical, which was any error which might result in the reporting of false negative results. The third rating was serious, which was any error that might result in poor precision, poor accuracy or the introduction of bias into the results. In the quality assurance plan, all critical and subcritical errors were treated as unacceptable. Serious errors were acceptable if they occurred within the limits of the adopted quality control criteria. Table II gives a listing of the risk assessment for the analytical methodology.

QUALITY CONTROL

Following the risk assessment study, a quality assurance plan was drafted and quality control procedures were implemented to eliminate or minimize the probability of the errors occurring. Each type of error was addressed and whenever possible the procedure was modified to eliminate the possibility of that error occurring as is shown in Table III. Procedural quality control was the most effective type

TABLE II
IDENTIFICATION OF POSSIBLE ERRORS AND PROBABLE CONSEQUENCES

	ERROR TYPE	PROBABLE CONSEQUENCES	RATING
1)	STORAGE/TAMPERING	– TOTAL FAILURE OF ANALYSIS	CRITICAL
2)	TRANSCRIPTION	– ERROR IN SOME SAMPLES	CRITICAL
3)	IMPRECISE SAMPLE HANDLING	– POOR PRECISION	SERIOUS
4)	CONTAMINATED REAGENTS	– FALSE POSITIVES OR FALSE NEGATIVES	CRITICAL
5)	CROSS CONTAMINATION BETWEEN SAMPLES	– FALSE POSITIVES	CRITICAL
6)	**CROSS CONTAMINATION FROM STANDARD TO SAMPLES	– FALSE POSITIVES	CRITICAL
7)	SAMPLE DEGRADATION (IMPROPER STORAGE AND HANDLING)	– BIAS (POSITIVE OR NEGATIVE) – POSSIBLE FALSE NEGATIVE	SERIOUS SUBCRITICAL
8)	DEVIATION FROM WRITTEN PROCEDURE	– TOTAL FAILURE OF ANALYSIS – SYSTEMATIC BIAS (POS OR NEG)	CRITICAL SERIOUS
9)	MATRIX INTERFERENCE	– FALSE POSITIVES OR FALSE NEGATIVES	CRITICAL SUBCRITICAL

TABLE III
IDENTIFICATION OF SPECIFIC QUALITY CONTROL TO BE USED TO ADDRESS EACH TYPE OF PROBABLE ERROR

ERROR TYPE	QUALITY ASSURANCE IMPOSED
1) STORAGE OR TAMPERING	— STRICT ENFORCEMENT OF CHAIN OF CUSTODY
2) TRANSCRIPTION	— MAKE UP ALL LABELS FOR CONTAINERS IN ADVANCE — CHECK ALL LABELS THREE TIMES — HAVE INDEPENDENT CHECK OF LABEL ACCURACY (SUPERVISORY)
3) IMPRECISE SAMPLE HANDLING	— USE OF INTERNAL STANDARD — IMPOSE ACCEPTANCE CRITERION ON I.S. RESPONSE
4) CONTAMINATION OF REAGENTS	— FREQUENT USE OF QC SAMPLES AT 0, 20 AND 100 ppb LEVELS — USE OF DISPOSABLE GLASS-WARE WHENEVER POSSIBLE — USE NEW BATCH OF REAGENT WITH EACH BATCH OF SAMPLES
5) CROSS CONTAMINATION BETWEEN SAMPLES	— ALWAYS RUN QC SAMPLES BETWEEN UNKNOWNS — NEVER USE SAME CONTAINER OF REAGENT/SOLVENT MORE THAN ONCE DURING THE PROCEDURE — SEQUENCE ALL OPERATIONS — USE DISPOSABLE GLASSWARE WHENEVER POSSIBLE
6) CROSS CONTAMINATION FROM STANDARD TO SAMPLE	— USE SEPARATE ROOM FOR HANDLING STANDARDS WHERE SAMPLES ARE NEVER ALLOWED — USE STANDARD COCKTAIL OF TRICHOTHECENES AT KNOWN RELATIVE CONCENTRATION FOR ALL QUALITY CONTROL AND STANDARD SAMPLES. — NEVER USE LIQUID HANDLING DEVICE FOR MORE THAN ONE OPERATION IN A BATCH RUN — SEQUENCE ALL OPERATIONS
7) SAMPLE DEGRADATION (IMPROPER STORAGE AND HANDLING)	— DETERMINE STABILITY OF SAMPLES DURING METHOD DEVELOPMENT — INSERT POSITIVE QC SAMPLES UPON RECEIPT OF SAMPLES — ADHERE STRICTLY TO SAMPLE STORAGE PROCEDURES
8) DEVIATION FROM WRITTEN PROCEDURE	— CLOSE SUPERVISORY CON-TROLS — INSERTION OF FREQUENT QC SAMPLES WHICH ARE "BLINDS" TO ANALYST — KEEP RESEARCH CHEMISTS AWAY FROM ROUTINE ANALYSIS
9) MATRIX INTERFERENCES	— DEVELOP RUGGED ANALYTICAL METHODOLOGY — DEVELOP AND ENFORCE STRICT RULES FOR REJ/ACCEPT. OF SELECTED ION MONITORING MASS SPECTROMETRY DATA

to implement in this study. Quality control samples were included in the sample stream to test for the occurrence of critical and subcritical errors and to measure their magnitude.

The Quality Control Coordinator (QCC) randomly divided the samples into groups of four unknown samples. One of the four samples was chosen randomly for replication within the same group. Quality control samples of the trichothecene mixture were added to the group at the 0, 20 and 100 ppb level. The QCC arranged samples such that quality control samples bracketed each unknown sample, thereby eliminating the possibility of cross contamination between unknowns if the procedural quality control were followed. The samples were assigned new numbers and internal standards were added. The analyst was given the concentration of one of the non-zero quality control samples in each group so the quantitative accuracy could be monitored vs the calibration data in real time.

PERFORMANCE OF THE ANALYTICAL METHOD

Figure 4 shows a single ion chromatogram of the base peak from the mass spectra of the heptafluorobutryl ester of scirpentriol resulting from the extract of a 5 part-per-billion standard in blood. Figure 5 shows typical data generated for an unknown sample which was verified to have T-2 toxin. This confirmation run was determined to be positive by comparison with the responses of standards at the same level. Figure 6 is a typical regression curve of concentration vs the peak height ratio of the base peak to the internal standard, Deoxyverrucarol DOV (1,22-24) for the extractable trichothecenes. Table IV lists the experimentally determined limits of detection, verification and quantification for this analytical technique when the above criteria were applied. Analytical chemists often face an apparent dilemma in how to report a value which falls above the criteria for detection but below the level required for verification, especially when the penality for reporting false positives is high. One responsible and accurate answer is that the analyte was indicated above the minimum detectable concentration but could not be verified.

COLLABORATIVE STUDY

The Center for Disease Control (CDC), Atlanta, GA was requested to prepare a set of properly stabilized and deactivated human blood samples containing four simple trichothecene mycotoxins with two toxins each from the type A and B groups. A statistician from CRDEC, in collaboration with CDC, established a sample numbering system and an experimental design for preparation of the sample set, which included at least four concentrations of each toxin. This experimental design had checks for dilution error and measurement error in making up the samples and provided a statistically significant number of replicate and blank samples to determine accuracy, precision and the expected rate for reporting false positive values. The samples were shipped to CRDEC under refrigeration. After CRDEC completed the analysis of the samples and reported the results to an independent organization, CDC released the sample key. The results were evaluated by the Ballistics Research Laboratory, Aberdeen Proving Ground. The results are summarized in Table V.

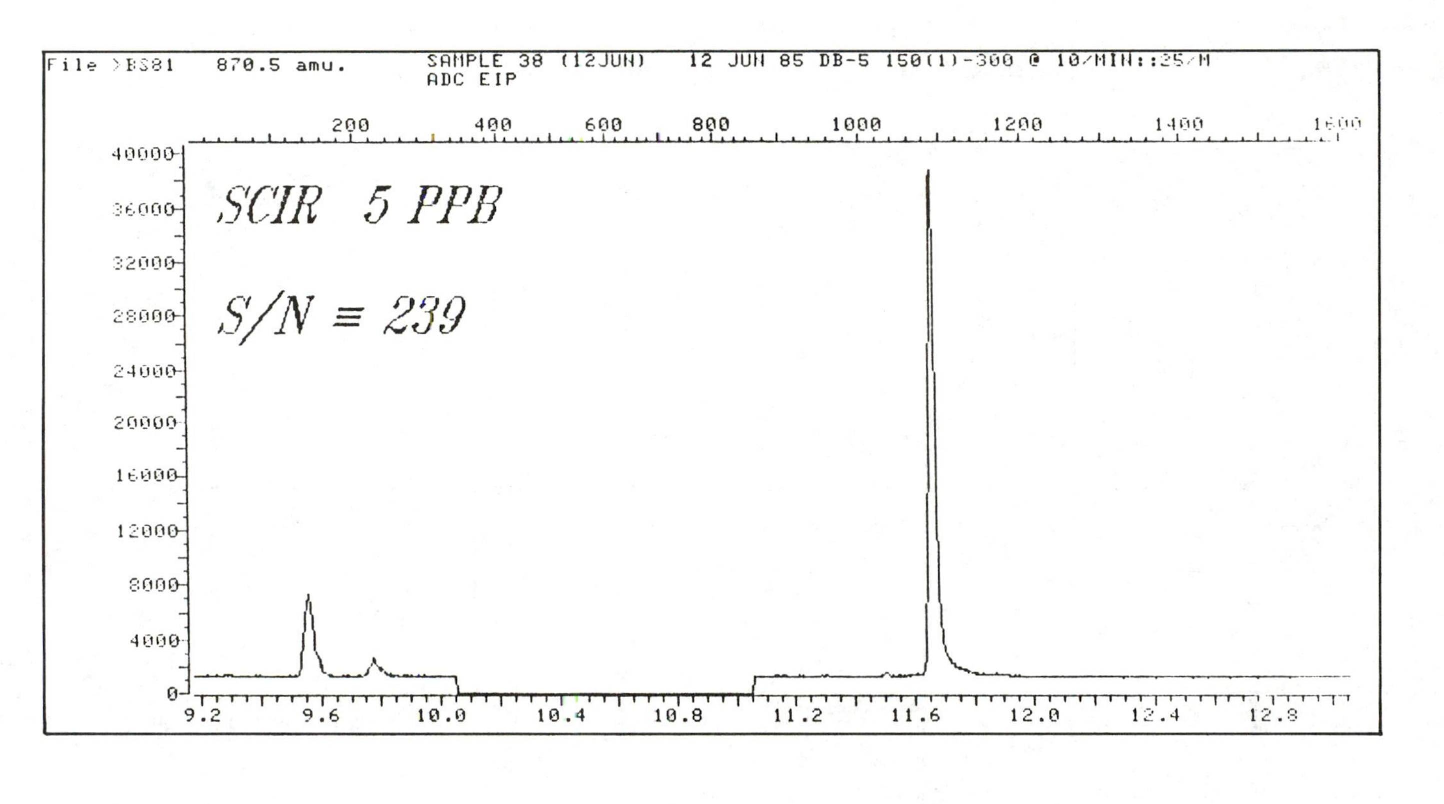

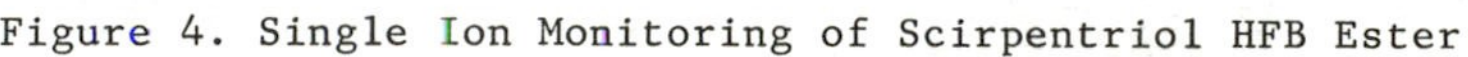
Figure 4. Single Ion Monitoring of Scirpentriol HFB Ester

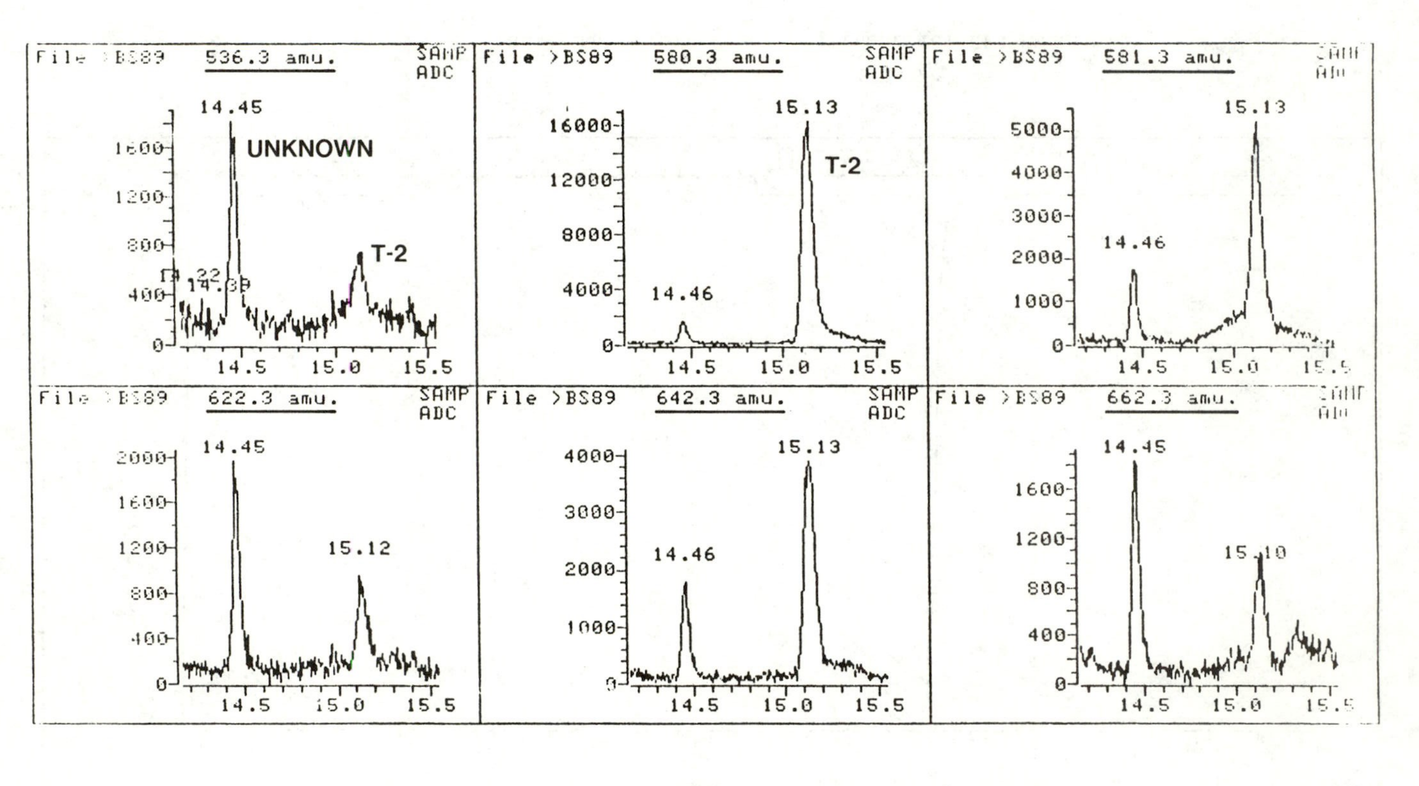

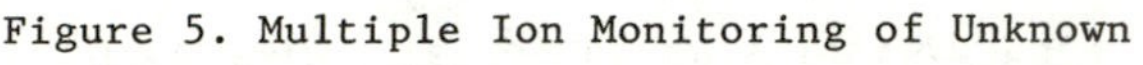
Figure 5. Multiple Ion Monitoring of Unknown

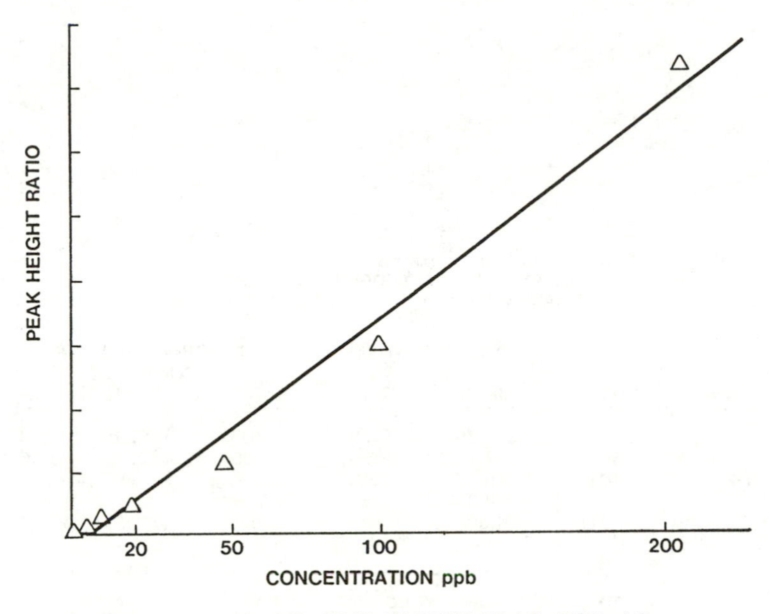

Figure 6. Calibration Curve for Verrucarol

TABLE IV
EXPERIMENTALLY DETERMINED LIMITS OF ANALYSIS FOR TRICHOTHECENES IN HUMAN BLOOD

TOXIN	DETECTION LIMIT* ppb	VERIFICATION LIMIT ppb	QUANTIFICATION RANGE ppb
T-2	0.5	3	1-200
HT-2	0.5	3	1-50
DAS	0.2	4	1-50
VER	.02	0.5	1-20
DON	.04	1	1-100
FUS-X	0.1	2	1-100
SCIR	0.1	2	1-100
T-2 TETRAOL	.01	0.1	.05-20**

* EXTRAPOLATED FROM LOWEST S/N VALUE.
**DETERMINED AFTER CDC COLLABORATIVE STYDY.

TABLE V
MEANS AND STANDARD DEVIATIONS OF THE MEASURED DATA IN ORIGINAL UNITS

		COLLABORATIVE STUDY		INTERNAL QUALITY CONTROL		
TOXIN	TARGET	MEAN	S.D.	TARGET	MEAN	S.D.
	200	218.75	9.78	100	102.4	28.4
	40	27.25	8.54	20	25.6	10.5
T-2	10	9.88	2.63			
	2	2.38	.50			
	0	bdl	0	bdl		0
	150					
	30	NOT				
NIV	10	ANALYZED		N/A		
	2					
	0					
	150	177.25	13.05	100	81.6	7.7
	30	28.75	2.87			
VER	10	15.00	0.00	20	18.6	2.4
	2	7.38	0.13			
	0	bdl	0	0	bdl	0
	200	113.25	24.41	100	82.8	14.4
	40	21.25	28.87			
DAS	10	4.25	0.50	20	18.6	4.2
	2	4.25	3.77			
	0	bdl	0	0	bdl	0

bdl = BELOW DETECTION LIMIT
n = 4
S.D. = Standard Deviation

The results from the quality control samples which were analyzed as part of the CDC collaborative study are shown in Tables VI and VII. Figures 7 and 8 are examples of control charts which show the precision of the analysis to be approximately 30% relative standard deviation for any concentration other than 0 regardless of the identity of the trichothecene. (On a response vs. concentration chart, the one standard deviation error bars would be seen as diverging from the line representing the mean as the concentration increases. If the same data were plotted as log response vs log concentration, the first standard deviation error bars would appear to parallel the calculated mean.) Figure 9 shows the results from plotting the first and second replicates of CDC samples containing T-2 toxin. This replication study demonstrates that reasonable estimates of the precision for unknown samples can be determined from control charts.

CONCLUSION

Analytical methodology was developed for accurate quantitative analysis of trichothecenes at low part-per-billion levels in blood. Although this methodology was arduous and lacked the ruggedness normally demanded of an analytical procedure which must have a low failure rates it proved to be both qualitatively reliable and quantitatively accurate when it was combined with a well planned quality assurance program. An indispensable part of developing the quality assurance plan was a formal risk assessment which specifically took into account the possibility of human error. This procedure was validated by collaborative study with independent laboratories.

TABLE VI
RESULTS OF QUALITY CONTROL SAMPLES
TARGET CONCENTRATION VS. REPORTED CONCENTRATION

SAMPLE #	TARGET CONCENTRATION (ppb)	REPORTED CONCENTRATION (ppb)		
		T-2	VER	DAS
QC 2	0	ND*	ND	ND
QC 6	0	ND	ND	ND
QC 9	0	ND	ND	ND
QC 12	0	ND	ND	ND
QC 13	0	ND	ND	ND
QC 1	20	21	17	19
QC 5	20	21	20	16
QC 7	20	16	16	14
QC 11	20	27	18	19
QC 14	20	43	22	25
QC 3	100	72	91	67
QC 4	100	82	80	67
QC 8	100	95	82	93
QC 10	100	140	85	93
QC 15	100	123	70	94

*ND = NOT DETECTABLE

TABLE VII

QUALITY CONTROL SAMPLES
Summary Report

TARGET CONC. (ppb)	X (MEAN)	STANDARD DEVIATION	X @ 95% CONFIDENCE LEVEL*	RELATIVE ERROR
T-2 20	25.6	10.5	12.6 - 38.6	28.0%
T-2 100	102.4	28.4	67.1 - 137.7	2.4%
DAS 20	18.6	4.2	13.4 - 23.8	− 7.0%
IDAS 100	82.8	14.4	64.9 - 100.7	−17.2%
VER 20	18.6	2.4	8.4 - 36.5	−25.7%
VER 100	81.6	7.7	72.0 - 91.2	−18.4%

*USING THE FORMULA FOR CONFIDENCE LIMIT FOR $\mu = X \pm ts/\sqrt{N}$
DEGREES OF FREEDOM = 4.

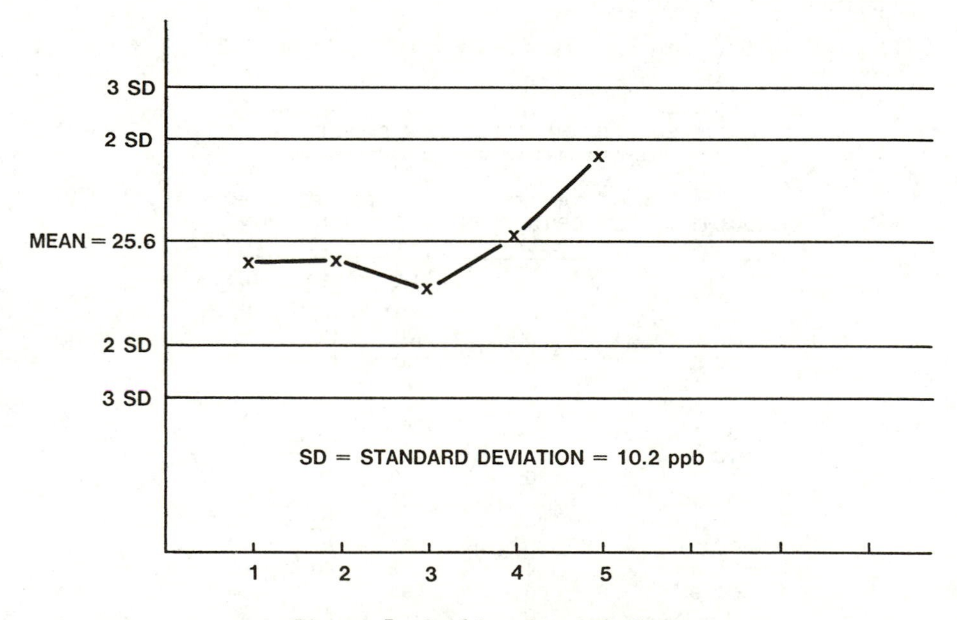

Figure 7. Quality Control Chart

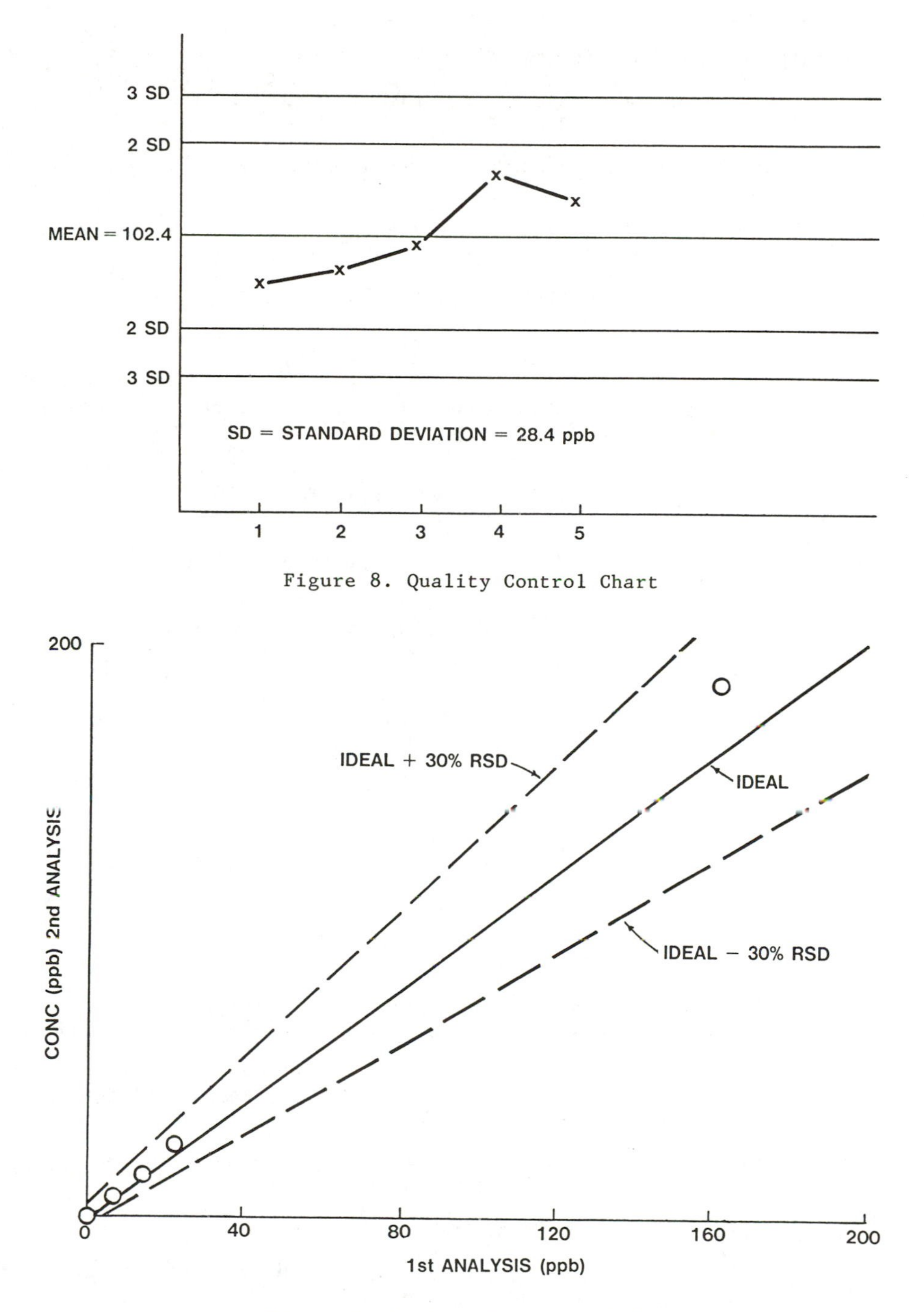

Figure 8. Quality Control Chart

Figure 9. Replication Control Study

ACKNOWLEDGMENT

The authors gratefully acknowledge Mrs. D. A. Paterno, Mr. M. A. Wasserman and Dr. T. Krishnamurthy for their support in developing the analytical procedure and Mr. L. Sturdivan for supplying statistical support to the project.

1. Krishnamurthy, K.; Sarver, E.W. J. Chromato. 1986, 335, 253-264

2. Kurata, H.; Ueno, Y. Toxigenic Fungi--Their Toxins and Health Hazard; Elsevier, New York, 1984.

3. Frank, B. Angew. Chem., Int. Ed. Engl., 1984, 23, 493.

4. Fontelo, P. A.; Beheler, J.; Bunner, D. L.; Chu, F. S., Appl. Environ. Microbiol., 1983, 45, 640-3.

5. Pare, J. R. Jocelyn; Greenhalgh, R.; Lafontaine, P.; Apsimon, J. W., Anal. Chem., 1985, 57, 1470-2.

6. Stahr, H. M.; Lerdal, D.; Hyde, W.; Pfeifer, R., Appl. Spectrosc., 1983, 37, 396-400.

7. Visconti, A.; Bottalico, A., Chromatographia, 1983, 17, 97-100.

8. Heyndrikx, A; Sookvanichsilp, N.; Van den Heede, M. Arch. Belg. Med. Soc., Hyg., Med. Trav. Med. Leg., 1984, Suppl. 143-6

9. Tanaka, T.; Hasegawa, A.; Matsuki, Y.; Ishii, K.; Ueno, Y., Food Addit. Contam. 1985, 2, 125-37.

10. Smith, R. D.; Udseth, H. R.; Wright, B. W.; J. Chromatogr. Sci., 1985, 23, 192-9.

11. Gore, J.; Rougereau, A.; Person, O., J. Chromatogr., 1984, 291, 404-8.

12. Wilson, C. A.; Everard, D. M.; Schoental, R., Toxicol. Lett., 1982, 10, 35-40.

13. Protection Against Trichothecene Mycotoxins, National Research Council, Report, 1983, Order No. AD-A137115, pp238.

14. Rodricks, J. V. ; Hesseltine, C. W.; Mehlman, M. A, Ed., Mycotoxins in Human and Animal Health, Pathotox Publishers, Park Forest, IL, 1977.

15. Mirocha, C. J.; Pawlosky, R. A.; Chatterjee, K.; Watson, S.; Hayes, W.; J. Assoc. Off. Anal. Chem., 1983, 66, 1485.

16. Kaiser, H., Anal. Chim. Acta, 1978, 33B, 551.

17. Fetterolf, D. D.; Yost, R. A.; Int. J. Mass. Spec. and Ion. Proc. 1984, 63, 33-50.

18. Johnson, J. V.; Yost, R. A.; Anal. Chem. 1983, 57, 758a.

19. Army Regulation 70-1, 15 March 1984, Headquarters, Department of Army, Washington, D. C.

20. Crouch, E. A. C.; Wilson, R.; Risk/Benefit Analysis, Ballinger, Cambridge, MA, 1982, p9.

21. McKean, K. Discover 1986, 7, 38

22. Schuda, P. F.; Potlock, S. J.; Wannamacher, R. W. Jr., J. Nat. Prod., 1984, 47, 514.

23. Jarvis, B. B.; Stahly, G. B.; Pavanasasivam, G., Med. Chem. (Academic), 1980, 23, 1054.

24. Jarvis, B. B.; Pavanasasivam, G., Appl. Environ. Microbiol., 1983, 46, 480.

RECEIVED September 1, 1987

Chapter 13

Evaluating the Impact of Hypothesis Testing on Radioactivity Measurement Programs at a Nuclear Power Facility

R. A. Mellor and C. L. Harrington

Yankee Atomic Electric Company, Rowe, MA 01367

NUREG 4007, "Lower Limit of Detection: Definition and Elaboration of a Proposed Position for Radiological Effluent and Environmental Measurements" proposed the application of hypothesis testing to nuclear power plant Radiological Effluent Technical Specifications (RETS). Although hypothesis testing has been implicit in the current RETS, the application of the NUREG proposal will require a more detailed knowledge of the measurement system uncertainties, establishment of a basis for assumed boundary conditions for these uncertainties and a heightened degree of control over measurement processes. This paper applies the requirements of the NUREG to three types of radiometric measurement systems routinely used for compliance with the RETS. The methods used to meet the requirements will be detailed.

Over the past several years Radiological Effluent Technical Specifications (RETS) have been implemented at nuclear power facilities in the United States. These RETS control the evaluation and release of radioactive material to the off-site environs. Radiological measurements of these releases are conducted using gamma spectroscopy, liquid scintillation, gross counting techniques and continuous on-line monitoring.

The Technical Specifications also contain requirements for Environmental Monitoring Programs (EMP). These programs are primarily designed to provide the determination of dose, assessment of trends of radioactivity in the environment and public reassurance (1). Many of the same radiological measurement techniques are used for environmental monitoring but with the intent of detecting radioactivity orders of magnitude less than effluent release source terms.

The concept of hypothesis testing has been applied to off-line radiological measurement techniques used in the RETS for both in-plant and environmental monitoring. Statistically valid multipliers, the random variation of the background, and

0097-6156/88/0361-0244$06.50/0

measurement protocol (MP) variables have been combined in standard radiometric formulations to provide an estimation of the lower limit of detection (LLD) for the MP. Numerical LLD values have been assigned as requirements which the MP must achieve for various radionuclides. Generally these requirements are set to ensure the release of radionuclides below these detection limits would have a negligible effect on the health and safety of the public.

NUREG 4007 (2) (hereafter called the NUREG) represents a definitive treatment of the LLD concept. The basic change from the current RETS approach recommended in the NUREG is the incorporation of upper bound relative systematic uncertainty terms in the determination of the LLD. This approach would allow the LLD determination to include the influence of any potential systematic uncertainty in either the background determination or MP variables. Explicit within this approach is the establishment of actual upper bound values for these relative systematic uncertainties. Currie, in the NUREG, has provided what he considers reasonable values for these systematic uncertainty bounds which represent "routine state-of-the-art" in radiometric measurements. However, there are instances within the NUREG where the substitution of site specific values are recommended and where more careful evaluation of these bounding conditions by the Nuclear Regulatory Commission (NRC) is left open to consideration.

This work evaluates the systematic errors possibly present in three MPs, compares these uncertainty bounds to those used in the NUREG and applies the results of these evaluations in a comparison of current and proposed LLD formulations. The measurement protocols chosen for this evaluation are: tritium (H-3) analysis in power plant liquid effluents, low level I-131 analysis in milk, and gamma spectroscopy of power plant liquid effluents. These protocols exhibit routinely achieved LLDs which are ten to twenty percent of the regulatory requirements.

Hypothesis Testing and LLD Formulations

The use of hypothesis testing to define the LLD has been evaluated previously (1-4). Two states of any measurement system composed of normally distributed random uncertainties are considered: the null hypothesis state in which the samples contain no net radioactivity and the distribution about the net count of zero is characterized by the Mean (μ_0) and the standard deviation (σ_0); and, the LLD state (Currie's "alternate hypothesis" in the NUREG) in which the distribution about the net counts at the LLD is characterized by the mean (μ_D) and the standard deviation (σ_D). The ultimate question for any sample data is "which state is most consistent with the data?". In making this decision, there is a chance that we will falsely conclude the data is part of the distribution about the LLD or that we will falsely conclude the data is part of the distribution about the net count of zero. These risks are defined by the probabilities α and β respectively. These risk probabilities may be chosen at any level but have been chosen for the purposes of regulatory LLD requirements at five percent. This leads to the regulatory definition of LLD.

> LLD - The smallest concentration of radioactive material in a sample that will yield a net count, above system blank that will be detected with at least a 95 percent probability with no greater than a 5 percent probability of falsely concluding that a blank observation represents a "real" signal.

Based upon this definition the current RETS formulation of LLD can be derived (2,4) and is shown below.

$$LLD = \frac{4.65\ S_b}{E*V*2.22*Y*Exp\ (-\lambda*t)} \qquad (1)$$

Where:

- E = detection efficiency
- V = sample size (mass or volume)
- Y = radiochemical yield
- 2.22 = dpm per picocurie
- λ = radioactivity decay constant for the particular radionuclide
- t = elapsed time between collection and analysis
- S_b = standard deviation of the background counting rate or the counting rate of a blank sample as appropriate

The NUREG formulation of the LLD (shown below) incorporates systematic uncertainty boundary conditions for E, V, Y, the counting time (T) and the background (B). The reader is directed to the NUREG for the complete rationale behind the manner of incorporation of these uncertainty bounds.

$$LLD = \frac{f(2\Delta_B B + 2.71 + 3.29\sigma_o)}{2.22\ (YEVT)} \qquad (2)$$

- f = an amplification factor providing conservative bounds for systematic uncertainty in Y, E, V or T (and = $1 + \Delta_A$)
- Δ_B = relative systematic bound in the blank B (B is in counts)
- σ_O = standard deviation of the true net signal

Assigning f = 1.1, $\Delta_B = 0.05$ and equating $\sigma_O = \sigma_B\sqrt{\eta}$ yields the following:

$$LLD = 0.11*BEA + \frac{1.1\ (2.71 + 3.29\sigma_B\sqrt{\eta})}{2.22\ (YEVT)} \quad (3)$$

Where:

σ_B = standard deviation of the blank

η = 1+1/b where b = ratio of the blank and sample counting times

BEA = Blank Equivalent Activity
= B/2.22 (YEVT)

NOTES: 1. If B>70 counts the constant of 2.71 is not required

2. $\Delta_A{}^2 = \Delta_Y{}^2 + \Delta_E{}^2 + \Delta_V{}^2 + \Delta_T{}^2$

The relative systematic uncertainty terms are the primary focus of this evaluation.

Methods of Evaluation

In preparing to determine the relative systematic uncertainty bounds associated with an a-priori LLD, it was decided to use methods which, although might not be rigorous from a statistical perspective, would allow the estimation of these upper bounds in a relatively easy manner. Four methods of evaluation have been used in this study.

Method I is a "theoretical" approach (5) which uses systematic boundary conditions for each of the parameters of the MP coupled with appropriate propagation techniques to arrive at an overall systematic uncertainty. In establishing certain of these boundaries, the central tendancy of the parameter was estimated, wherever possible, by determining the mean of a minimum of sixty observations of the parameter in question. The systematic uncertainty bound was established by determining the confidence interval for the 95.45 percent confidence coefficient of the distribution (2 s_x) and expressing this value as a percentage of the mean. The use of the confidence interval at infinite degrees of freedom is estimated to induce only a two percent uncertainty in the determined quantity versus using the student-t statistic.

In Method II, the relative systematic uncertainty in other parameters such as the yield were determined by propagating the uncertainties quoted from manufacturer literature for balances and automatic pipetting equipment with the experience estimates for uncertainties of carrier content, stable element contributions and variations in final precipitate weights. Estimation of the boundary conditions for efficiency determinations were, in some instances, based upon the maximum deviation from the mean response.

The relative systematic uncertainty in the background was estimated (Method III), where possible, from background distributions having a minimum of 120 observations. The background distributions were evaluated for normality using the standard chi-square test. The determination of the relative systematic uncertainty was made by assuming a perfectly normal distribution and calculating the difference between the observed standard deviation and the theoretical standard deviation. Two times the absolute value of the fractional difference divided by the mean $[2 \mid (S_n - \sigma_n/\mu_o \mid]$ was considered to be the upper bound for the systematic uncertainty in the distribution.

Method IV is empirical and involves the evaluation of intra-laboratory knowns (5). The systematic uncertainty may be estimated by comparing the mean experimental result with the known value in accordance with the equation below.

$$\hat{\Delta} = e \pm \delta_M$$

Where:

e = the total uncertainty (mean-known)

δ_M = two times the standard error of the mean.

As much independance from the original protocol as possible was maintained during these evaluations. It is recognized, however, that the laboratory dependance of these measurements may limit their usefulness for evaluation purposes. These techniques were applied, as appropriate, to each of the three MPs under consideration. The detailed evaluations of each technique are presented below.

Tritium (H-3)

The tritium measurement protocol is relatively simplistic and consists of neutralization of the sample, single plate distillation, removal of an appropriate size aliquot (usually eight mL.) via reproducible automatic pipets and the addition of 15 mL. of dark adapted scintillation cocktail under incandescent lighting conditions. After shaking to ensure a uniform gel, the sample is allowed to settle and dark adapt for up to twenty but no more than sixty minutes. The liquid scintillation unit efficiency is determined daily on a previously prepared standard and background measurements are determined at least daily. Quench corrections are not applied to the system due to the lack of an external standards ratio capability and an effort to minimize the amount of hazardous waste which would be generated if an internal standards approach were adopted.

Over a five month period, approximately 60 data points were collected for the efficiency and 150 data points for background determinations. The frequency distribution for the efficiencies is shown in Figure 1. The relative standard deviation for this distribution is 0.011 and the relative systematic uncertainty bound determined in accordance with Method I is 0.022.

The effect of the lack of a quench correction was evaluated by reviewing the same technique accomplished at a different laboratory within the Corporation but with the added protocol step of efficiency determination via the internal standard method. A total of 66 data points comprise the distribution depicted in Figure 2. This data is considered normally distributed with $2S_Q = 0.071$. These two relative systematic uncertainty boundary conditions for the efficiency have been propagated in quadrature to yield a relative systematic uncertainty of 0.074.

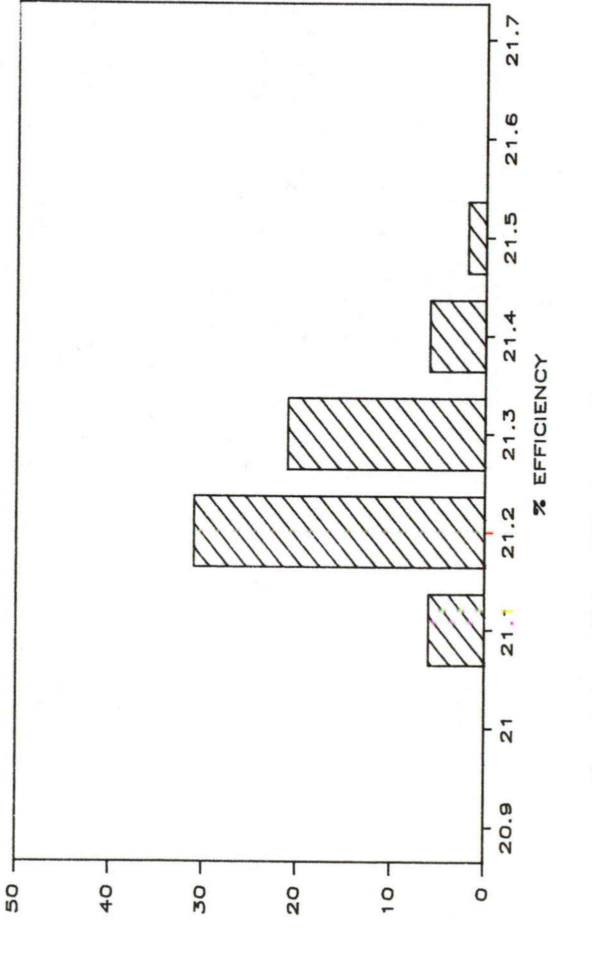

Figure 1 Tritium Efficiency Distribution for a Liquid Scintillation Counter.

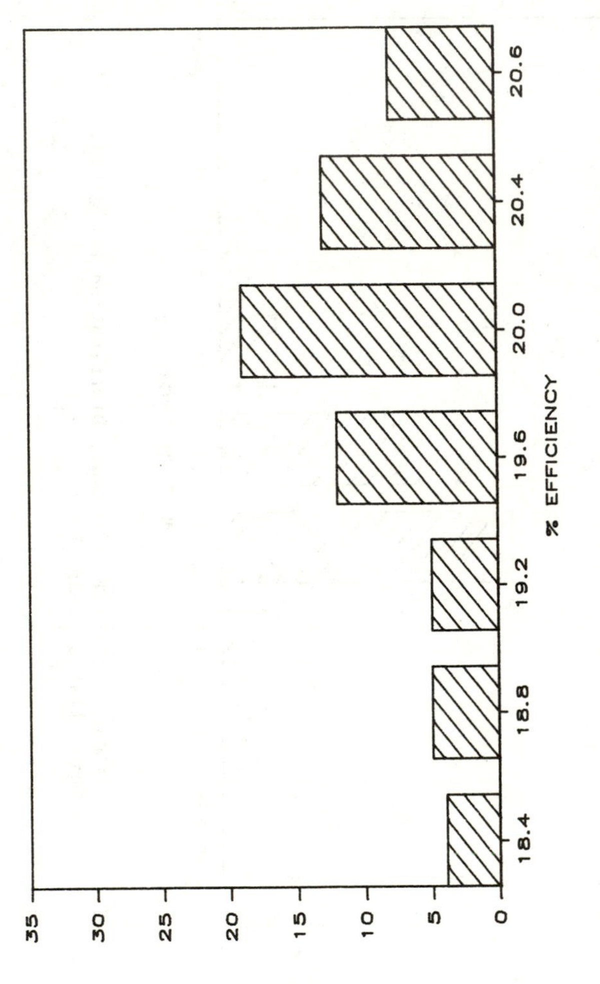

Figure 2 Tritium Quench Corrected Efficiency Distribution.

The effect of the sample standing (dark adapting) time on the efficiency of detection of H-3 in a sample has been evaluated via a series of three experiments. Samples prepared in the routine manner were immediately placed in the well of the liquid scintillation counter and a one minute analysis initiated. The sample was then re-analyzed at approximately five minute intervals and the ratio of the observed counts at time t=x to those at time t=0 were determined and plotted versus the elapsed time x. The results, illustrated in Figure 3., indicate a 5.3 percent average bias (n=17) after thirty minutes which remains relatively constant. The bias averages (n=13) 3.8 percent after twenty minutes. The relative standard error of the mean (RSEM) for the zero to twenty minute data (0.004) is equal to the RSEM for the thirty to sixty minute data. The ranges of the two means do not cross indicating the difference is real. Since samples are routinely held for twenty minutes prior to analysis, the difference between the two means is considered the maximum systematic uncertainty for this portion of the protocol. It has been combined in quadrature with the relative systematic uncertainty previously determined to obtain the relative systematic uncertainty bound for the efficiency of 0.076.

At the end of 55-60 minutes, the maximum deviation of the individual trial ratios was two percent. This minor deviation has been assigned to variability in sample preparation and has been added in quadrature with the automatic pipet literature precision estimate of one percent to arrive at the relative systematic uncertainty bound for the volume of 0.022. The total relative systematic uncertainty bound for the efficiency-volume term in the denominator of the LLD equation will therefore be 0.079. This was calculated by adding the relative systematic errors in quadrature.

The background frequency distribution for the liquid scintillation unit is graphically portrayed in Figure 4. The distribution was determined to be normal. The value of $2 \mid s_B\,(H\text{-}3) - \sigma_B\,(H\text{-}3) \mid / \mu_B$ (Method III) was determined to be 0.004 based upon μ_B = 22.59 cpm and S_B = 1.46 cpm for a ten-minute analysis. The results of the evaluations of relative systematic uncertainty bounds have been tabulated in Table I.

I-131 Low Level Analysis

Low level (LLD < 1 pCi/Kg) I-131 measurements have been conducted at the Yankee Atomic Environmental Laboratory (YAEL) for over eight years. Chemical separation, precipitation and mounting of CuI, and analysis on beta/gamma coincidence systems form the basis of the technique (6). The stable iodide concentration is determined by an ion selective electrode method. The large volume milk samples are weighed on top-loading balances and stable element iodide carrier is added via control checked automatic pipets. After equilibration of the carrier, the sample is processed through an anion ion exchange column. The stable iodide present on the column is subsequently removed; and through a

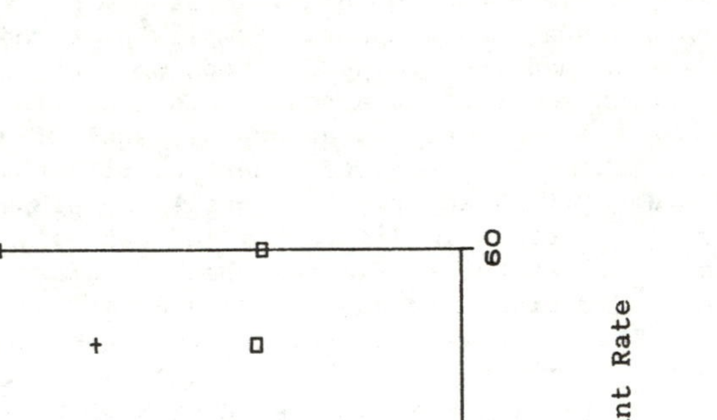

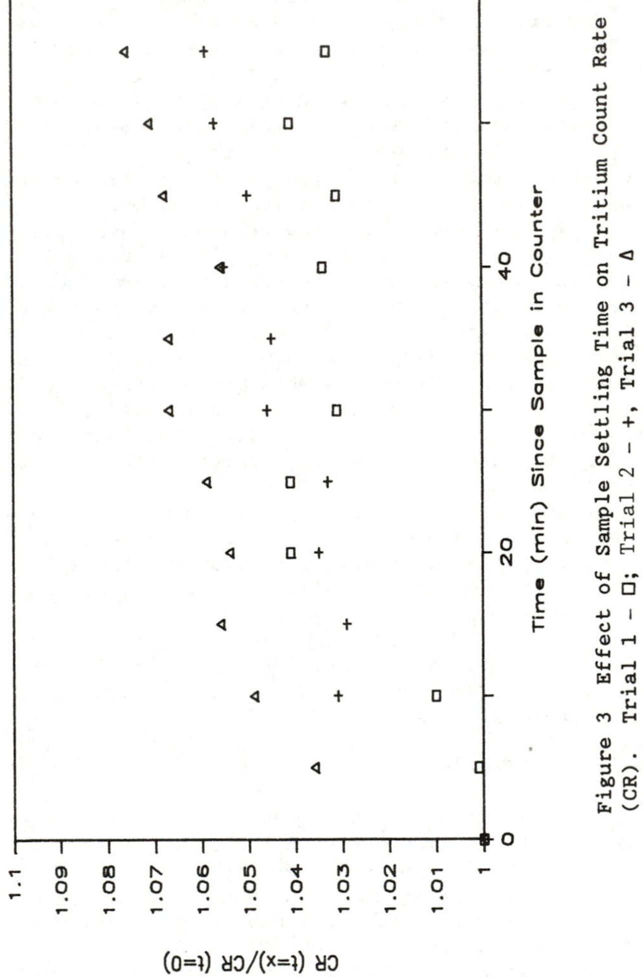

Figure 3 Effect of Sample Settling Time on Tritium Count Rate (CR). Trial 1 - □; Trial 2 - +, Trial 3 - Δ

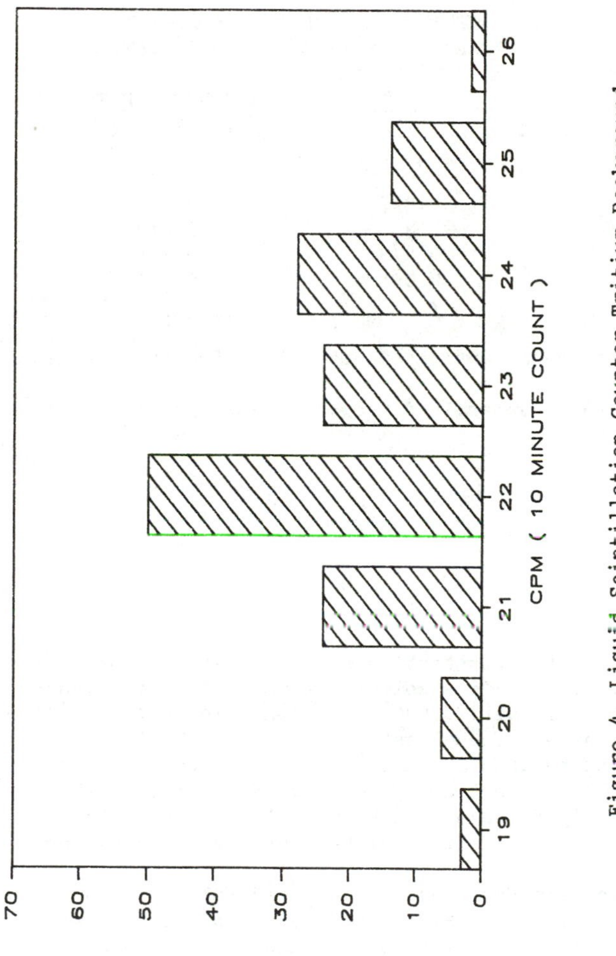

Figure 4 Liquid Scintillation Counter Tritium Background Distribution (CPM = Counts per Minute).

TABLE I

ESTIMATES OF THE UPPER BOUNDS
OF RELATIVE SYSTEMATIC UNCERTAINTIES

UNCERTAINTY	MEASUREMENT PROTOCOLS		
TYPE	H-3	I-131 (LL)	GAMMA
Δ_E	0.076	0.075	0.068
Δ_V	0.022	0.010	0.014
Δ_Y	0	0.074	0
Δ_A	0.079	0.106	0.070
Δ_B	0.004	0.039	*

NOTE: Δ_T has been considered negligible for these evaluations.

* Not able to be determined from available gamma spectroscopy information.

series of oxidation/extraction/reduction steps, the iodine present is isolated as an iodide solution and CuI is precipitated by the addition of an acidified CuCl solution. The CuI is filtered, mounted and dried to constant weight. The chemical yield is determined based upon the total weight of stable iodide in the sample (stable present in the milk plus carrier). The purified sample is analyzed on a beta/gamma coincidence analysis system which detects coincident I-131 beta and gamma emissions from the sample.

A large data base has been amassed relative to the analysis of milk samples for I-131. The data are presented here along with the "theoretical" estimates of systematic uncertainty in each of the parameters of concern in order to provide the reader with a review of the type of data required for experiment redesign and optimization. This historical data might be used to establish an _a-priori_ LLD estimate of a measurement system.

The units of reporting environmental I-131 concentrations in milk at the YAEL are pCi/Kg and thus, the sample mass is determined for each sample processed. The historical mass data are portrayed in Figure 5 and the spread of data shown is primarily dependent on the exact amount of milk submitted for

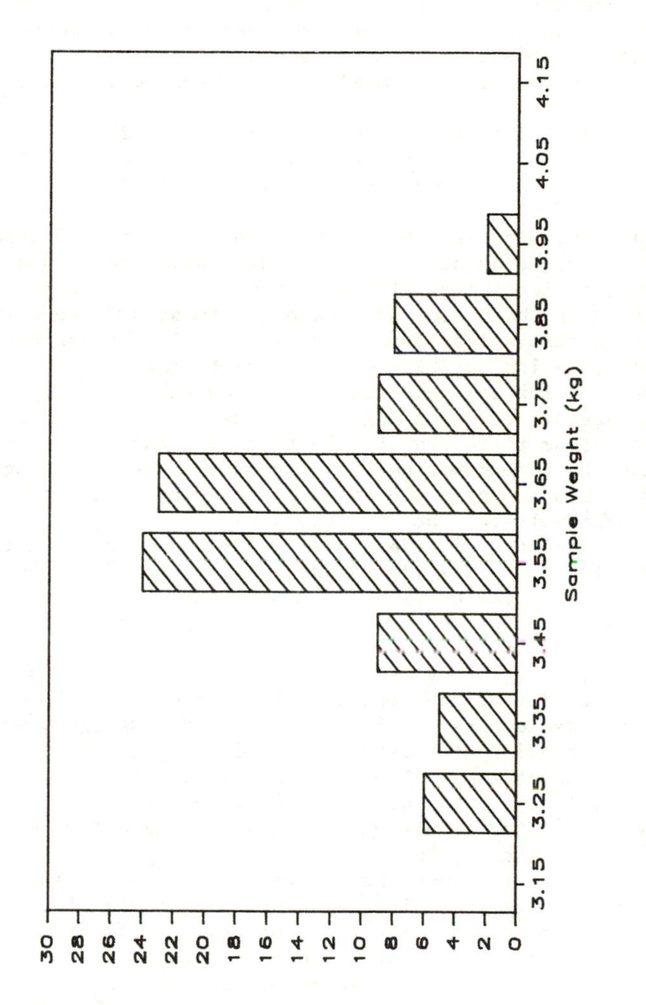

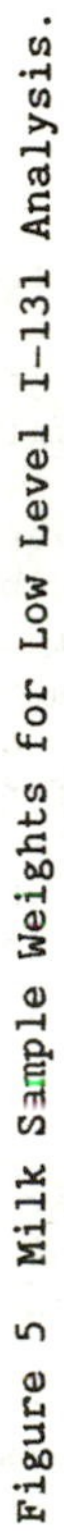
Figure 5 Milk Sample Weights for Low Level I-131 Analysis.

analysis. The evaluation of this data determined a mean (μ_W) = 3.598 Kg and a standard deviation (σ_W) = 0.165 Kg. For any individual measurement, a maximum uncertainty in the weight would be 0.5 percent based upon experience.

The chemical recovery of CuI for this system is determined gravimetrically. Each of the weight terms has an associated uncertainty which are listed below. These uncertainties are based upon the precision allowed (e.g. in a carrier calibration), the maximum uncertainty in measuring the stable iodide concentration and an estimate of the effect of variation of the CuI precipitation solution pH causing the co-precipitation of CuCl.

[mg. of iodide added (20.0)] = 0.02
[mg. of iodide present (5.0)] = 0.10
[mg. of CuI recovered (29.41)] = 0.02

The values in parentheses indicate the maximum boundary condition for the quantity under consideration based upon the average historical chemical recovery of 78.42 $\pm$ 5.38 percent. Propagation of the listed uncertainties in accordance with established statistical methods yields a relative systematic standard deviation (σ_R/R) = 0.037. The frequency distributions for the chemical recoveries exhibited by two chemists are portrayed in Figure 6. The main point to be remembered from these two distributions is that no two individuals will have the same distribution of results. Establishing an <u>a-priori</u> LLD based upon the results of a single individual may not be applicable to other individuals.

The efficiency of detection for any individual sample is a function of the calibration in use at the time of sample processing. Any calibration based upon the preparation of sources from one stock solution, one carrier preparation and/or one series of instrument settings must, by its very nature, be considered to be a new baseline for systematic uncertainty (2). Thus, only by comparing the differences between calibrations can the upper bound of the systematic uncertainty be determined. Data was available for the calibration of a beta/gamma coincidence unit over a five-year span of time. The maximum fractional deviation of any single efficiency at the average precipitate weight from the mean efficiency is 0.075. This will be used as the upper bound of the systematic uncertainty in the efficiency.

Two times the volume and recovery relative systematic uncertainties were added in quadrature with the maximum deviation in the efficiency to provide the boundary condition for the quantity Δ_A with a result of 0.106. This value is consistent with the value assumed in the NUREG.

The background frequency distribution for a beta/gamma system is depicted in Figure 7. This distribution is composed of 175 800-minute determinations of the system background spanning five different calibrations. Any effect due to the different calibrations has been assumed negligible. The data are normally distributed. The four highest data points were rejected based upon Chauvenet's criteria [K_{172nd} = 3.62 > 2.98]. The

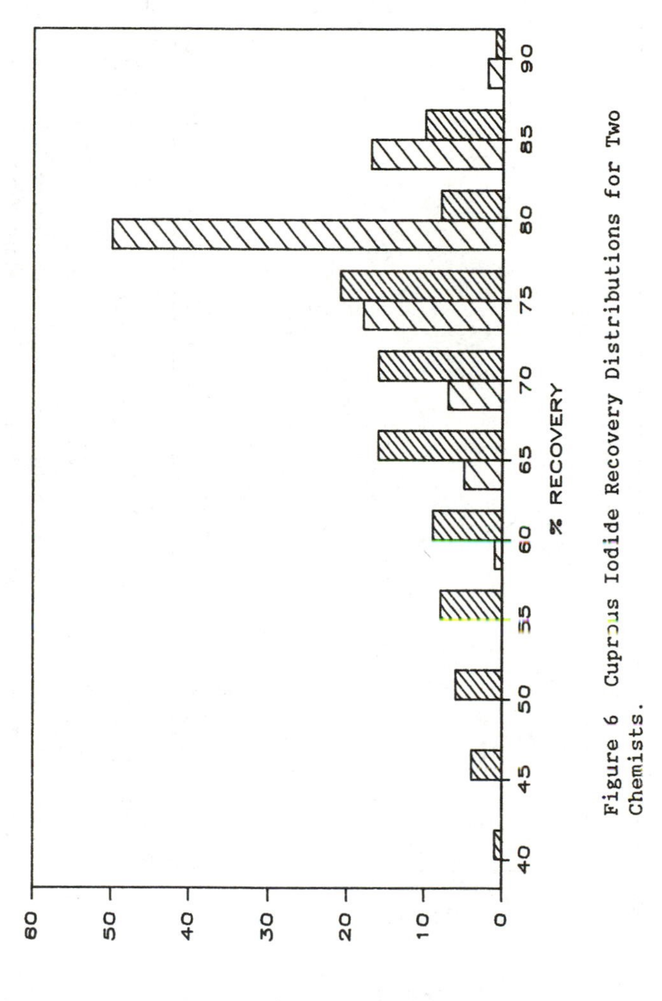

Figure 6 Cuprous Iodide Recovery Distributions for Two Chemists.

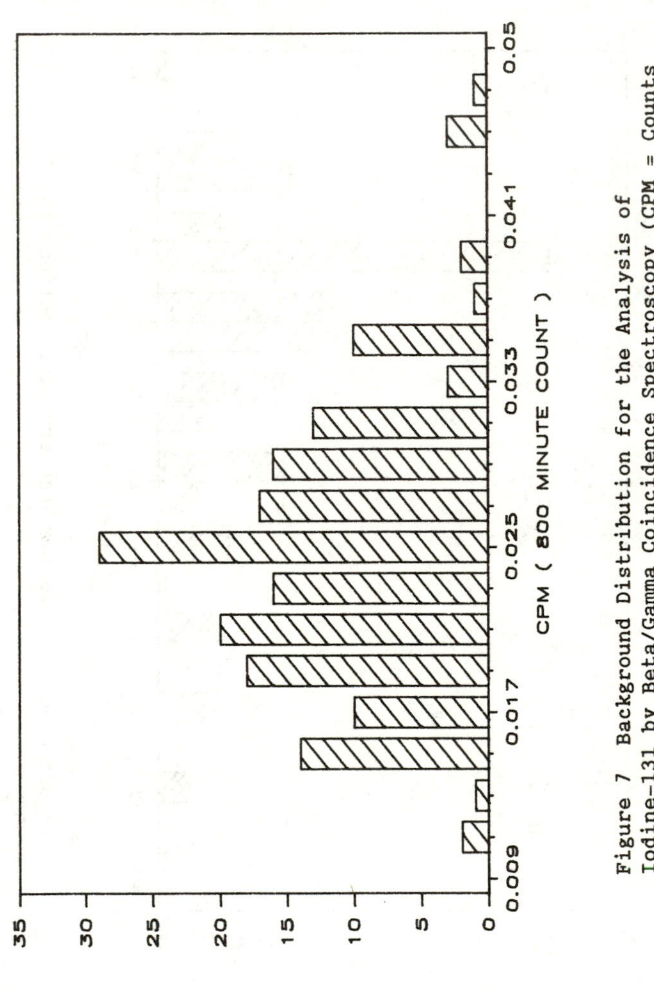

Figure 7 Background Distribution for the Analysis of Iodine-131 by Beta/Gamma Coincidence Spectroscopy (CPM = Counts per Minute).

estimation of the relative systematic uncertainty bound in the background was determined as:

$$2 \mid s_B \text{ (I-131)} - \sigma_B \text{ (I-131)} \mid / \mu_B = 0.039.$$

This value was based upon μ_B = 0.0236 cpm and s_B = 0.0059 cpm for an 800-minute count. The systematic uncertainty boundary conditions for the low level iodine-131 method are provided in Table I.

Gamma Spectroscopy

This section deals with the process of determining the photon-emitting components of liquid radiological releases from pressurized water nuclear power facilities. The method of performing liquid effluent evaluations at the Yankee Nuclear Power Station (YNPS) involves the collection of the sample in question followed by immediate analysis for photon emitters in a 3.5 L geometry. The system used for analysis is a "state-of-the-art", commercially available, gamma spectroscopy system including 13-15 percent relative (to NaI at Co-60) efficiency high purity germanium semi-conductor photon detectors housed in poured four-inch thick lead shields, high resolution spectroscopy amplifiers coupled to 500 MHz analog to digital converters (ADC) and a multi-channel analyzer for data acquisition. Spectra are analyzed via vendor software to reduce the observed signals to μCi/mL concentrations.

Estimates of batch release concentrations are performed without decay correction since the exact time of the release is not known at the time of the measurement, the release is in the near future (usually less than six hours from analysis) and failure to incorporate a decay correction maximizes the reported activity for any isotope. Samples are analyzed for a length of time (usually 5,000 seconds) which will ensure the LLD requirements are achieved for each individual sample. This concept must be incorporated since the YNPS RETS require notification of the NRC, on a yearly basis, of all samples for which the LLD has not been achieved. This concept is tantamount to establishing the last spectral evaluation of the sample in question as the latest a-priori LLD estimate for the sample under investigation (i.e. the analysis incorporates all of the interferences specific to that sample at that time) (2).

The volume analyzed for each individual sample is a tightly controlled variable. Although the four-liter marinelli beakers (a water containing volume having a center cavity which will house the photon detector) used in the analysis are currently filled to a specified height (14.1 cm), the allowed variation does not (by protocol) exceed 0.2 cm. For chemically pure water specimens, this translates to a maximum allowable difference of 1.4 percent by volume. This maximum allowable uncertainty has been used for the final estimation of the relative systematic uncertainty in the volume term of the activity or LLD equation.

The efficiency of detection of photon emitters in the four L marinelli beaker geometry has only been determined three times since the system has been functional. The maximum relative variation (difference) in the observed efficiencies has been utilized as the relative systematic upper uncertainty bound for the efficiency. A total of 21 points were involved in the evaluation with the maximum fractional difference for any three points at the same energy being 0.068.

The combined (in quadrature) relative systematic uncertainty bounds for E and V have been determined to be 0.070. This relative systematic uncertainty is within the boundary condition established within the NUREG. It should be stressed that the areas evaluated represent only a portion of the analytical evaluation performed by the current "state-of-the-art" software systems. Peak search and complex spectral fitting algorithms have not been addressed directly to date in this evaluation. An attempt will be made to address some of these items in a later section through evaluation of samples containing added isotopes of known quantity.

The estimation of the relative systematic uncertainty of the background is a much more difficult task for those facilities which do not have full knowledge of the algorithms employed or the exact background count rates achieved for various areas of the photon spectrum during LLD estimations. If, however, the measurement protocols under investigation in this study are rigorous with respect to the required LLDs then they should be able to reliably predict the concentration of an analyte present at the required LLD. This approach has been applied to the gamma spectroscopy system at the YNPS. A review of the systematic uncertainty boundary conditions for gamma spectroscopy is provided in Table I.

A typical liquid release was chosen at random to be evaluated in detail with regard to the LLD for two specific isotopes - Ce-144 and Ru-106. Cerium -144 and Ru-106 were analyzed not to be present in the sample at one-tenth the LLD requirements. The intention of this experimental study was to evaluate the gamma spectroscopy system response when presented with the presence of Ce-144 and Ru-106 at the estimated LLD and at the required LLD. In order to determine the level at which the isotopes should be added, an experiment was conducted to determine the average LLD for both isotopes over a relatively short, three day, period of time. The isotopes determined to be present in this release and their approximate concentrations (μCi/mL) at time t = 0 are I-131 (7E-8), Cs-134 (5E-8), Cs-137 (7E-8), Co-60 (2E-8), Kr-85 (7E-6) and Xe-133 (2E-4). One potential problem which was noted in the NUREG is interestingly portrayed in Figure 8. As can be observed from an evaluation of the LLD versus days from sampling, the LLD decreases for Ce-144 but remains (relatively) stable over the time period of the measurement for Ru-106. This reduction in observed LLD is possibly a function of the decreasing I-131 concentration in the sample which would, most probably, produce the largest variation in the Ce-144 LLD.

After repeated evaluation of the sample, a known amount of Ce-144 equivalent to the average observed LLD value of Ce-144 was added to the sample. This amount was approximately equal to 1.2 times the system LLD at the midpoint of subsequent evaluations. Subsequent to ten determinations (results reported on Table II) an additional amount of Ce-144 was added to the sample to bring the

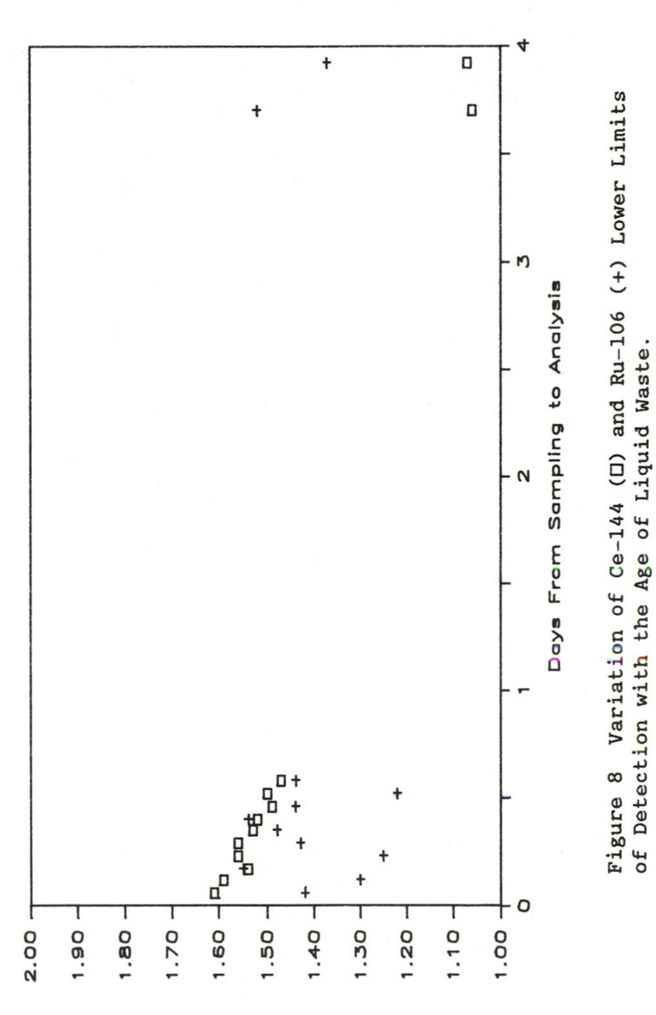

Figure 8 Variation of Ce-144 (□) and Ru-106 (+) Lower Limits of Detection with the Age of Liquid Waste.

TABLE II

EXPERIMENTAL RESULTS OF KNOWN ADDITIONS TO A ROUTINE RETS LIQUID MATRIX

	Ce-144		Ru-106	
	At 1.2xSystem LLD	At Required LLD	At System LLD	At Required LLD
Known Value (μCi/ml)	1.37E-7	5.00E-7	1.34E-7	5.00E-7
Number of Measurements	10	25	10	13
Number Detected*	10	25	9	13
$\bar{X}$ (μCi/ml)	1.28E-7	5.15E-7	1.46E-7	5.06E-7
s (μCi/ml)	0.24E-7	0.37E-7	0.24E-7	0.70E-7
Number Within $\bar{X}$+ - 2s	10	25	10	12
RSD Ranges for Individual Measurements	0.15-0.24	0.06-0.07	0.19-0.26	0.10-0.13
$\bar{S}$ (net counts)	115	454	36	123
s (counts)	20	30	5.9	17

* Detected by peak search algorithm at routine operational settings.

NOTE: At the system LLD $S > 3.3s$ indicating a possible bias of the LLD algorithm.

known level of Ce-144 to the required LLD concentration (5E-7 µCi/mL). A total of 25 more measurements of the activities in this sample were obtained. The pertinent results are portrayed in Table II. During the measurements of Ce-144, Ru/Rh-106 was added to the sample. First the addition was made at the system quoted LLD and later (after ten subsequent measurements) an addition was made to bring the level up to the required LLD for other isotopes (5E-7 µCi/mL). The point should be made that the measurement protocol and sample count time remained the same throughout all of the measurements. Table II depicts results which are indeed encouraging. The maximum deviation of the mean from the known was nine percent with a minimum of ten measurements. The use of this methodology at the Yankee Atomic Electric Company (YAEC) should not be taken as condoning the use of known value "samples" at the quoted LLD of a licensee. This approach should not be undertaken on a routine basis and could only result (generally) in a determination of detected or not detected. However, analysis of a sample having a known concentration of a particular isotope at the required LLD and typical interferences should be detected greater than nine out of ten times if the requirement to utilize an initial evaluation of the sample matrix as an a-priori LLD estimate is imposed.

Regardless of the regulatory approach, it has been demonstrated that analysis of two of the most problematic nuclides in gamma spectroscopy, at levels well below those deemed harmful to the public can be accomplished and can provide protection to the public without recourse to a revised formulation of the LLD equation.

Potential Impact of the Proposed NUREG

The effort extended to establish the appropriateness of the NUREG assumed (or assigned) relative systematic uncertainty bounds has been substantial for the three techniques under evaluation in this study. Several man-months have been expended in collecting, analyzing and portraying the data for the three techniques. These three techniques represent only 15-20 percent of the techniques which may require validation under the proposed RETS changes. A definitive estimate of the time involved would, perhaps, predict one-two man years of effort per facility. This estimate may indeed be placed higher if the number of on-line monitors is substantial in any single utility. However, this effort might be considered worthwhile if the health and safety of the public were indeed improved as a result of the efforts undertaken by each of the utilities (estimated from above as approximately two hundred man-years of effort).

In order to define the potential improvement to the health and safety of the public, the actual LLDs calculated under the current RETS, proposed RETS with the assumed relative systematic uncertainty bounds and the proposed RETS with calculated relative systematic uncertainty bounds have been determined for two of the robust techniques under consideration: tritium analysis and low level I-131 analysis. The third technique, gamma spectroscopy, has been shown to be robust with respect to the LLD requirements via the method of "known addition". As can be seen from Table III even under the worst set of circumstances, the calculated LLD is increased only to within a factor of two of the required LLD. It must be remembered that the calculations are based upon the boundary conditions and may not be truly indicative of the routine conditions of the measurement process. If a significant change in boundary conditions is noted, a strict application of the NUREG concepts might require re-evaluation of the LLD parameters.

TABLE III

COMPARISONS OF CALCULATED LLDs FOR VARIOUS FORMULATIONS OF THE LLD

	CURRENT RETS	PROPOSED RETS (CASE 1) (CASE 2)	REQUIRED LLDs
H-3 µCi/ml	1.79E-6	2.60E-6 2.03E-6	1E-5
I-131 pCi/Kg	5.50E-2	7.30E-2 7.30E-2	1

Case 1: Based upon the NUREG-4007 relative systematic uncertainty boundary conditions.

Case 2: Based upon the estimated relative systematic uncertainty boundary conditions from this study.

The tritium analysis case was further evaluated in accordance with Method IV by preparing a tap water sample (from well water) with a tritium standard so the resulting concentration was equal to the required LLD (1E-5 µCi/mL). The sample has been analyzed a total of twenty-six times under the standard protocol with the following results: $\mu_s = 1.08 \pm 0.06E-5$ µCi/mL. Each sample analysis would have predicted (based upon the experimentally determined NUREG LLD) the presence of tritium at the calculated critical level (1.3 E-6 µCi/mL). Although more analyses may be required to demonstrate with absolute assurance that the tritium methodology is capable of detecting (and quantitating) tritium at the LLD requirement, the implication from the data is clear: the public is well protected by these protocols if the achieved LLDs truly reflect the concentrations which have been established to protect the public (e.g. H-3 LLD = 1E-5 µCi/mL).

Acknowledgments

The authors gratefully acknowledge the help of M. W. Thisell and Mark J. Smith of the YNPS for their help in collecting and portraying data and the staff of the YAEL for their support of this project.

Literature Cited

1. Upgrading Environmental Radiation Data, HPSR-1 (1980), Watson, J. E., Chairman, August 1980.
2. Currie, L. A., NUREG/CR-4007, Lower Limit of Detection: Definition and Elaboration of a Proposed Position for Radiological Effluent and Environmental Measurements, September 1984.
3. Altshuler, B. and Pasternak, B., Health Physics, 1963, 9, 293-298.
4. Currie, L. A., Anal. Chem., 1968, 40, 585-593.
5. Currie, L. A. and DeVoe, J. R., Systematic Error in Chemical Analysis, American Chemical Society, Washington, DC, 1977, 114-139.
6. McCurdy, D. E.; Mellor, R. A.; Lambdin, R. W.; and McLain, M. E., Health Physics, 1980, 38, 203-213.

RECEIVED May 19, 1987

Chapter 14

Radioactivity Analyses and Detection Limit Problems

of Environmental Surveillance at a Gas-Cooled Reactor

James E. Johnson and Janet A. Johnson

Department of Radiology and Radiation Biology, Colorado State University, Fort Collins, CO 80523

The Lower Limit of Detection (LLD) values required by the USNRC for nuclear power facilities are often difficult to attain even using state of the art detection systems, e.g. the required LLD for I-131 in air is 70 fCi/m^3. For a gas-cooled reactor where I-131 has never been observed in effluents, occasional false positive values occur due to: counting statistics using high resolution Ge(Li) detectors, contamination from nuclear medicine releases and spectrum analysis systematic error. Statistically negative concentration values are often observed. These measurements must be included in the estimation of true mean values. For this and other reasons, the frequency distributions of measured values appear to be log-normal. Difficulties in stating the true means and standard deviations are discussed for these situations.

The Fort St. Vrain High Temperature Gas-cooled (HTGR) power reactor, operated by Public Service Company of Colorado is located approximately 40 miles north of Denver at the confluence of the South Platte river and St. Vrain creek. It is the only gas-cooled power reactor in the country and while it has had operating difficulties, the nuclear aspects of the design have great promise for the following reasons:

1. Net electrical efficiency is 39.2%
2. Extremely low in-house worker radiation dose rates
3. Extremely low radioactivity release to the environment

The last two characteristics are primarily due to the unique HTGR fuel element design and the rather innocuous environment of the core. Table I illustrates this point with a comparison of HTGR effluents with those from Boiling Water Reactors (BWR) and Pressurized Water Reactors (PWR).

0097-6156/88/0361-0266$06.00/0

Table I: Radioactivity Released by Reactor Type

	H-3 (Ci/MWe-year)		Cs-137 (mCi/MWe-year)
	Gaseous	Liquid	Liquid
BWR	0.05	0.1	25
PWR	0.2	1.2	1.7
HTGR	0.1	6.5	2.8 E-4

Since the radioactivity release is so low, the radioactivity concentration in all environmental sample types is essentially all due to natural radiation background sources with fallout and releases from other industries principally medical, superimposed upon it. The background source term due to primordial radionuclides is essentially constant and the frequency distribution can be normally or log-normally distributed. The fallout source term is time dependent and highly variable and generally log-normally distributed. Since the objective of a reactor environmental monitoring program is to document the presence or absence of radioactive materials due to reactor effluents, it is imperative that background in the reactor environs be documented. For Fort St. Vrain the preoperational period occurred during periods of significant Chinese weapons test fallout and essentially all radioactivity data shows lower values after reactor start-up. This makes comparison of operational periods to preoperational periods of no value. Thus it is necessary to compare site data near the reactor or in the predominant wind directions to control data.

Figure 1 shows gross beta particulate concentrations in air. Gross beta measurements include both background (due to K-40 and the U-238 and Th-232 series) with fallout superimposed. Shown are the half yearly arithmetic means for the four stations located in the predominant wind directions from the reactor and the means for the reference (control) air sampling stations. There is no difference in the means between the two station sets. The extremely large temporal variation is due to fallout from the Chinese atmospheric weapons tests during the period. (Chernobyl fallout measured during the spring of 1986 reached weekly maxima of 560 fCi/m^3).

Figure 2 shows that fallout Cs-137 concentrations in milk are log-normally distributed. (The geometric mean (median) multiplied by and divided by the geometric standard deviation includes 68.1% of the area under the frequency distribution.)

Figure 3 shows gross beta particulate concentrations in precipitation at the two facility site collectors. The total data (solid circles), if assumed to be log-normally distributed, would be very misleading. There are clearly two separate frequency distributions with very different geometric means and geometric standard deviations. The high activity concentration distribution is assumed to be fallout and the lower one due to natural background.

Iodine-131 is certainly the key radionuclide in reactor environmental monitoring. It is an indicator of release since the radionuclide is almost always gaseous and it is the predominant fission product contributor to radiation dose to the general public due to its food chain mobility. Reactor license requirements therefore put great emphasis on I-131 measurements in air and

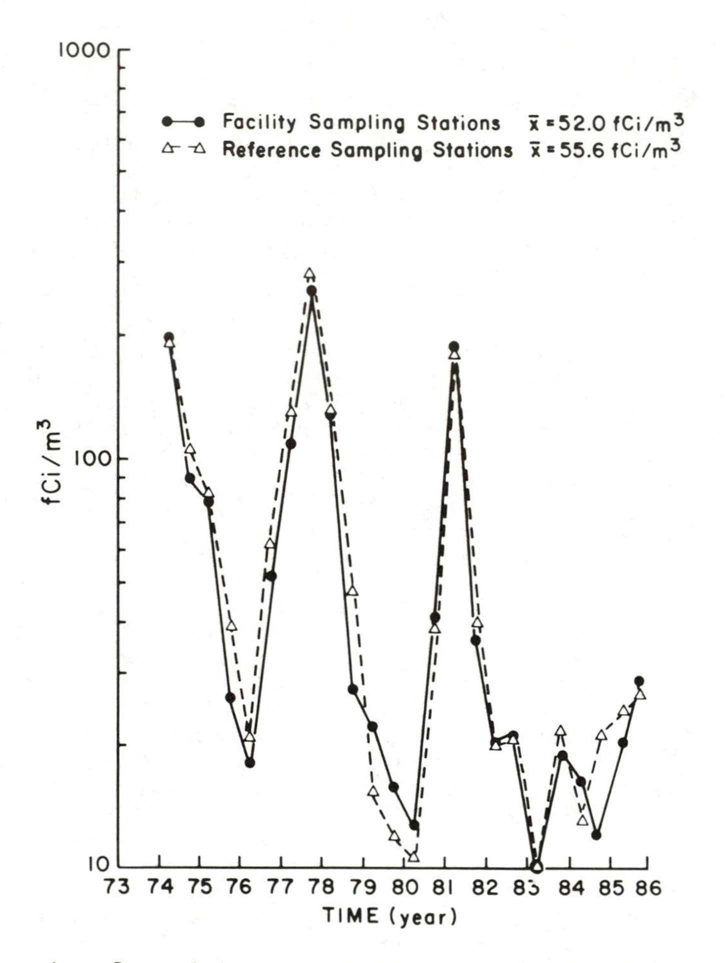

Figure 1. Gross beta concentrations in air for 1974 through 1985.

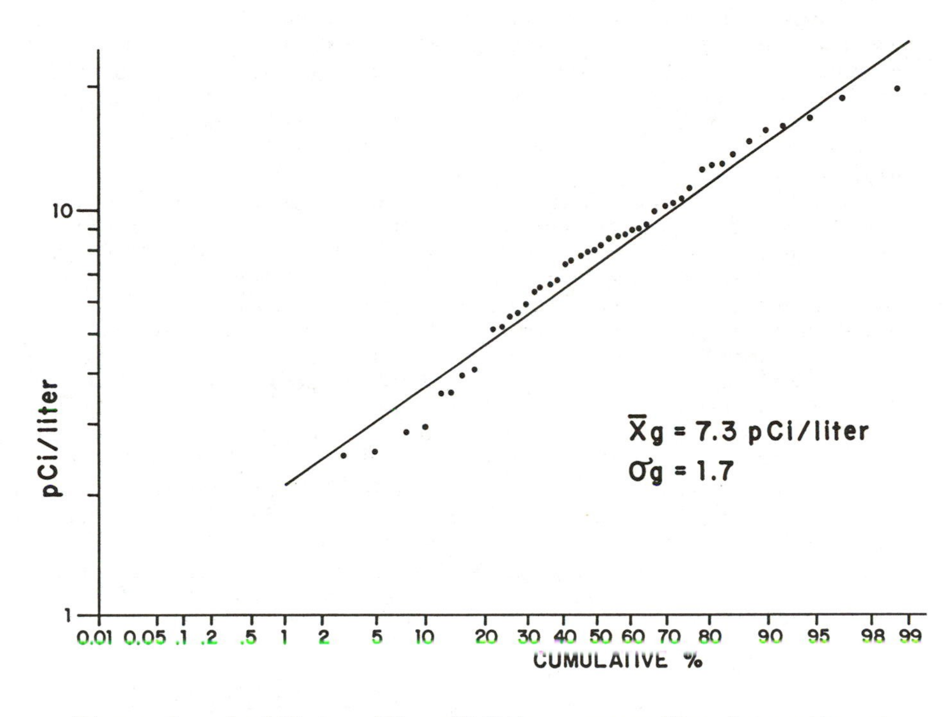

Figure 2. Cs-137 in milk, pCi/liter, composites from adjacent area, 1974 and 1975.

milk. The detection limit (LLD) for I-131 in air required by the USNRC is 70 fCi/m³. Using activated charcoal as the collection medium and a fixed air sampler flow rate, it is possible to achieve this value with Ge(Li) spectroscopy and counting times of approximately 300 minutes. (The net count rate due to I-131 is obtained by subtracting the sum of the count rates in four channels adjacent to the peak on each side from the sum of the count rates in the eight channels in the peak.)

During all of 1985 the Fort St. Vrain reactor was shut down, therefore no I-131 could possibly have been released. I-131 is released from local hospitals after patient diagnostic procedures and treatment with I-131 but this release is confined to surface waters. Figure 4 shows the measured I-131 concentration at all seven of the air sampling stations for all of 1985. If, in fact, there was no source term for I-131 in air, the expected frequency distribution of the measured values would be due only to methodological uncertainty and would be expected to be normally distributed. The arithmetic mean would be expected to be zero. Figure 4 shows the observed mean was actually -1.5 fCi/m³ indicating a negative systematic error, probably due to bias in the subtraction method used to obtain net count rate. The frequency distribution was indeed normal. This illustrates the necessity of including all negative values in data analysis. If only positive values were averaged, the mean would obviously be biased toward a false higher value. It also illustrates a method of determining systematic error. Using the simplified expression of currie (1) for S_c (S_c = 2.33 s_b) the a-priori, S_c for our measurement parameter was 33 fCi/m³. For the data shown in Figure 4 (n =356), only 1.2% of the values exceeded the S_c value where 5% false positive values would be expected. If the negative bias is taken into account and the distribution normalized to a mean of zero by adding 1.5 fCi/m³, more of the values would be greater than 33 fCi/m³ bringing the false positive percentage closer to the expected 5%.

The a-priori S_c determined from the sum of the count rates in channels adjacent to the peak can be compared to the S_c based on σ_o determined for the net peak count rates for the 356 trials:

$$\sigma_o = \text{S.E.M. } (n)^{1/2} = (0.63)\,(356)^{1/2} = 12 \text{ fCi/m}^3$$

$$S_c = 1.65\,\sigma_o = (1.65)\,(12) = 20 \text{ fCi/m}^3$$

As stated above, detection of effluent releases depends upon comparing mean activity concentrations over a period of time with those in a reference or control area, as close in characteristics as possible to the reactor area. To properly compare means the appropriate variances must be used. When reporting standard deviations therefore, it is important to remove the fraction of the total uncertainty which is due to the method. Since the precision (random uncertainty) to be attached to the method is commonly determined with spiked samples, only the coefficient of variation may be used. Figures 5, 6 and 7 illustrate an approach to subtracting the coefficient of variation of the method from Sr-90 measurements in soil. Figure 5 shows that the observed total

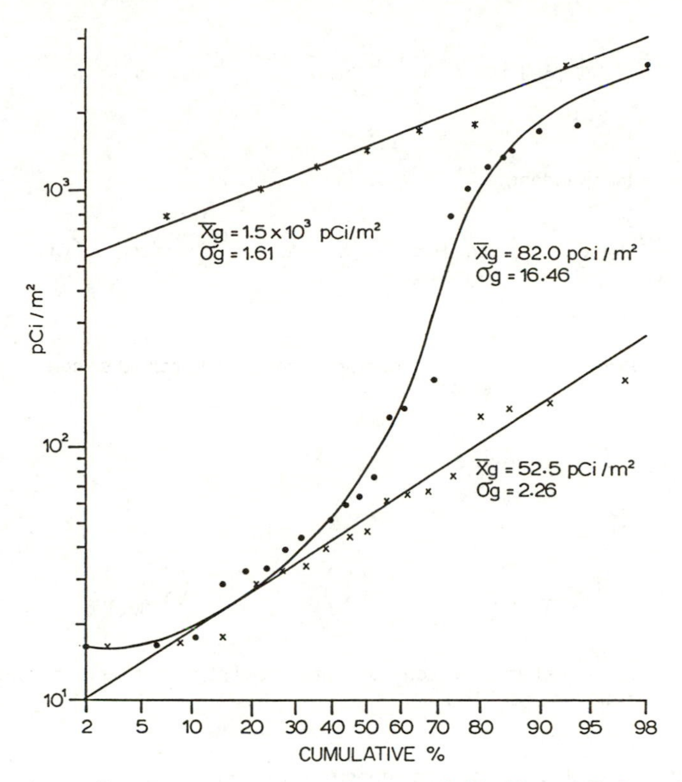

Figure 3. Total gross beta in precipitation, F-1 and F-4 combined July 1976 through June 1977 (F-1 and F-4 were the sample locations for two large precipitation collector funnels).

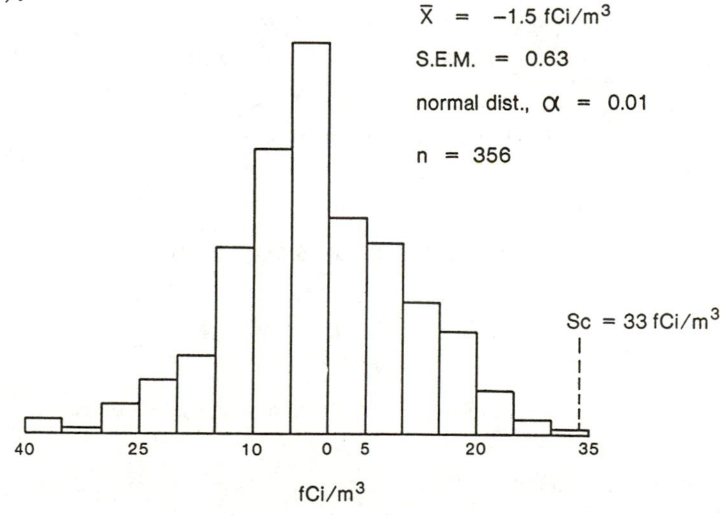

Figure 4. I-131 concentrations in air, 1985.

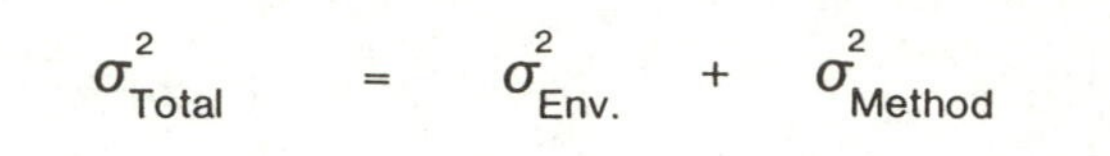

$$\sigma^2_{Total} = \sigma^2_{Env.} + \sigma^2_{Method}$$

If Independent,

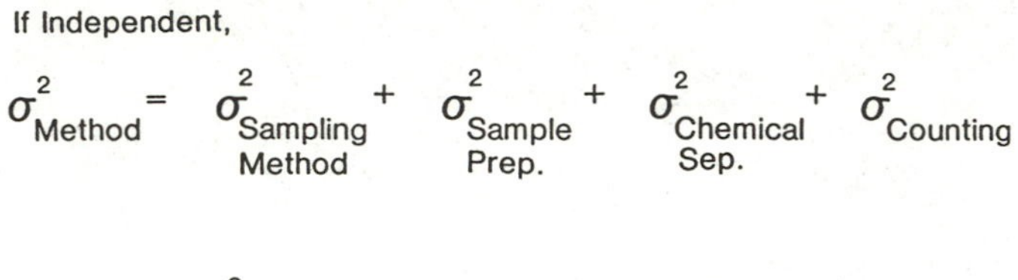

$$\sigma^2_{Method} = \sigma^2_{\substack{Sampling\\Method}} + \sigma^2_{\substack{Sample\\Prep.}} + \sigma^2_{\substack{Chemical\\Sep.}} + \sigma^2_{Counting}$$

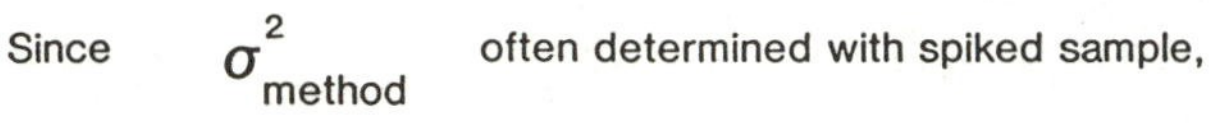

Since σ^2_{method} often determined with spiked sample,

Use:

$$\left(\frac{\sigma}{\bar{X}}\right)^2_{Total} = \left(\frac{\sigma}{\bar{X}}\right)^2_{Env.} + \left(\frac{\sigma}{\bar{X}}\right)^2_{Method}$$

Figure 5. Determination of uncertainty due to method and environmental variation.

From a single homogenized sample

$n = 10$

$$\left(\frac{\sigma}{\bar{X}}\right)_{method} = 0.22 \qquad \left(\frac{\sigma}{\bar{X}}\right)^2_{method} = 0.05$$

This includes counting uncertainty

Method: $^{89,90}Sr$ by $SrCO_3\downarrow$, $SrNO_3\downarrow$

^{90}Sr by $^{90}Y(OH)_3\downarrow$

Sr chemical yield by AAS of carrier.

Figure 6. Precision (random uncertainty) of Sr-90 method, 1979.

$$n = 38$$

$$X = 220 \text{ pCi/kg}$$

$$\sigma = 140 \text{ pCi/kg}$$

$$\left(\frac{\sigma}{\bar{X}}\right)_{Total} = 0.65, \quad \left(\frac{\sigma}{\bar{X}}\right)^2 = 0.42$$

$$\left(\frac{\sigma}{\bar{X}}\right)^2_{env.} = \left(\frac{\sigma}{\bar{X}}\right)^2_{total} - \left(\frac{\sigma}{\bar{X}}\right)^2_{method}$$

$$\left(\frac{\sigma}{\bar{X}}\right)^2_{env.} = 0.42 - 0.05 = 0.37$$

$$\left(\frac{\sigma}{\bar{X}}\right)_{env.} = 0.61$$

Figure 7. Sr-90 measurements in surface soil, 1980, Ft. St. Vrain environs.

coefficient of variation term squared is equal to the coefficient of variation of the method squared and the square of the coefficient of variation due only to environmental factors. Since it is only this last term that should be used to describe the results for a given environment, the uncertainty due to the method must be subtracted. In this case 10 replicate subsamples from a homogenized large sample were analyzed identically and the coefficient of variation determined to be 0.22 (Figure 6). This is not large considering the number of chemical and counting steps involved in the method. Figure 7 shows that if the method uncertainty is subtracted (as the coefficient of variation squared) from the total coefficient of variation squared of a set of 38 soil samples analyzed for Sr-90, the resulting coefficient of variation was 61%. Although it is true that often the methodological uncertainty is small compared to biological or environmental variation, the approach must still be used. When the uncertainty frequency distribution of the method is normal and the observed distribution is log-normal, this approach is not rigorous.

In conclusion, it is obvious that environmental measurements at radioactivity levels near, or in the case of Fort St. Vrain, below preoperational backgrounds present a complex problem. Attention must be given to determination of the proper frequency distribution of the data and to the proper statistical test to compare facility area data to that from background locations. Statistical methodology cannot always be rigorously applied and must be combined with a common sense approach to the available data. Negative values must always be included in the calculation of means and standard deviations in order to avoid biasing results towards high values.

Literature Cited

1. Currie, L. A. Anal. Chem. 1968, 40, 586-693.

RECEIVED May 19, 1987

Chapter 15

Detection Limits for Amino Acids in Environmental Samples

P. E. Hare[1] and P. A. St. John[2]

[1]Geophysical Laboratory, 2801 Upton Street, N.W., Washington, DC 20008
[2]St. John Associates, Inc., 4805 Prince Georges Avenue, Beltsville, MD 20705

Liquid chromatography techniques using fluorescent derivatives and laser-induced fluorescence can separate and detect sub-femtomole levels of amino acids. The widespread distribution of amino acids and proteins from living organisms produces nanomolar and higher levels of amino acid material in most environments of the earth's surface. Contamination usually occurs to some extent during sample collecting and processing and must be recognized and addressed before meaningful amino acid concentrations and distributions can be obtained from environmental samples.

Amino acids are distributed ubiquitously throughout much of the earth's crust, including the atmosphere (1). Their occurrence and important role in living organisms are well known, but amino acids have also been found in fossils and rocks hundreds of millions of years and even billions of years old (2). Amino acids have been reported at parts-per-billion levels in extra-terrestrial samples such as the Apollo moon rocks as well as in several meteorites (3,4). Even distilled water, reagent-grade HCl and other chemicals frequently contain trace amounts of amino acids. Figure 1 summarizes the levels of amino acids found in several samples of environmental interest.

Current analytical techniques for the analysis of the common amino acids are able to detect less than femtomole (10^{-15}) amounts using liquid chromatographic methods with fluorescent derivatives. However, this level of sensitivity may be extremely difficult to utilize because of the widespread presence of amino acids from living organisms. For example, the relatively high levels of amino acids in human body tissues and fluids make human fingerprints or even a person's breath a potential serious contaminant in detecting amino acids in specific environmental samples (5). Sample collecting and preparation are major concerns to amino acid trace analysis. Even with careful handling during sample collection and preparation it is always possible that the sample may have been contaminated in situ before its collection.

0097-6156/88/0361-0275$06.00/0

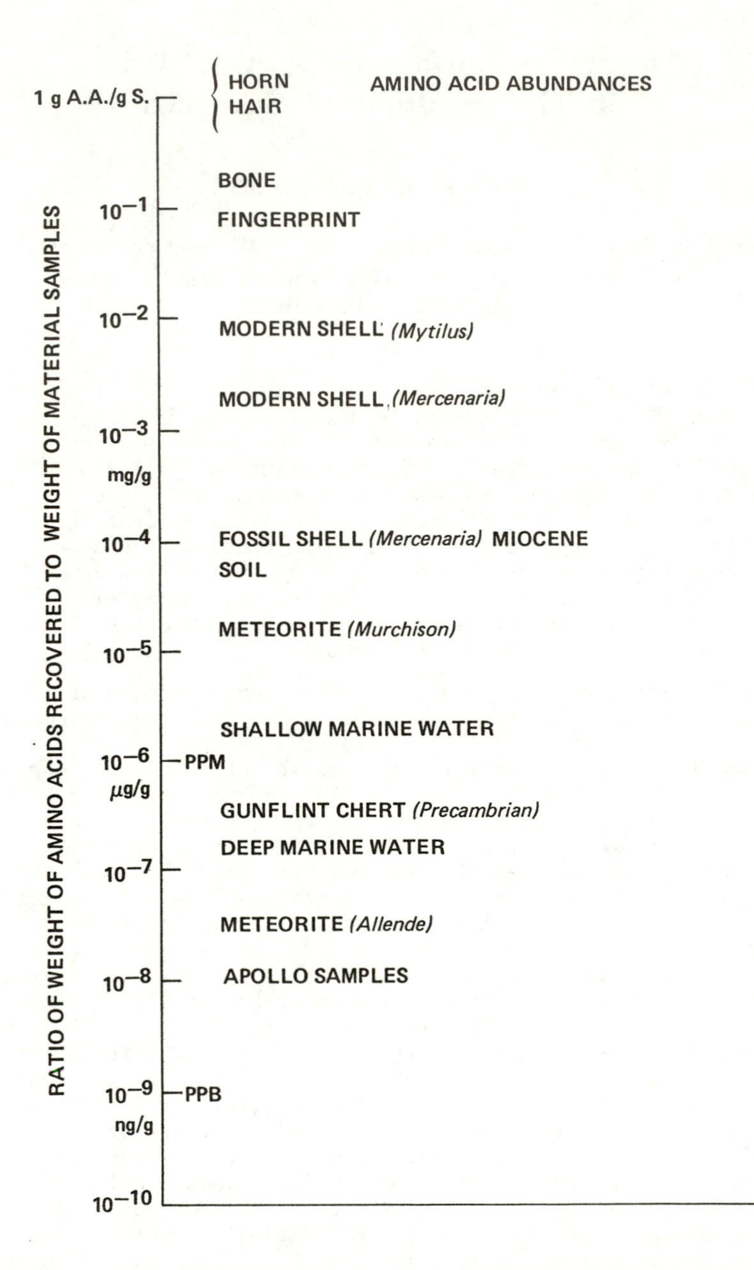

Figure 1. Distribution of amino acids in selected components of the Earth's crust. Ratio of weight of recovered amino acids to weight of sample. Pure proteins plot as 1 gram per gram.

Because of the widespread distribution of living organisms it is often difficult to interpret the significance of finding amino acids at trace levels (<10^{-6} g/g) in geological samples. The presence or absence of unstable amino acids such as serine and threonine as well as D-amino acids from racemization can be helpful in distinguishing older in situ fossil amino acid assemblages from modern amino acid contaminants. Stable isotope ratios of nitrogen and carbon can also be helpful in interpreting sources and pathways for amino acids found in environmental samples. Micromole levels of the amino acids are needed for isotope studies, and this level may not be available for some samples.

In spite of the difficulties encountered with contaminants from living organisms, several studies have shown the usefulness of determining the amino acid compositions and concentrations in geological samples ranging from nearly modern samples to samples billions of years old (1). As techniques for high-sensitivity amino acid analysis are further developed, studies of the distribution of amino acids in various environments will be increasingly helpful in tracing pathways of amino acids and proteins from living organisms into the environment. Food chains as well as airborne and waterborne diseases are research areas that should benefit from data on the pathways of distribution of amino acids and proteins throughout the crust of the earth.

Approaches to Amino Acid Analysis by Liquid Chromatography

The analysis of amino acids by liquid chromatography follows one of three main approaches depending on how the amino acids or their derivatives are chromatographed and detected. The first and most straightforward method is to separate and elute the free amino acids from a column while monitoring the effluent at around 210 nm wavelength.

The second approach involves post-column derivatization of the amino acids, which improves the sensitivity for detection by creating either a highly fluorescent derivative or one with a strong chromophore. The traditional approach for amino acid analysis uses post-column derivatization in which the reagent ninhydrin mixes with the column effluent and reacts with the amino acids to form a derivative that can be detected by absorbance at 570 nm and 440 nm.

The third approach is to make the amino acid derivatives before injection into the column. This pre-column derivatization method has potentially the highest sensitivity and is increasing in popularity in analytical laboratories. By choosing appropriate chiral (optically active) reagents for derivatization and/or chromatography, it is possible to make diasteroisomers that can be separated during chromatography or diastereomeric interactions that allow determination of the optical configuration of the amino acids in the sample. The distribution of D and L amino acids among living organisms is not well known and a systematic study would be helpful in interpreting the geological and environmental occurrences of D amino acids.

UV Detection of Underivatized Amino Acids

All of the common amino acids, including hydroxyproline and proline, show some absorption of UV light at around 200-210 nm (6). This non-destructive detection method is particularly useful for a preparative system for amino acids using ion-exchange resins and HCl elution. Adequate separation is achieved starting with an isocratic (constant composition) elution of 0.6M HCl followed by a linear gradient to 2.5M HCl to elute phenylalanine and the basic amino acids: histidine, lysine and arginine. Work on nitrogen and carbon stable isotopes in individual amino acids has used this method of separation successfully. Evaporation of the volatile HCl eluent facilitates sample preparation for isotope analysis. Considerable isotopic fractionation of the order of 1% occurs during chromatography, making it essential to achieve complete resolution and recovery of the entire chromatographic peak. This method could be developed into a suitable analytical technique for amino acid analysis, with sensitivity in the low nanomole or even picomole levels. It detects the secondary amino acids proline and hydroxyproline and needs no pre-column or post-column derivatization, which all other current methods require.

Post-Column Derivatization of Amino Acids

Modern amino acid methods of analysis started with the paper by Spackman, Stein and Moore in 1958 (7). Chromatographic columns were packed with ion-exchange resin particles of uniform size. Sodium citrate buffers were pumped through the columns eluting the amino acids. A 150 cm by 0.9 cm column separated the acidic and neutral amino acids, while a 15 cm by 0.9 cm column separated the basic amino acids from a second sample aliquot. Ninhydrin reagent was pumped continuously into a tee and mixed with the column effluent containing the separated amino acids. The mixture was pumped through a length of teflon tubing heated in a boiling water bath for 15 minutes to effect the reaction of the amino acids with the ninhydrin. The mixture was monitored in a flow-through dual-channel photometer at 570 and 440 nanometers. A multipoint recorder printed out a continuous record of the photometer output. The limit of detection (taken as signal to noise ratio of 2) was around 50 nanomoles (10^{-9}) for most of the amino acids. The total analysis time was 24 hours.

In 1987, using a pre-column derivatization method, the common amino acids can be analyzed in less than 20 minutes with the limit of detection less than 0.5 femtomole (10^{-15}) (8). This represents an increase in sensitivity of at least eight orders of magnitude and a decrease in analysis time by around two orders of magnitude! A brief description of how these changes have come about follows.

The first significant improvement came as a result of using smaller particle-size resins to reduce band-spreading. This permits the use of smaller volume columns, which require less volume of mobile phase to elute each amino acid (9). This in turn increases the concentration of amino acids in the column eluent. With better resins and smaller volume columns alone, it became possible to improve the detection limits by two orders of magnitude. At the same time it was possible to shorten analysis times to around a few hours. Another improvement was a flowcell design that increased sensitivity by another order of magnitude by increasing the pathlength while decreasing the volume of the flowcell. Further improvement in sensitivity has resulted from the development of still smaller column volumes with yet finer size particle packings (5-micron resin beads), and in particular the development of fluorescent derivatives of amino acids (10,11).

In post-column derivatization systems, the derivatization reagent is continuously mixed with the column effluent; consequently, the presence of trace amounts of amino acids or interfering components such as ammonia in the column eluent limits the ultimate sensitivity attainable. Figure 2 shows the effects of buffer contaminants on the baseline of a typical high sensitivity analysis (12). The basic contaminants are held up on the column until elution by the final buffer. Detection of the basic amino acids below 10-picomole levels is difficult using this approach. The use of a shorter second column with isocratic elution for the analysis of just the basic amino acids is a good solution to the problem but does require the use of a second aliquot of sample.

Chromatography of Amino Acid Derivatives

Pre-column derivatization and subsequent chromatography of the amino acid derivatives potentially provide the highest sensitivity available for amino acid analysis ($<10^{-15}$ molar). The important advantage in this approach is that trace amino acid contaminants in the column eluents do not interfere as in the post-column systems. Only amino acids in the sample that have been derivatized are detected, and consequently the baseline is not influenced by possible contaminants contained in the eluents.

There is a wide choice of amino acid derivatives that can be used. Phenylisothiocyanate (PITC) reacts with both primary and secondary amino groups to form moderately stable phenylthiocarbamyl (PTC) derivatives that are separated on a reversed-phase column and can be detected in a UV detector at 254 nm wavelength (13). Detection limits are at about the picomole level. Sample derivatization with PITC takes around 20 minutes and requires close attention to details if consistent results are to be obtained. The presence of salts such as NaCl in the sample interferes with the derivatization of several of the amino acids, and care must be taken in processing samples containing salts.

The development of fluorescent derivatives of amino acids and their chromatography on reversed-phase columns yield a significant gain in sensitivity. Many fluorescent derivatives of amino acids are available that greatly enhance the sensitivity of detection. O-phthaldialdehyde/mercaptoethanol (OPA) reagent reacts with most of the common amino acids (but not proline) to form fluorescent derivatives (14). Because OPA derivatives are not very stable, it is essential to chromatograph the OPA derivatives within a few minutes. In order to achieve consistent analytical results, it is necessary to automate or to time accurately the derivatization step. Amino acid analysis with pre-column OPA takes less than 15 minutes including the derivatization step. It has become a popular technique wherever proline values are not necessary.

The search for improved pre-column derivatives for amino acid analysis is a continuing process. Dansyl derivatives are stable and have been used for amino acid analysis, but hydrolysis products (dansyl OH) of the dansyl reagent are difficult to eliminate and can interfere with the dansyl amino acid peaks (15). FMOC-Cl (9-fluorenylmethyl chloroformate) is another derivative that has been used and that shows promise for the analysis of the secondary amino acids proline and hydroxyproline as well as the primary amino acids (16).

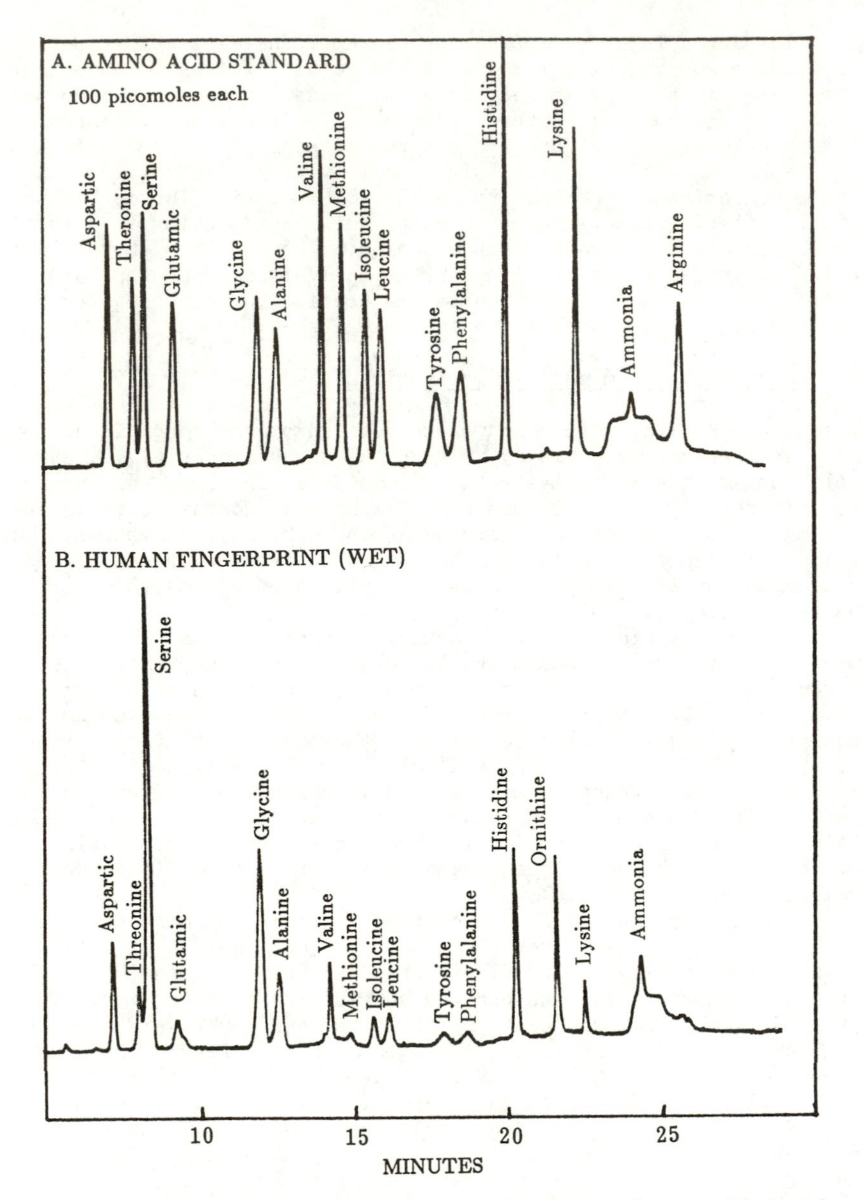

Figure 2. Post-column amino acid chromatograms showing limitations to detection limits due to contaminants in the mobile phase of the liquid chromatography system. Ion-exchange column with sodium citrate buffers and OPA reagent for post-column reaction. Vertical axis is fluorescence with the same sensitivity for each run. A. Amino acid standard with 100 picomoles of each amino acid injected. B. Human fingerprint showing high levels of serine, glycine, and ornithine. Continued on next page.

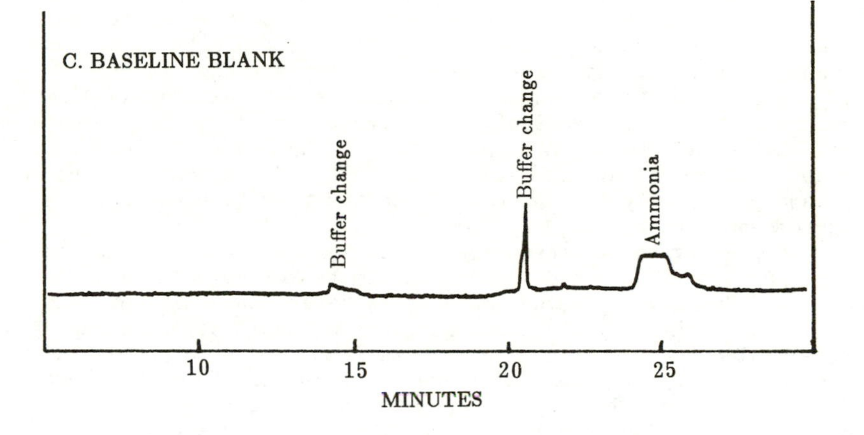

Figure 2.--Continued. C. No sample injected. Mobile phase contaminants accumulate on the column and elute with change in mobile phase.

The ideal pre-column derivative for amino acids, when it is found, should react with both secondary and primary amino acids and also make it possible to separate enantiomers during chromatography.

Detection Using Laser-Induced Fluorescence (LIF)

Fluorescence intensity is directly proportional to excitation light intensity. The use of laser excitation would seem to be a logical way to improve detection limits. The potential is great: monochromatic radiation, sharply collimated beam geometry, and high intensity at the laser wavelengths. These features match well with the requirements for fluorometric liquid chromatography detectors, in particular the conflicting requirements for small-cell volumes, long optical path lengths, and efficient minimization of stray light. Unfortunately, several aspects of laser light sources limit their utility as fluorescence excitation sources. Stability of output intensity is poor, and noise level is relatively high in both continuous wave and pulsed lasers as compared to alternative light sources. Wavelength availability is limited to a few choices and is very limited in the ultraviolet region of the spectrum. Wavelength-tunable lasers are available, but the cost and complexity of these systems poses a significant barrier to their routine use in fluorescence detectors.

Complexity and expense notwithstanding, laser excitation has been shown to produce good detection limits for amino acid analysis. Roach and Harmony (8) have demonstrated detection limits of 5 to 15 femtomoles of hydrolysate amino acids in standard solutions using O-phthaldialdehyde pre-column derivatization and 334-363 nm argon-ion laser excitation. These limits are on the order of tenfold better than most commercial fluorescence detectors under similar conditions. These limits are lowered tenfold to 0.2 to 0.5 femtomoles using 2,3 naphthalenedialdehyde as the pre-column derivatizing agent when a 457.9 nm argon laser line is used as the excitation source. The better detection limits using visible-light excitation are attributed to the lower inherent noise level of the visible-laser line and the avoidance of high background fluorescence induced by ultraviolet excitation. An equally important contribution to the achievement of a low dectection limit appears to be due to the use of the flow cell described by Sepaniak and Yeung (17). It utilizes a quartz fiber light pipe as an efficient optical couple to a small-volume cell.

Novotny (18) has pointed out an alternate trend in the effort to utilize the potential power of the laser for HPLC detection, namely the "tuning" of the chemistry of the system to match the available laser wavelengths (19). The use of naphthalenedialdehyde as discussed above is a good example. An even more striking example is the work of Sauda, Imasaka and Ishibashi (20), who used indocyanine green as a fluorescent tag for protein labeling. The dye/protein complex absorbs at about 765 nm and emits at 820 nm. This couples well with the output of a pulsed semiconductor laser diode (780 nm, 15 mW). The resulting equipment is capable of detecting 90 nanograms of bovine serum albumin. The authors stress that blank fluorescence is "completely negligible" in near-infrared fluorometry; consequently, the low blank is the dominant factor in such low limits of detection. These authors quite candidly point out that such detection limits are of little practical use in the absence of reagent specificity.

Chemiluminescence Detection

Chemiluminescence detection of amino acids is another promising approach that has the potential for very high sensitivity. Replacing electromagnetic excitation of analyte molecules with chemical excitation eliminates the scattered light component of the usual blank signal associated with a light source. Furthermore, the specificity requirements of the chemi- or bioluminescence detection reactions should reduce the overall level of the blank resulting from chemical contaminants.

Chemiluminescence results from energy transfer from peroxyoxalate reaction systems to fluorescent analytes. This approach to high sensitivity amino acid analysis was taken by Kobayashi and Imai (21) and Miyaguchi et al (22). Dansylated amino acids were detected in this manner using bis (2,4,6-trichlorophenyl) oxalate and H_2O_2 as the detection reagent. The use of such aryl oxalate systems has been reviewed by Imai and Weinberger (23). A procedure utilizing the well-known luminol-peroxide chemiluminescence system has been presented by MacDonald and Nieman (24). They chose a quenching mode of operation whereby the Co(II)-enhanced luminol/peroxide luminescent solution was continuously pumped and mixed with the effluent of the HPLC column. Complexation of the Co(II) by analyte amino acids effectively removes the Co(II) from interaction in the chemiluminescence reaction sequence and results in decreased light emission. This type of procedure is dependent on the formation constants of the amino acid-Co(II) complexes. Limits of detection range from about 4 picomoles for good complex formers such as histidine and cystine to as high as 20 nanomoles for poor complexers such as aspartic acid. These results are not directly comparable with the detection limits quoted for the peroxyoxalate procedure because of the relatively inefficient bonded cation exchange column used to separate the amino acids.

Discussion

It is possible to detect sub-femtomole levels of amino acids and their enantiomers with state-of-the-art liquid chromatography systems. Under most circumstances, these levels of detection are unattainable due to the presence of amino acids and proteins in virtually all environments. In order to realize the potential limits of detection, it is necessary to utilize new approaches to sample and reagent preparation. To approach the limits of detection possible with the technique, a properly used clean room facility is a must. A serious and seldom recognized source of contamination in amino acid chemistry is the use of 6N HCl for the hydrolysis of peptides and proteins to free amino acids. Even reagent-grade HCl frequently contains amino acid contaminants, but even more serious is the storage and handling of the reagent. Whenever the HCl reagent bottle is opened, the HCl can become seriously contaminated by airborne contaminants such as dust and pollen. Laminar flow hoods and special handling reduce the problem but do not eliminate it. Vapor phase hydrolysis with 6M HCl is another promising approach to reducing contaminant levels. Analysis even at picomole and low nanomole levels needs careful controls during sample preparation. Controls should include samples heated to remove possible amino acids as well as frequent reagent blanks processed and analyzed to determine background contaminant levels. It is embarrassing to find out that the sample contained the same levels of amino acids as the reagent control blanks!

In geological samples the problem of in situ contamination by living organisms and their metabolic products limits the usefulness of amino acid geochemistry in older geological samples. The original levels of proteins in shells, bones and teeth decrease with time due to chemical breakdown and leaching. Ground water and other sources contain levels of amino acids that can sometimes exceed the levels of amino acids left from the original organic matter in the fossil. When this happens, amino acids can diffuse into the fossil and show higher than expected levels.

This frequently encountered situation is usually recognized by the presence of excessive levels of serine. Serine is one of the least stable amino acids found in modern organisms and decreases rapidly in progressively older, uncontaminated fossils. Amino acids in contaminants are generally of the L-configuration, again pointing to living organisms as the source. The original protein in fossils undergoes hydrolysis and racemization with increasing levels of D-amino acids. It may not always be possible to distinguish racemization-derived D-amino acids from cell-wall D-amino acids in living bacteria.

In the earlier days of amino acid geochemistry it was necessary to extract the amino acids from relatively large samples, frequently entire shells or bones. As limits of detection have improved it has been possible to study the distribution of amino acids within a single shell and even detect gradients from the surface to the interior of a shell or bone. Frequently the outside of a contaminated fossil will show much higher levels than the interior as well as a different composition. An uncontaminated fossil shows less concentration of amino acids on its surface and a greater concentration in its interior. Only with improved limits of detection have studies on distribution of amino acids in single fossils--and even in microfossils such as diatoms and forams--been feasible.

With attention to sample collecting and handling it will be possible to study effectively the occurrences and distribution of amino acids throughout the crust of the earth and even in regions (submarine vents, volcanic areas and extraterrestrial regions, etc.) where abiotic synthesis of amino acids has been suggested. Prospects are exciting!

Literature Cited

1. Hare, P. E. In Organic Geochemistry, Methods and Results; Eglinton, G.; Murphy, M.T.J., Eds.; Springer-Verlag: New York, 1969; p 438.
2. Schopf, J. W.; Kvenvolden, K. A.; Berghoorn, E. S. Proc. Natl. Acad. Sci. 1968, 59, 639-648.
3. Kvenvolden, K. A.; Lawless, J. G.; Ponnaperuma, C. Proc. of Natl. Acad. Sci. 1971, 68, 486.
4. Harada, K.; Hare, P. E. In Biogeochemistry of Amino Acids; Hare, P. E.; Hoering, T. C.; King, K., Eds.; John Wiley & Sons, Inc.: New York, 1980; p 169.

5. Hamilton, P. B. Nature 1965, 205, 284.
6. Schuster, R. Anal. Chem. 1980, 53, 617.
7. Spackman, D.; Stein, W. H.; Moore, S. Anal. Chem. 1958, 30, 1190.
8. Roach, M. C.; Harmony, M. D. Anal. Chem. 1987, 59, 411.
9. Hamilton, P. B. Anal. Chem. 1963, 35, 2055.
10. Hare, P. E. Methods in Enzymol. 1977, 47, 3.
11. Benson, J. R.; Hare, P. E. Proc. Natl. Acad. Sci. 1975, 72, 619.
12. Hare, P. E. In Chemistry and Biochemistry of the Amino Acids; Barrett, G. C., Ed.; Chapman and Hall: New York, 1985; p 415.
13. Bidlingmeyer, B. A.; Cohen, S. A.; Tarvin, T. L. J. Chromatogr. 1984, 336, 93-104.
14. Hill, D. W.; Walters, F. H.; Wilson, T. D.; Stuart, J. D. Anal. Chem. 1979, 51, 1338.
15. Perrett, D.; In Chemistry and Biochemistry of the Amino Acids; Barrett, G. C., Ed.; Chapman and Hall: New York, 1985; p 426.
16. Einarsson, S.; Josefsson, B.; Lagerkvist, S. J. Chromatogr. 1983, 282, 609-618.
17. Sepaniak, M. J.; Yeung, E. S. J. Chromatogr. 1980, 190, 377-83.
18. Novotny, M. V. Chrom. Forum 1986, 1, 19-25.
19. Gluckman, J. C.; Shelly, D.; Novotny, M. J. Chromatogr. 1984, 317, 443.
20. Sauda, K.; Imasaka, T.; Ishibashi, N. Anal. Chem. 1986, 58, 2649-53.
21. Kobayashi, S.; Imai, K. Anal. Chem. 1980, 52, 424-27.
22. Miyaguchi, K.; Honda, K.; Imai, K. J. Chromatogr. 1984, 316, 501-05.
23. Imai, K.; Weinberger, R. Trends Anal. Chem. 1985, 4, 170-75.
24. MacDonald, A.; Nieman, T. A. Anal. Chem. 1985, 57, 936-40.

RECEIVED March 2, 1987

PANEL DISCUSSIONS OF SOME CRITICAL ISSUES

Chapter 16

Real-World Limitations to Detection

David A. Kurtz[1], John K. Taylor[2], Larry Sturdivan[3], Warren B. Crummett[4], Charles R. Midkiff, Jr.[5], Robert L. Watters, Jr.[6], Laura J. Wood[6], W. William Hanneman[7], and William Horwitz[8]

[1]Department of Pathology, Pennsylvania State University, University Park, PA 16802
[2]National Bureau of Standards, Gaithersburg, MD 20899
[3]CBM Branch, Chemical Research and Development Center, U.S. Army Chemical Research, Aberdeen Proving Ground, MD 21010
[4]Analytical Laboratories, Dow Chemical Company, Midland, MI 48667
[5]National Laboratory Center, Bureau of Alcohol, Tobacco, and Firearms, Rockville, MD 20850
[6]Center for Analytical Chemistry, National Bureau of Standards, Gaithersburg, MD 20899
[7]Center for Technology, Kaiser Aluminum and Chemical Corporation, 6177 Sunol Boulevard, Pleasanton, CA 94566
[8]Food and Drug Administration, Washington, DC 20204

Data obtained at or near the limit of detection is often improperly prepared by scientists and improperly interpreted by the receiving public. This panel offers advice to scientists preparing data to have statistical support for the data produced. Four examples from the real world illustrate interpretation problems: limit of detection reporting, the role of matrix effects and matrix variability in forensic work, the effects of the blank in ICP work, and federal regulatory problems.

Transferring carefully prepared data from the laboratory to the general public can result in problems and difficulties often not considered beforehand. Data are created by analysts, usually at the request of others. These data are of varying quality but are used in many areas of life. Though the general purpose is to solve problems, problems are often created because of the varying qualities of the data and the varying backgrounds of the users of the data. We have convened a panel to discuss some of these problems suggesting in some cases solutions that could alleviate them as problems.

There are four groups of people who use analytical information in daily life. These are the non-experts (everyone except the analysts), the governmental regulators and those regulated, the rulers of the courts, and the analysts themselves. Each of these has

0097-6156/88/0361-0288$08.25/0

different backgrounds, different areas of thought, and different purposes in using analytical information. The interaction between these groups on such issues can lead to discussion, fear, and threats. This panel will show how this can come about.

The writers each address specific areas where there have been problems when diverse people are using analytical information. Dr. Taylor addresses the analysts because this is of the first importance. Good information must come first. Dr. Sturdivan picks up this theme in further discussing how to obtain good information in the narrower field of calibrating instrumental response to amounts or concentrations in samples. When analytical reports are made public, misunderstandings may occur amongst the "non-experts" as outlined by Dr. Crummett. He discusses problems in such areas as the presence of dioxin in human fat and contamination of water supplies with toxic solvents. Dr. Crummett discusses the risks of false negatives and false positives. The interpretation and calculation of the limit of detection when taking the risks into account has not been readily available. Accordingly, a short section is included here, perhaps in simplified form, to acquaint the reader with the term, limit of detection. Further real-world problems in the forensic field are discussed by Mr. Midkiff. He describes some examples at the legal interface. In this area the analyses are even more difficult because of the unusual matrices found and the lack of sample to work with. The governments, of course, are asked by the people to become active in this area because of their concerns and lack of knowledge of both the areas of analysis and toxicology. Dr. Hanneman has shown that broadly written regulations can be difficult to comply with. Here is also an example where researchers have not read the Federal Register, the central regulation document, carefully enough to get its true meaning. Hence, they can get the wrong impression and support an erroneous plan of action. Finally, Dr. Horwitz takes the other side of the coin and defends the government in putting out good documentation about their well planned regulations. He also recommends some basic points for analysts to remember as rules of the thumb to follow...as long as they first know and follow good analytical practice.

The following text discusses problem areas. Here, analysts are working in trace amounts or at the limit of detection. Frequently, problems occur due to the interaction between the analysts, courts, and non-experts. Key questions that might be answered are:

*** Can a quality label be put on data?

*** How can we teach the public the concept of error?

*** Can we discuss data, taking error into account, without inducing an alarmist reaction?

*** How do we understand the special difficulties of obtaining analysis data relating to forensic problems.

*** Can we interact with federal regulators in the making of suitable regulations?

Basic Data Quality (Dr. J. K. Taylor)

Have we always had in the laboratory a state of statistical control on a given problem? Dr. Taylor says in this regard, "Statistical control of the measurement process is the first requirement for meaningful detection, since without it all data are meaningless. This means that a limiting mean must be realized and a stable standard deviation must be attained and maintained." By limiting mean we have in mind a mean that approaches the true mean as the sample size becomes infinite. A stable standard deviation means one that does not change in size with the addition of additional samples. In other words an analyst must be able to reproduce his results at all times. Otherwise, any result at any time is meaningless.

"Statistical control applies to all parts of the analytical system - sampling process, the calibration, the blank, and the measurement. Statistical control is attained by the quality control of the entire system and involves maintenance of realistic tolerances for all critical operations. A system of control charts is the best way to demonstrate attainment of statistical control and to evaluate the appropriate standard deviations. In the simplest form, the results of measurement of a stable check sample, obtained over a period of time, are plotted. Statistical control is demonstrated when the values are randomly distributed around their average value." Control limits are often taken as $\pm$ 2 or 3 standard deviation units of these replicates. Dr. Taylor also adds, "Even the ranges of duplicate measurements of the actual samples tested can be plotted in a similar manner to demonstrate a stable standard deviation. In either case, the statistics of the control charts are the best descriptors of the variability of the measurement process."

Dr. Taylor continues his general advice: "Later speakers will discuss how measurement variability quantitatively defines the limits of detection and quantification. Due to the nature of measurement, each laboratory (analyst) will have somewhat different measurement uncertainties, and hence different limits of detection. Published values of LOD's (Limits Of Detection), MDL's (Method Detection Limits), or what-have-you for methodology are typical, at best; hence they have no predictable quantitative relation to those obtained by any laboratory or analyst. Each must evaluate them for itself and will make somewhat different decisions concerning precision and detection when analyzing the same samples. Each has the professional obligation to obtain all information necessary to support the quality of its data, which must be technically sound and defensible.

"Measurement uncertainty becomes critically important as it influences the decision process. In high-accuracy analysis, this influences the last significant figure. In trace analysis, it influences the first (which is also the last) significant figure. No figure can be significant without statistical support. Luckily, many measurements are made with more significant figures than are needed for a decision process. Analysts must remember that they cannot use statistics obtained by any one else to support their own data and hence in making decisions using it.

"A fact that is not always appreciated is that the use of appropriate methodology is necessary but not sufficient for reliable measurement data and/or for attaining specified limits of detection. In the EPA Love Canal studies, for example, the contract laboratories varied by more than a factor of ten in their detection limits while utilizing the same methodology. Accordingly, it is fallacious to assign quality labels to data based on typical or even collaborative test results. Data from various laboratories may be of better or poorer quality than such indicators. Only reliable estimates based on the performance of the producer should be used.

"The wide differences that can occur in the quality of data have serious implications for data compilations. Unless there is some way to code and/or to associate data with its uncertainty, poor data can be unduly influential in subsequent data analysis. In detection situations imprecise data leads to large detection limits; hence non-detection can have a different meaning, depending on how the measurement was made.

"In this speaker's opinion the precision attained in measuring real samples [ed.: as opposed to measurement of standards only or standards dissolved in substitute background matrices] is the only reliable basis for decisions on detection. In large measurement programs, the use of duplicate-sample control charts is the most feasible way to establish the precision parameters needed and to defend limits of detection. Otherwise, a sufficient number of replicate measurements must be made on the samples tested for this purpose. Without documented demonstration of precision, the data are meaningless."

Calibration Errors (Dr. L. Sturdivan)

Another aspect of data quality that needs to be addressed is the error that is produced during the calibration step in analysis. While, strictly speaking, calibration errors are slightly outside the realm of detection limit errors, it is also true that much analytical work is done at the detection limit and the calibration errors that are produced can be proportionately much larger at the detection limit than elsewhere on the graph. Hence, it is most appropriate to discuss the matter at this time.

Most quantitative analytical data results from a calibration step. Whether the data comes from a gas chromatograph, liquid chromatograph or from a spectrometer, calibration is required. Calibration relates instrumental response with a specified amount of substance to be measured. Calibration requires standards to cover and enclose the entire concentration range of interest. In preparation of these standards, however, Dr. Sturdivan comments: "Often the standard samples are made by serial dilution. They're treated as though they are independent samples," which they aren't in that case. "Now there is nothing wrong with making serial dilutions, but if that is done, it would be desirable to make two series of serial dilutions, side by side." What he is referring to comes from the following discussion: The best method for the preparation of standards is to prepare each concentration directly (or with serial dilutions) from neat (100% pure) compounds (with the same number of dilution steps). This can be costly in both time and materials. The next best method, and one that can be shown to be statistically nearly correct is to start from two different neat standards and dilute them in parallel tree-like steps. Tree structure dilutions would be to prepare AA and AB from A; then prepare AAA, AAB, and AAC from AA and ABA, ABB, and ABC from AB, etc.

Another approach in sample preparation, and one having some bias, is to start from one neat weighing measurement assuring that each standard has the same number of dilution steps in its preparation. The dilution steps would be of a tree structure. Thus each dilution standard will have the same preparation variance. The most common standards preparation method is one where each standard is serially produced from the proceeding one. In this case a large bias is developed as dilutions proceed. The error variance of dilution gets larger and larger with each dilution step.

Dr. Sturdivan continues, "There may be dilution errors. These tend to occur more frequently at the lower end, down there where the detection limit is. If there is adsorption to the walls of the container, the lowest standards resemble blanks."

Dr. Sturdivan emphasizes "Good confidence limits around the calibration graph must be done in equal variance space." What he is saying is the following: Calibration standards are prepared at various concentration levels to cover the desired analytical range. The square of the standard deviation, the variance, at each level must be approximately equal. A quick look at any chromatographic calibration data covering a wide range will show that this condition is not met. Some sort of treatment must be done with the data in order to have an equal-variance condition. Transformation of response data has been found to satisfy the equal variance requirement (1)

He continues: "Often the calibration graphs are linear over a particular region but not over the region of interest. Therefore, one should check the linearities and use a non-linear calibration graph as well." If fitting to a linear function is desired (for simplicity's sake), the transformation function used for the amount variables need not be the same as that used on the response variables for equal variance. In this manner data can often be brought into conformity with the simpler linear function (1).

"Okay, what are the better solutions? Use equal variance space. Put confidence limits around that graph in equal variance space. You may have to do at the beginning a little more calibration, make a few more standards, and put a little more work into establishing those calibration graphs. Don't try to extend the curve beyond your region of interest. It is not reasonable to associate the same variance with the blanks that you associate with the curve. And therefore, one would have to do a little more work with the blanks and determine what the decision limit is based on. If the dilution error is significant, determine the 'decision limit' on the basis of blanks alone.

"What would the ideal system consist of? Well, first of all we would determine what the cost of the Type I error (false positive) is and what the cost of a Type II error (false negative) is. Then we would determine the probability of encountering those particular errors. The third step is to determine the distribution of future samples (whether uniform, normal, or what). Finally one can minimize the probable cost of future errors."

These laboratory problems as related in this first part of the panel discussion are important to solve in order to obtain good working data. However, when the data gets put into the public arena, look out for more problems. A few are described in the next sections.

Limit of Detection and Public Trust (Dr. W. B. Crummett)

"Over the past several years great progress has been made by the scientific community working largely through scientific societies to define and understand the meaning of the 'limit of detection' and the 'limit of quantification'. Thus, the American Chemical Society has issued guidelines (2) and principles (3) of environmental analysis. The Association of Official Analytical Chemists (AOAC) and the American Society for Testing Materials (ASTM) have continued to emphasize collaborative studies, cooperating with the International Union of Pure and Applied Chemistry (IUPAC) in holding symposia and studying best ways to conduct such studies," writes Dr. Crummett.

He continues: "In spite of this extraordinary effort, however, analysts in the real world continue to present their results in forms which sometimes cause the credibility of the data to be questioned or the meaning to be misinterpreted. The problem becomes much more serious as the 'limit of detection' is approached." Either the credibility of science is put into question or the wrong impression is given to the public in one manner or another.

The biggest problem with interpretation by the public is the general public lack of understanding about uncertainty. If I have three apples in my basket, then the person on the street knows there are three apples there because he can count them. However, if a measurement of 2 ppm for an analysis of pesticide "x" in my drinking water is reported, he expects that not only was there exactly 2 ppm in the sample but that there is exactly and always 2 ppm there, now and anytime later. Our schools simply do not teach uncertainty and change, but they should...way down in the lower grades.

The first example Dr. Crummett talks about refers to analyses of dioxin compounds in human fat. Ten years ago analyses were done at the ppm level, and there was uncertainty in those figures, both in identification and in quantification. More recently, levels were being reported in the high ppb range. At this time quantities of dioxin in the medium and low ppt (parts per trillion, picograms/g) ranges are being reported. These are 1000 to 1,000,000 times more sensitive than the best analyses previously done. Here is the first example:

"Look at the results in a study sponsored by the Environmental Protection Agency, EPA, and conducted by Phil Albro of the National Institute of Environmental Health Sciences, NIEHS (4). Eight internationally known laboratories participated in the study, each using their favorite method.

Human fat was spiked with various concentrations of three PCDD's and three PCDF's at low parts per trillion levels. As part of the data set, the number of unusually low and unusually high values were reported. These were the number of values that deviated by 50% from the spiked amount in the sample." A total of 54 samples were reported in this example from each laboratory. The data are found in Table I.

The variability of analysis in this case is quite pronounced, but it may be understandable in view of the extremely low level of

Table I. EPA/NIEHS Collaborative Study of Dioxin in Human Fat. Number of Values Above and Below 50% of Spiked Amount in 54 Analyses

	Analytical Method, Laboratories							
	1	2	3	4	5	6	7	8
No. of Low Values	0	0	0	1	10	8	7	0
No. in Mid Range	50	54	52	47	42	43	31	51
No. of High Values	4	0	2	6	2	3	16	3

analyte present and the number of analysis steps the process required. Dr. Crummett concludes in this case: "Although such performance is sufficient for some scientific studies in which trends are the goal, it is not quantitative enough for decision-making when regulation is indicated."

Data of this sort, however, can be handled in regulatory cases if the proper statistical calculation for the decision limit and the limit of detection is performed (5). In regulatory cases attention is given to the standard deviations of the blank and the analyte detected in the region of the limit. Similar risks for false positives and false negatives (α and β) should be specified. As it turns out, however, in the example above cited, there were probably laboratory procedures affecting the variability that should have been examined before the data were submitted for regulatory purposes.

Analysis problems for dioxin at the parts per trillion level in human fat would be similar to parts per quadrillion (1000 times more remote) in water because with water one works usually with a much larger sample size. Since few work at the parts per quadrillion level in water, it is expected that the dioxin problem should be extremely severe. Another study quoted by Dr. Crummett involved analysis at this level in five private water wells performed by four different laboratories. As Dr. Crummett relates:

"Some of the consequences and confusion of using such methodology can be seen from a study the EPA attempted in 1984 to determine if 2,3,7,8-TCDD had contaminated potable water in the Midland, MI area. The results of this study are shown in Table II.

Table II. 2,3,7,8-TCDD Analysis in Five Water Wells performed by Four Laboratories over a Period of Time. Analysis in pg/L (limit of detection, pg/L)

Sampling Date	Lab 1	Lab 2	Lab 3	Lab 4
November, 1984	(a)			
December, 1984	ND(5-50)			
June, 1985		ND (10)	ND (8)	ND (10) (a)
August, 1985			ND (1)	ND (10)
September, 1985		ND (2)	ND (1)	ND (3)

(a) Laboratory sample contamination present

"The first laboratory was reported to have contaminated the first set of samples, and it obtained out of control detection on the second set. A third sampling was sent to three different laboratories, using superior methodology. Still, some contamination and variability was experienced, and the wells had to be sampled twice more. The following front page headline in the Midland Daily News offered one resolution to erudite people who may have seen the variable limits of detection for these analyses: 'EPA Says those who Don't Trust Well Water Should Buy it in Bottles.' (6)

"It can be shown that people can have a great concern when variable sampling and analytical work is known to them. However, this example also shows a lack of understanding on the part of the media and the public concerning the uncertainty and difficulty of such an analysis."

Dr. Crummett concludes his discussion, "The use of inappropriate sampling analytical methodology or faulty technique brings false numbers into the data set of the laboratory report. These may then be communicated to the public. Misinterpretations and faulty reading of the laboratory report may also cause headlines in newspapers which convey a message contrary to the data in the initial laboratory report.

"All false negative results thus generated lead to a false sense of security. False positive results, on the other hand, lead to an expensive course of investigation which is sure to involve one or more governmental agencies and produce unwarranted fear in the general public. Are we analytical chemists totally responsible for this unnecessary activity and fear? We are probably not, although we are often accused of it."

Limit of Detection Revisited (Dr. D. A. Kurtz)

At this point, perhaps some effort should be made in simplified terms to describe a method of calculating a limit of detection in statistical terms for typically generated calibration data. The basic problem of this limit, LOD, is that it is not being dealt with in the overall measurement. An LOD is a simple number containing stated risks so there is no danger in reporting it to the public. Its α and β have to be shown, and this could be done in terms of its risk. Too frequently the detection decision level ("critical level") and the detection limit are calculated to be one and the same with the result that the false negative ends up having a risk of 50%, infra.

Dr. Currie has described this in simple terms (5): "Our basic task is to distinguish the blank or background from a true signal at the detection limit. [That] can be done, provided that the signals are random, independent, and stationary. To completely specify the false positive (α) and false negative (β) risks, we must know the form of the [signal] distribution and its parameters. For most analytical situations we assume the distribution to be normal (Gaussian), and the dispersion parameter is simply the imprecision (standard deviation)."

Following this argument we must first accept the levels of risk, α and β. Conventionally, we choose equal risks at the 5% level. Hence, $\alpha = \beta = 5\%$. If we select Student's t-statistics in place of z-statistics because of limited sample size, we define the decision level (L_C) and detection limit (L_D) as shown in the following sketch:

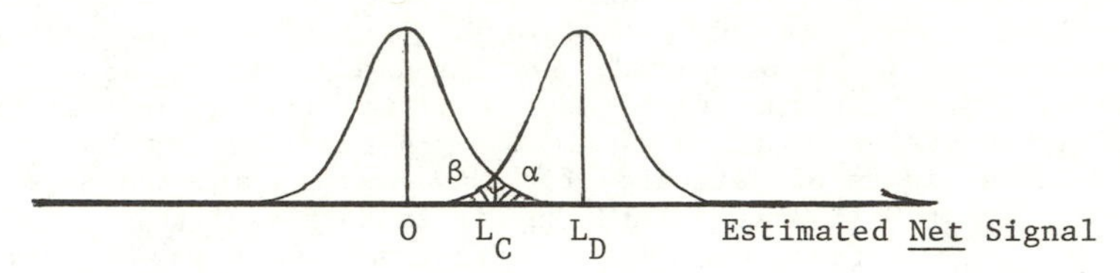

The equations of the two limits are as follows:

$$L_C = t_{1-\alpha}\sigma_o \quad (1a)$$

$$L_D \cong L_C + t_{1-\beta}\,\sigma_o \quad (1b)$$

The standard deviation of the blank, σ_B, is related to the standard deviation of the net sample estimate (σ_0) near the limit of detection by the relation:

$$\sigma_0 = \sqrt{1 + 1/n}\;\sigma_B$$

where n is the number of replicates used to estimate the blank ($\bar{B}$). Since the risks α and b are chosen to be of equal value, the calculation for L_D is simply $2L_C$ for the case when the scale refers to the net (blank-corrected) signal (i.e., has a mean of zero). Observed values that are less than L_C are simply reported as being not detected because the false positive risk has been exceeded from the chosen amount. True values less than L_D are not detectable because the false negative risk exceeds β.

Limitations on the Interpretation of Forensic Laboratory Results (Mr. C. R. Midkiff)

The value of the limit of detection in forensic work is shrouded in details not usually present in other trace analysis efforts. In this area the performance of the analyst relates to samples submitted as legal evidence, perhaps for some crime. The sample may often be in such short supply that only limited analyses can be performed on it. In these cases the limit of detection can be greatly affected.

Mr. Midkiff introduces this topic in the following manner: "Recently, increasing concerns have been addressed about the reliability of results from forensic laboratories. One question of particular concern which may be asked is the extent to which these poor results are attributable to attempts to push detection limits beyond the limits of 'reasonable scientific certainty', making interpretation of results tenuous.

"The forensic chemist is concerned not only with conduct of an analysis and preparation of a report, but also with the presentation of the results in a court of law. Conclusions and opinions of the analyst may have economic impact on individuals and organizations ranging from denial of initial employment to loss of a current position or the necessity for payment of sizeable monetary damages. For individuals accused of a crime, the potential effects are even more serious, involving loss of liberty or life itself."

In the courtroom such analytical questions as the following can be asked of the analytical chemist during cross-examination procedures: "Were the values which you claim to have obtained close to the detection limit for the method used? What is the detection limit for (material) in (sample type) using (method)? Is that detection limit based upon examination of a pure sample or one like the one which you examined in this case? Are you familiar with the IUPAC method (7) for the determination of the detection limit?"

The Sample in Forensic Work. "In the context of conventional analytical chemistry," Mr. Midkiff relates, "a method is designed to optimize the determination of a particular analyte in a defined matrix. Although the composition of the matrix may vary, the variation is normally within an expected range, and the overall composition of the matrix is reproducible. As a result, the effects of the matrix on the determination are considered in the design of the analytical method. Once a determination of detection limit is made, the analyst may be confident that if the material of interest is present in the sample above a minimal level, it will be detected and measured with an acceptable level of precision and accuracy.

"In an analytical situation typical of a quality or process control laboratory, adequate sample is provided for all tests to be conducted More sample may be obtained if needed. In addition, the sample matrix is usually known and its effects understood.

"By contrast, in the forensic laboratory, the quantity of sample is often limited and additional material is unobtainable. As a result, there is little opportunity for analytical methodology modification and matrix adjustment or optimization. Further, the matrix itself is variable as in cases involving arson.

"Additionally, current legal trends require that not all the evidence be consumed in the testing process. The courts have, in recent years, taken a dim view of situations in which the entire sample of evidentiary material was consumed during analysis and unavailable for examination by opposing experts."

Matrix Effects. Signal Suppression and Enhancement. Major effects on the LOD can be found due to the unavoidable presence of the sample matrix. These effects can take the form of suppression, enhancement, or masking of the analyte signal. Suppression and enhancement will be discussed here and masking in the next section. Without taking these effects into account, major errors can be made in the conclusions drawn from the evidence.

Mr. Midkiff writes: "Signal suppression refers to either a decrease in the overall analytical signal or suppression of the signal at selective analytical wavelengths. Matrix-related signal suppression is a common problem in mineral analysis by atomic spectroscopy. In a study using Inductively Coupled Plasma, ICP, atomic emission, it was found that while most analyte sensitivities were depressed by matrix effects -- some up to 30%, lithium could be either suppressed or enhanced depending on small changes in conditions (8). In a similar study of low-power ICP, all metals studied were suppressed by increasing sodium concentration except arsenic which was enhanced (9).

"Self-absorption and saturation effects also lead to suppression of the analytical signal. In gamma or X-ray analysis, self-adsorption is a function of sample thickness. Significant self-absorption in samples analyzed by X-ray emission could result in failure to detect elements present as traces or indicate their concentration to be below the actual level.

"For non-destructive gamma measurements, a mock-up of the object being examined is often used for calibration purposes. Standards of the elements measured are incorporated into the mock-up which simulates the geometry and chemical composition of the samples as closely as possible."

Although the analyst may be unaware of its existence in the sample, signal suppression serves to decrease system sensitivity and raises the working detection limit. Spectral enhancement, on the other hand, increases the system sensitivity and lowers the working detection limit. "Enhancement is a well known problem in atomic absorption spectroscopy (10). A variety of approaches, such as Zeeman effect correction have been proposed for its elimination (11). To avoid artificially high results, calibration standards must contain concentrations of the enhancing species equivalent to those in the sample. Ordinary standard solutions are not representative of the analytical situation."

Spectral overlap is a special case of enhancement. "Spectral overlap can cause detection and measurement of a material not present in the sample (10) and give 'false positive' results. Detection limits must be recognized as being inseparably linked to selectivity because reliable detection limits cannot be established unless there is certainty in what is being measured."

Matrix Effects. Signal Masking. Signal masking raises the effective detection limit and seriously complicates interpretation of the analytical data. One cause of signal masking is elevation of the background signal leading to unfavorable signal/noise ratios. "An example is the examination of swabs collected from the hands of a suspected shooter. Antimony and barium are relatively uncommon in nature or manufactured products but are present in the primer composition of most types of modern ammunition. During weapon discharge or handling, these elements are deposited on the hands and can subsequently be collected with acid-moistened cotton swabs. Unfortunately, other materials present on the hands, such as, dirt, grease, oil or blood are also collected by the swabbing process.

"In the laboratory, the swabs are leached with nitric acid to extract the barium and antimony for FAAS analysis. Light swabbing results in ineffective extraction of these elements from the swabs. The analyst will obtain a 'false negative' result for the presence of firearms discharge residues. If longer leaching times or agitation of the swab are used, contaminants cause a background elevation to an extent not readily noted or correctable, even with sophisticated instrumentation.

"Masking can also occur when blindly relying upon sophisticated instruments. For example, many modern atomic absorption instruments provide for display of only one signal at a time, either the background alone or the background corrected signal. The analyst relying on automatic background correction may be unaware of the magnitude of the background signal being subtracted (12,13).

Background subtraction, regardless of system used, does not change the signal/noise ratio in the system. It cannot, and should not, be relied upon to convert a sample unfit for analysis into one from which important information and conclusions will be drawn."

Matrix Variability. Limits of detection of various analytes are dependent not only on the matrix but also on the changeable qualities of the matrix due to the incident. This latter quality is seen in addition of analytes from pyrolysis and the subtraction of analytes from adsorption onto pyrolyzed material.

Mr. Midkiff continues, "When a sample from a suspected arson is examined by gas chromatography, additional peaks from materials present at the scene, as for example, plastics, in the sample may be observed. These additional peaks make difficult pattern recognition, normally relied upon for detection/identification of flammable liquids in the debris. Similar problems may be encountered in the analysis of samples from a bomb scene where chemicals in soil or debris from the bomb crater complicate the detection and identification of explosive components.

"When the material sought is not detected in an adsorbent-collected sample, as in the case of a charcoal sample produced by fire, partial or total retention by the adsorbent may be a factor. Conversely, the adsorbent may have varying effectiveness in retaining the analyte because of surface inactivation, water saturation, or some other cause (14)."

Conclusions. Mr. Midkiff concludes with these remarks. "In the forensic laboratory, to ensure a reasonable interpretation of the analytical data, the analyst must have certainty in what is being measured. There must be a focus on analytical fundamentals, and the analyst must be cognizant of factors impacting detection and measurement. These factors include matrix effects in the actual sample; potential contamination or loss during sample handling, storage, and analysis; and instrument sensitivity. The analyst must be sufficiently familiar with the material to determine the probative value of similarities and the discrimination value of minor differences. The analyst should recognize that preset criteria may negate the need for finer data and avoid a tendency to attempt to use lower and lower values. The next decimal place may be uncharted ground and just because it can be measured, does not give it probative value. While improved detection limits offer advantages in the examination of physical evidence, the additional data requires care in interpretation. The results may have a major impact affecting an individual's fortune, freedom, or life. To ensure fairness and overcome questions about their credibility, forensic scientists must consider the problems that are created by improved detection limits and the interpretation of results in the forensic laboratory."

The Role of the Blank in the Measurement of Detection Limits in Atomic Spectrometry (Dr. R. L. Watters, Jr. and L. J. Wood)

Some of the problems in forensic analysis related to sample matrix and detection limits can be found in other analytical laboratories as well. Besides the enhancement and suppression effects on analyte signals described by Mr. Midkiff in the previous section, Dr. Watters and Ms. Wood of the National Bureau of Standards provide some specific information regarding matrix effects on spectral background. Background level and spectral structure comprise the chief source of measured signal when blanks are being measured for detection limit estimation. Their example is taken from inductively coupled plasma (ICP) spectrometry.

Dr. Watters writes: "A detection limit (DL) is usually used for comparison, and the specific type of comparison often dictates how the DL is measured or calculated. When valid blank measurements and calculations are made, such as those described by Currie (15), the resulting DL is applicable only to that specific analytical system and conditions used for the DL determination. Erroneous conclusions frequently arise when this same DL is used as a comparison for other systems or conditions."

One way to ensure that DL's are used only for valid comparisons is to examine the specific reason for a particular DL estimation. Watters and Wood continue: "To examine the validity of DL comparisons, we can pose two questions that are supposed to be answered by a DL estimate: (1) Is one instrument better than another for determining a trace analyte? (2) Can an analyte be determined in a specific sample matrix using a specific method?

"The usual approach to answering either of these questions is based on establishing the instrument response with respect to analyte concentration and measuring the variability of this response when no analyte is present. Some analysts prefer to call the measured response the 'blank'. The solution used for the measurement is also called the 'blank', and it is the composition of this blank that often leads to the problems mentioned above. For simplicity of discussion, we will consider samples and blanks to be solutions.

"Although the distinction is seldom made, there are at least three different types of 'zero-analyte' samples that are called 'blanks'. The first is the solution that contains only reagents mixed in quantities that are used in the calibration standards or the final sample dilution. This 'solvent' or 'reagent' blank is sometimes used as the 'zero standard' to define the y-intercept of the calibration curve.

"When samples are prepared using dissolution methods, the true analytical blank consists of all reagents and steps used in the method. The only analyte present in this second type of blank is caused by contamination from any reagent or contact with laboratory environment and apparatus. The level of analyte in this 'analytical' blank and its variability are key quantities to be evaluated in accurate trace analysis (16). The content of the analytical blank is more method dependent than that of the reagent blank.

"The third type of solution blank is one that contains every component of a specific sample except the analyte of interest. Rarely does one encounter such a solution in the normal course of analysis, and indeed it is difficult to make such a solution by design. The usefulness of having this type of 'matrix' blank will be discussed in terms of detection limit measurements using inductively coupled plasma (ICP) spectrometry as an example."

The example chosen to demonstrate the significant differences in detection limit estimates obtained by measuring different types of blank solutions is taken from ICP spectrometry. However, analogous situations arise in any form of spectrometry where net absorption or emission peaks are measured. Dr. Watters describes the relationship between the questions posed above and the type of blank to be measured for DL estimation:

"If one wishes to answer question 1 above, replicate ICP measurements are made with or without spectral background correction, while aspirating a solution of dilute pure acid in distilled water. The choice of whether or not to use background correction may be dictated by the ICP spectrometer vintage, i.e. some older instruments are not capable of background measurement on either or both sides of the analyte spectral peak. Accurate measurements near the detection limit require background correction. Detection limits should therefore be estimated with net intensity measurements of the blank. In this case, the question of instrument comparison is adequately answered by measuring the reagent blank, provided all measurement parameters are realistic and similar from instrument to instrument. On the other hand, if one has to answer question 2, DL estimates using the reagent blank can yield erroneously low DL's and use of an analytical or matrix blank is more appropriate. The reason more complex solutions result in higher detection limits is that certain components of variance are proportional to the level of signal being measured. Any increase in background level, spectral line caused by contamination, or spectral line feature caused by an interfering matrix element line can result in an increase in the detection limit." This effect may be offset if the increase in blank measurement variance is compensated by matrix suppression of the calibration curve as previously mentioned by Mr. Midkiff.

Ms. Wood describes how certain types of DL estimates were made using ICP spectrometry and different kinds of blank solutions: "Regardless of the mathematical expression one uses to estimate a DL, the term common to many approaches is the standard deviation of the background, σ_b. This term is estimated by replicate measurements of the background as s_b. We have examined the effects of blank type on the value of s_b for a number of spectral analysis lines using the ICP. Where differences occur from one blank type to another, the underlying reasons for the differences are examined. Five different blanks were prepared as examples of reagent, analytical, and matrix blanks. The reagent blank consisted of 1% V/V HNO_3 in distilled water. An analytical blank was prepared using a dissolution method for steel samples. Another analytical blank was prepared using a procedure suitable for dissolving geological samples by lithium metaborate fusion. Matrix blanks were prepared using 99.999% pure iron in one and 99.999% pure aluminum in the other. In each case, 1 g of metal was dissolved in 100 mL of 10% V/V 12M HCl. A sequential ICP spectrometer was used to measure net signals for the elements and spectral lines listed in Table III. Ten measurements of each of the five solutions were made and the standard deviations of these replicate measurements at each wavelength are also listed in the table. In cases where two wavelengths are listed for a given element, the second wavelength has been identified as one with no spectral interference from iron (17).

"The reagent blank can be used as an indication of the baseline instrumental detection limit, when its standard deviation is multiplied by the appropriate constant. Data from this solution could be used to answer question 1 above, but in a number of cases cannot be used to answer the second question. The analytical blanks can be used as indications of method detection limits, wherein wet ash digestion or alkali fusion procedures are employed to dissolve the sample. In the case of the fusion blank, this analytical blank also serves as a specific type of matrix blank since the principal constituents in solution are lithium and boron. The concentrated iron and aluminum solutions represent solution matrices that contain the major constituent when the determination of trace elements in these matrices is of concern. These solutions can be used for answering question 2 above. In addition, these blanks also serve as analytical blanks since dissolution procedures were employed in their making. In some cases the differences in calculated detection limits among these solutions is significant."

Dr. Watters points out specific examples from the data: "Several differences in standard deviation for replicate net intensity measurements between the reagent blank solution and the other blank solutions can be found in Table III. In general, one can ascribe an increase in the standard deviation for analytical and matrix blanks compared to the reagent blank to three possible causes: These are contamination from the dissolution procedure, broadband shifts in the spectral background caused by a matrix element, and spectral line interference from a matrix element. Table III contains examples of all three of these and their occurrence is indicated in the Table. Specific examples can be understood by inspection of wavelength scans in the region of the blank measurement.

"The differences in background standard deviation for the Ca measurements at 393.366 nm provide examples of contamination and spectral line interference effects. Wavelength scans for the five solutions are plotted in Figure 1. The effect of Ca contamination in the lithium metaborate fusion flux is most evident. Effects of trace amounts of Ca contamination in dissolving the aluminum and in the steel blank solutions are also visible in the figure. The wavelength scan of the iron solution clearly shows the spectral interference from the 393.3605 nm line of Fe. Although the most serious degradation in the Ca detection limit occurs for the fusion blank, this case is not likely to cause practical analysis difficulties. Fusion techniques are most often employed to dissolve geological samples in which Ca is usually a major constituent. Hence trace detection of Ca is not a concern.

"The detection of Cd at 214.438 nm is subject to interferences from Fe and Al. In Figure 2, the spectral line interference of the 214.445 nm line of Fe is apparent, as well as the large sloping shift caused by the presence of the high concentration of Al. This latter effect is even more pronounced for the detection of Cr at 205.552 nm. The scan plotted in Figure 3 shows the large continuum shift caused

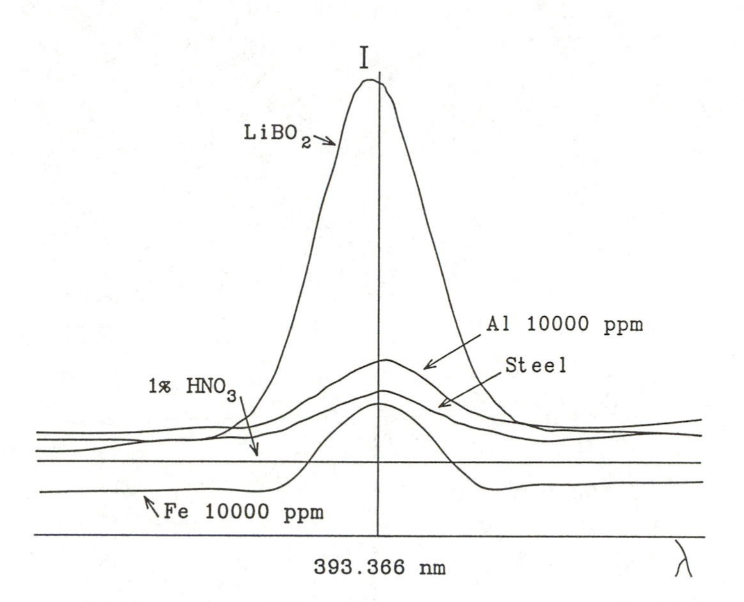

Figure 1. Spectral scan of ICP emission intensity versus wavelength in the region of the 393.366 nm line of Ca for various types of blank solutions.

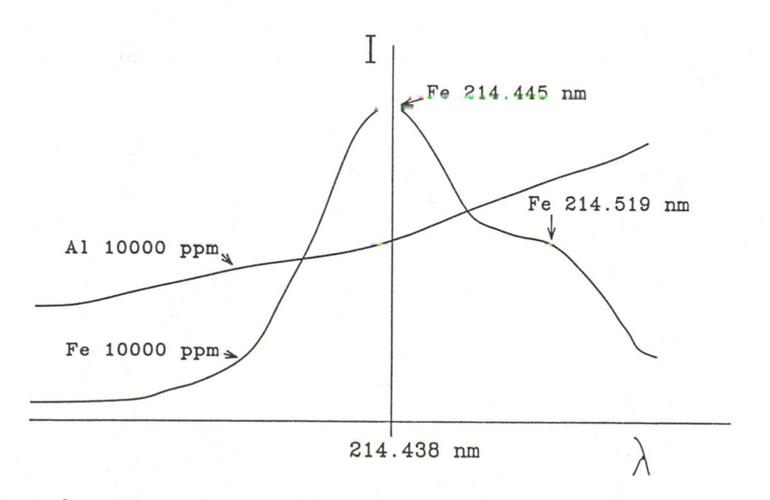

Figure 2. Spectral scans showing the effect of 10000 ppm Fe and Al on the background near the 214.438 nm line of Cd.

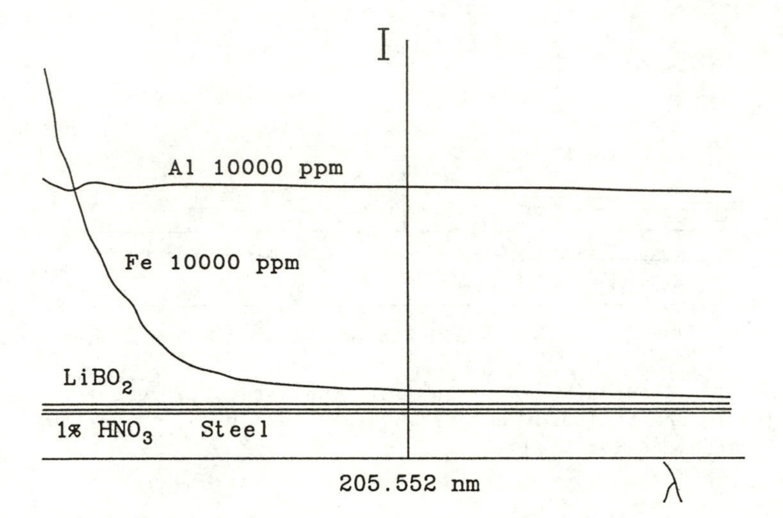

Figure 3. The effect of 10000 ppm Al and Fe on the spectral background of the 205.552 nm line of Cr.

Table III. Standard deviation of 10 replicate measurements of three types of blank solutions for analyte element detection using ICP spectrometry. Standard deviations are expressed in units of ng/mL

Element	λ (nm)	1% HNO_3	Fusion Blank	Steel Blank	Fe	Al
Ca	393.366	0.1	1.0[a]	0.1	0.3[b]	0.6[a]
Cd	214.438	9.6	9.4	6.5	38[b]	20[c]
Cr	205.552	21	35	19	40[b]	148[c]
Cr	427.480	22	18	23	18	20
Cu	327.396	4.5	3.3	2.3	86[b]	4.0
Mg	279.553	0.2	0.2	0.1	0.9[b]	1.0[d]
Mg	280.270	1.5	3.0	0.9	1.8	3.3
Mo	202.030	21	28	21	82[b]	76[c]
Mo	281.615	14	7.0	8.5	8.9	322
Ni	231.604	8.0	5.1	8.6	21[b]	13.9
Pb	220.353	81	70	57	148[b]	250[c]
Si	251.611	7.5	13	24[a]	274[b]	26
Si	288.158	13	36	18	36	13
V	311.071	3.2	2.1	3.2	13[b]	3.7
Zn	213.856	6.2	5.5	5.1	36[b]	8.2

[a] Contamination from reagents or procedures
[b] Matrix spectral line interference
[c] Matrix continuum background shift
[d] Unidentified spectral line interference

by Al. This phenomenon has been explained by Larson and Fassel (18), who described the background shift from recombination continuum in the spectrum of Al. The recognition of the importance of this continuum shift led to the general acceptance of the need for background correction near the ICP analyte emission line. Consequently, ICP spectrometers have the ability to measure background adjacent to a spectral line to correct for systematic errors of quantification due to background shifts. It is evident here that the same phenomenon also affects the random error component related to detection.

"It is interesting to note that the secondary lines for Cr, Mg, Mo, and Si, chosen to avoid iron interference, do alleviate or reduce the effects of Fe line overlap. For Cr and Mo, standard deviation of the background for the secondary lines is lower for the Fe matrix blank than for the dilute acid blank. This is probably caused by matrix suppression of the calibration curve as described by Mr. Midkiff."

Dr. Watters concludes: "It is evident that the term 'detection limit' can have multiple meanings. The type of blank that is used for an experimental estimate of the detection limit is related to the specific type of detection limit desired. We have presented examples taken from ICP spectrometric data that indicate the need for clean laboratory sample preparation facilities, so that contamination effects can be reduced or eliminated from any type of detection limit estimation. Even under the best laboratory conditions, the problem of analyte detection in the presence of complex matrix elements at high concentrations can be especially difficult. Detection limit estimation related to a specific sample treatment and sample matrix requires the careful selection of the appropriate blank to ensure realistic estimates. So called 'instrumental detection limits' may be useful in comparing one instrument to another for samples with no interfering species. However, realistic analytical detection questions must be answered by realistic estimation procedures and blanks."

Problems with Regulatory Limit Settings (Dr. W. W. Hanneman)

[*Note: The quoted opinions expressed here are those of the author alone]

Federal administrative agencies regularly have the task of setting limits of detection as mandated by legislative Acts that require a no-detection level of a chemical to be classed as "clean". While a great deal of outside advice is sought, received, and used by the agencies, difficulties arise when an Act covers such a broad range of activities as chemical analysis. Requirements must cover organics, inorganics, organometallics, and biochemicals; they must handle gases, solutions, large molecules and small. We have already seen some examples in the section above. However, even in more common analytical areas, there are claims that problems have arisen when an agency has set a detection level that may be unattainable.

Method Detection Limit. In EPA, where much of this work was done, early attempts at setting detection limits were based on the signal/noise concept. In 1984, came the initial efforts to utilize statistical concepts when the Method Detection Limit (MDL) was brought into use. The next step was the Maximum Contaminant Level (MCL). Finally, the Practical Quantification Limit (PQL) has been proposed. In at least some cases these limits are unattainable in the real world, according to Dr. Hanneman, who describes these problems:

"Most toxicologists believe that the dose makes the poison. Nevertheless, the EPA and state regulatory agencies, abetted by analysts and the media, have promoted the zero risk concept. This concept offers some attractive advantages. It reduces the need for scientific judgment, and it simplifies the regulation of carcinogens. The logical consequence of accepting this premise is that exposure limits must be set to zero.

"The zero risk concept produced the greatest growth area for analysts in the history of mankind. New methods had to be developed, limits of detection had to be lowered, and compliance had to be demonstrated. All this was fueled by the linear extrapolation to zero.

"Let's examine the case of benzo(α)pyrene (BaP). We could do the same with PCB's or almost any suspected carcinogen. As promulgated in 1979, EPA Method 610 (19) is applicable to industrial discharges. It was designed to meet monitoring requirements of the National Pollution Discharge Elimination System (NPDES) permits. Its limit of detection for BaP was stated to be 40 parts per trillion using fluorescence detection (19). One of our plants had a NPDES permit which allowed a discharge of 50 parts per trillion of BaP. We had to regularly demonstrate compliance.

"The stated Detection Limit of Method 610 was determined from the signal/noise ratio measured in reagent grade water. Since actual detection limits of methods depend upon the level of interferences rather than instrumental limitations, there is no compelling reason to expect that we will ever be able to demonstrate compliance until our plant discharges only reagent grade water.

"In 1984 the EPA 'officially' recognized that one could not make measurements at 'precisely' the detection limit so they introduced the concept of MDL (20).

> 'Definition: The Method Detection Limit (MDL) is defined as the minimum concentration of a substance that can be measured and reported with a 99% confidence that the analyte concentrations are greater than zero.'"

EPA recognized that the MDL will vary depending on instrumental sensitivities and matrix effects. The reference included a table of representative MDL's determined from reagent water. For BaP the value listed was 23 ppt (21) in the HPLC method. The reference went on to say that "similar results were obtained using representative waste waters. The MDL actually achieved in a given analysis will vary depending on instrument sensitivity and matrix effects." Paragraph 15.3 (20) said: "This method was tested by 16 laboratories using reagent water, drinking water, surface water, and three industrial wastewaters spiked at six concentrations over a range of 0.1 to 425 µg/L. Single operator precision, overall precision, and method accuracy were found to be directly related to concentration of the parameter and essentially independent of the sample matrix." Reagent water was defined as water having no background peak in the vicinity of the analyte in question. Quality control acceptance data for BaP (21) at a 'detection limit' test concentration of 10 µg/L gave a standard deviation of 4 µg/L (n=4). This amounts to a Relative Standard Deviation (RSD) of 40%.

[Note: One of the biggest problems, however, with the MDL is that it ignores errors of the second type, accepting false negatives. The method puts a 50% confidence level on this type of error since the mean of the distribution of positive values is three standard deviations (to give 99% confidence) up from zero, the mean of the blank. In order to properly assess values at the 99 % level it would be necessary to locate the MDL another 3 standard deviations away from zero. In this position there is the 99% confidence that when the analyte is there, it is reported and when not there, it is not reported.]

Dr. Hanneman illustrates the real data, obtained for setting the MDL for BaP, in Figure 4 (22). "The histograms represent the raw data from Method 610 validation, at 200, 2000, and 12,000 parts per trillion. For BaP the RSD ranged from 40 to 53%. Even so, in order to bring it down to this level, the statisticians had to discard about 1/5 of the data. The results at all levels were the same. The method did not define a limit of detection."

And he continues, "This prompts the question, 'What are the performance criteria for a valid method?' The answer, in essence, is ...a validated method is one that has undergone the validation procedure. There is no requirement that it produce meaningful numbers...

and in the case of benzo(α)pyrene, the method is essentially worthless for measuring at these levels."

Maximum Contaminant Level. "In November of last year new measures were proposed. EPA established the Recommended Maximum Contaminant Level, RMCL (23), for carcinogens and suspected carcinogens." The RMCL is the maximum level of a contaminant in drinking water at which no known or anticipated adverse effect on the health of persons would occur and which includes an adequate margin of safety. RMCL's are non-enforceable. Maximum Contaminant Levels, MCL's, are enforceable standards. They are set as close as feasible to RMCL's, with the use of the best technology, treatment techniques, and other means, which the administrator finds are generally available (taking costs into consideration).

Dr. Hanneman continues, "While these MCL's are presently touted as drinking water standards, history teaches that NPDES permit writers will invoke them. Since our discharge ponds are exposed to the atmosphere, and urban dust is constantly contributing BaP to the surface, BaP will always be present. We are not presently in compliance and can never be in compliance."

Practical Quantification Limit. "As of now the LOD is a dead issue as far as suspected carcinogens are concerned. All the analytical tests have been reduced to nothing more than qualitative indicators. What is even more frightening is that there is no way to counteract a single 'false positive' result.

"EPA is now proposing the Practical Quantification Limit, PQL (24). The PQL is to be set somewhere between 5 and 10 times the MDL."

[Note: The PQL was set for volatile organic compounds, VOC, in the following manner: Seven of the most experienced EPA and contract labs reported MDL's in the range of 0.2 to 0.5 μg/L. When the same test was repeated in the same or higher ranges of concentration amongst 30-40 other laboratories, the number of laboratories reporting concentrations different from the true value was noted. There was a fairly large number of laboratories reporting figures different from the true value by more than 20%, but no more than 5 labs (6 in one case) reported analysis figures different from the true value by more than 40%. Hence, it was concluded that a high percentage of laboratories could determine these compounds within 40% of the true value. The PQL was determined by taking the high end of the 7-lab range, 0.5 μg/L, adding 40% of that to obtain 0.7 μg/L and then choosing the midpoint of 5 to 10 times that figure. This amounted to 5 μg/L. Thus, this figure was determined in this case without standard deviations and no RSD can be calculated.]

"I want to close by reiterating what Dr. Horwitz demonstrated five years ago. Figure 5 (25) shows the relationship between the interlaboratory relative standard deviations, RSD (or coefficients of variation) with concentration. The intersection of the heavy curve with the 34% RSD line indicates that there is no reason to believe that one can even define a true MDL below approx. 8 ppb, let alone at 50 ppt. The intersection with the 20% RSD line indicates that the PQL probably cannot be set below 250 ppb. This is not 5-10 times above the MDL, but rather 12,000 times that of the 17 ppt value

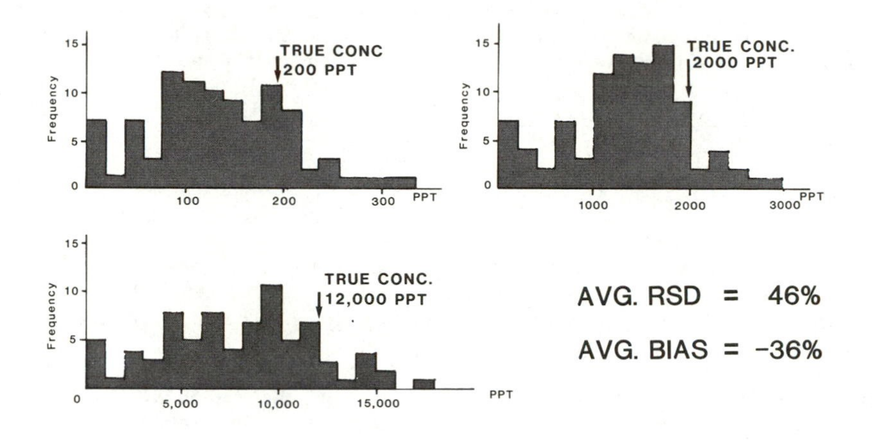

Figure 4. Histograms of raw data from Method 610 Validation Study.

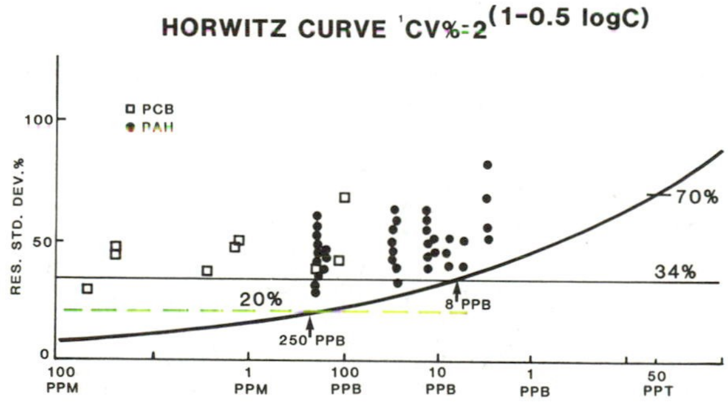

HORWITZ CURVE $CV\% = 2^{(1-0.5 \log C)}$

□ PCB
● PAH

RES. STD. DEV.%

70%
34%
20%
8 PPB
250 PPB

100 PPM, 1 PPM, 100 PPB, 10 PPB, 1 PPB, 50 PPT

Figure 5. Horwitz Curve and analytical results.

decreed for the non-carcinogenic polynuclear aromatic compound, PNA, benzo(k)fluoranthene.

"Finally, the data points shown are typical of those generated in the validation studies for PNA's and PCB's. Mind you, these results were obtained after eliminating about 20% of the results.

"There is little evidence that an RSD of 20% can even be reached. However, the greatest pity of all is that this activity will in no way contribute to the better health of the public."

Remarks on Detection Limits (Dr. W. Horwitz)

When all is said and done about the intricate details of correctly calculating limits of detection and the like, we need some advice to be able to stand back from what we are doing just to see the total effect. This section provides such advice and is given by Dr. Horwitz, long a mentor in the field. One of the most interesting and useful items of advice is to examine regularly the Federal Register and to respond to requests for input as a part of the regulation-setting process. The quotes below are all from Dr. Horwitz.

First, Analysis Purpose. "The first thing that should always be known in performing any analysis is its purpose. In almost all cases when dealing with a limit of detection or limit of determination, the primary purpose of determining that limit is to stay away from it. Almost by definition, the reliability of results near the limits of detection or determination is extremely poor and in fact ultimately reaches a point where a determination of whether an analyte is present or absent can be obtained just as reliably by tossing a coin as by making the so called measurement."

Second, Identify before Measure. "The second important point is a matter of definition. There are so many possible definitions of limits of detection that any statement using these terms should always be accompanied by a definition as to how the user means it. 'Limit of detection' can have three different primary meanings whether or not an analyte (1) was identified, (2) measured, or (3) identified and measured. Although measurements can be made without knowing what is being measured, they cannot be interpreted without this information. The property of knowing what is being measured is related to the attribute of method specificity. A complete analytical system of sampling, determination, and interpretation requires that we place identification before measurement. Often it takes considerably more material to identify the presence of an analyte than to measure its amount. Therefore, in speaking of limits of detection we are really measuring only a naked signal whose origin is obscure. Often we merely assume we know what we are measuring at these low levels. This obviously creates considerable ambiguity in our work."

Third, Read the Federal Register. "There has been considerable criticism with regard to how regulatory agencies utilize the concept of 'limits of detection' in various applications. As background for this point, I would like to mention that analytical chemists should become familiar with that eminent scientific journal called the

Federal Register. There is considerable technical material in the Federal Register. Its contents include numerous methods of analysis and justification for these methods. Extensive reviews of the toxicology of various chemical materials appear here, often because the bulk is too large to publish anywhere else. These reviews have taken many man years for compilation. They are well worth noting. The usual procedure when an agency wishes to promulgate a regulation is to first publish a proposal in the Federal Register accompanied by an invitation to comment." It should be noted that announcement of these invitations is printed regularly in Chemical and Engineering News. "Ordinarily 60-90 days are given for receipt of comments, but this can be extended by a formal request. Even so, agencies will consider comments that come in after the deadline. Regulatory agencies are influenced by these comments. For example, almost 10 years ago the Food and Drug Administration published a proposal known as the 'SOM Document', dealing with trace residues of carcinogenic compounds in tissues of food-producing animals. Although it dealt primarily with risk analysis, exposure calculations, and carcinogenesis, there was considerable analytical chemistry involved. As a result of extensive comments by the analytical community, that document was recently republished as a proposal with much of the objectionable analytical material removed. Consequently, it is the obligation of the professional analytical chemistry community to bring their views to the attention of the responsible regulatory agency to avoid the promulgations of regulations not based on sound science. Sometimes revisions are not possible, because of the way the law is written. An example is the case of the Delaney Clause which mandates a zero value. At the time it was enacted, it was not realized that attaining zero was impossible. Therefore agencies are attempting to do the best they can to adapt to the situation by developing practical zeros.

"In this connection you should be interested in a recent proposal made by EPA for setting a practical quantification limit for fluoride in drinking water. They recommend a regulatory limit of 4 ppm of fluoride and a practical quantification limit of 0.5 ppm. For those who are familiar with fluoride determination, the quarrel with this particular value would be that it is undoubtedly set too high. You would think that practically all laboratories could comfortably determine 0.5 ppm fluoride. Yet this value was set from actual interlaboratory studies that showed that there were still a certain number of laboratories that could not achieve even such a relatively liberal limit of determination. I suggest that EPA is to be congratulated in coming up with a value for a quantification limit that can be practically achieved by most laboratories involved in this particular determination and recognizing that there is no need to go any lower (26). This concept of a practical quantification was first proposed (24) in conjunction with volatile organic compounds (VOC). It should be examined by all who are interested in the 'limit of detection' concept."

Reporting Results. "Another important related matter is that of reporting low level results. Again we start with the purpose of the analysis. In some cases an unambiguous answer of yes or no is what is desired; in other cases, a good approximation to the average is needed. The Food and Drug Administration's total diet program is a specific example of the need to know the average intake of trace nutrients and toxic contaminants in the American diet. A specific number is required for interpretation by toxicologists with respect to safety as well as how that value changes with time. Statements of 'not detected' or below a certain limit of determination value are unsatisfactory because they cannot be handled by descriptive statistics in cases where the amounts present may be below any reasonable definition of 'limit of determination,' especially when a blank subtraction is involved. In such cases, we can take advantage of the fact that in many instances the average of a large number of very poor measurements gives a remarkably good average." This, of course, assumes that the errors are random and not biased. "That is in these cases we are dealing primarily with random errors of relatively large magnitude. In the long run random errors cancel each other out leaving the core of the basic value we are looking for.

"Although scientists may have an aversion to reporting negative values, these are a direct consequence of the utilization of normal distributions (Figure 6). Consider a contaminant present in a food matrix at 0.1 ppm with a relative standard deviation (RSD) of 100%. An RSD of 100% means that one standard deviation subtracted from the average gives zero and any value less than this, say two standard deviations below the average, is in the negative region. In fact, in this example 16% of the measurements automatically will be negative. Physically this means that the blank measurement was greater than the analyte measurement. This is a natural consequence of using the normal distribution. In fact if you do not get about 16% negative values in this situation, there is probably something wrong with your measurements. A sound objection may be raised on the grounds that measurements in this area are not normal and this is a good possibility. However, in almost all cases we never have enough data to determine whether our measurements are normal, log normal, or follow any other type of distribution. Because of the wealth of information known about the normal distribution and the many instances where measurement data have been shown to be normally distributed, it is generally best to assume normality and make any necessary modifications to the interpretation if there is evidence to indicate non-normality."

Summary. "To summarize the major points: (1) The usual reason we want to measure a detection limit is to stay away from it. (2) Report low level measurements as they come - positive, negative, or zero - since their average is usually a reasonably good approximation of actuality. (3) As we go down in our concentration measurements, precision degenerates into a question of false positives and false negatives, and at low enough levels a toss of a coin will give us just as good an answer as will our most elaborate instrumentation. (4) At these low levels, you probably will never have enough data to determine if you have a normal or non-normal distribution, and when your random error is so great, the type of distribution probably does not make any difference anyway."

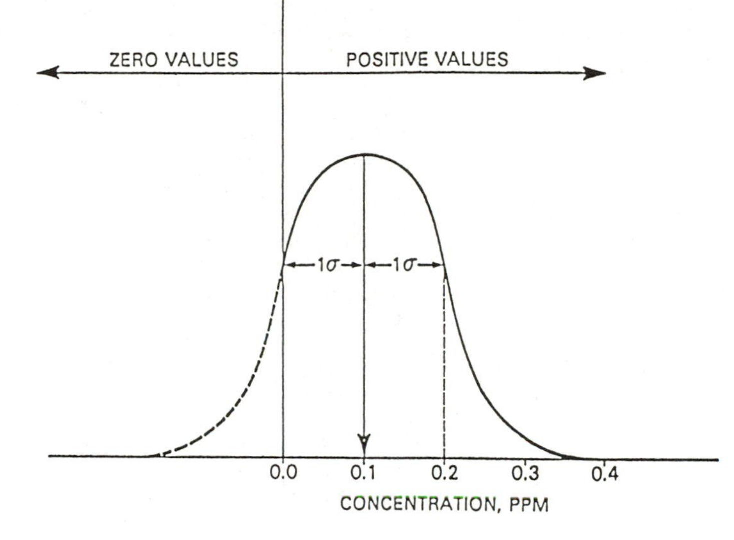

Figure 6. Normal distribution near zero showing the necessity to report negative values.

Epilogue

Real world problems resulting from analysis of small samples at the lowest levels are found in four arenas: perception by the real world people who are not analytical experts, governmental regulations, influences of the courts, and production of the analytical data by the expert analyst. Each of these arenas has its own characteristics, its own operating rules, and its own purposes. We must find, however, enough of a common base so that we can all live amicably in this one world.

The person on the street, the non-expert, wants a simple piece of information. He wants to believe what experts say. Problems in the past, however, put doubts on what experts have said. These doubts are clouded further by a lack of knowing about probability risks and relevance. Dr. Crummett has spoken to this point.

The government regulators (legislators as well as administrators) like to have neat packages outlined by rules that broadly cover the problem areas. The world of analysis, however, covers many completely different analytical areas - from radioactivity counting, to chromatography, and to emission and adsorption analysis - that this approach often becomes ineffective. Dr. Hanneman's experience suggests that regulations should be problem specific and physically tested under real conditions before putting them into use. Rules should, perhaps, be broad and understated. They would be made specific only where unique problems are identified that require attachment of specific numerical requirements.

The courts have a very difficult time handling analytical information, as Mr. Midkiff has explained. The courts look at information in a precise way, but they don't like error. Everything has a specific cause and it just can't be "possibly" there. Aside from educating all the judges and attorneys with statistical techniques, the best effort is to assure that the expert testimony provides very carefully determined information.

Most of the information presented in this panel touches on the role of the analytical technician and researcher. These individuals need to be more sensitive to the various amounts of error that are present in each part of each analysis. It goes along with the common quip in data handling, "garbage in, garbage out." If the analytical data coming from the laboratory is poorly controlled or lacks the estimations of error, it can't be used by others for information leading to meaningful decisions and action. The first suggestion, coming from Dr. Taylor is to always put control charts into each separate test. This tells how the progress of the analytical tests is going. Each person, each test, and each use of a test needs the control chart. Those using calibration graphs, as Dr. Sturdivan recommends, should handle the information so that there is constant variance at each level across the graph.

Dr. Crummett has suggested that the analyst must clearly state the risks of making a false determination when true and a true determination when false. This is especially important in state-of-the-art determinations. Researchers need to be terribly careful of the blank and its accompanying uncertainties, as pointed out by Dr. Watters. The risks of making both the false positive and the false negative should be clearly defined and when so done, signals found to be less than the decision limit (critical level) should simply not be reported as present. Blank signals should also be reported. Forensic work, as implied by Mr. Midkiff, requires a great deal of experience because of the wide sample types. The matrix in such work also has a profound effect on the signal - it may enhance and it may mask the signal - and is variable as well. Finally, Dr. Horwitz suggests: 1. know the purpose for an analysis before you measure and 2. be aware of what the government has said about the topic.

On the other hand, error calculations must be simplified. There is a utility in using simplified, practical approximations, but it must be certain that such approximations are clearly noted when used.

A restatement of the key questions as key actions is appropriate:

*** Put a quality label on data
*** Teach the public the concept of error
*** Publish data, taking error into account, and let the chips fall as they may
*** Be extremely careful in producing forensic data so that unequivocal actions can be taken on account of them
*** Interact with federal regulators in the making of suitable regulations.

The basic goal in trace analysis work is to be able to first report the numbers. The second goal is to indicate the error of each number. The third goal is for the user of the information, whether it be an agency, a newspaper reader, or the like, to accept the reports as having error, look for it, and then use those facts to give information and to form knowledge (27).

Acknowledgment

This paper is published as Journal Series Paper No. 7609 of the Pennsylvania Agricultural Experiment Station.

Literature Cited

1. Kurtz, D. A.; Rosenberger, J. L.; Tamayo, G. J. In "Trace Residue Analysis"; D. A. Kurtz, Ed.; ACS Symposium Series No. 284, American Chemical Society: Washington, D. C., 1985; pp. 133-166.
2. Keith, L. H.; Crummett, W. B.; Deegan, J., Jr.; Libby, R. A.; Taylor, J. K.; Wentler, G. Anal. Chem. 1983, 55, 2210.
3. MacDougall, D.; Crummett, W. B. Anal. Chem. 1980, 52, 2242.
4. Albro, P. W.; Crummett, W. B.; Dupuy, A. E., Jr.; Gross, M. L.; Hanson, M.; Harless, R. L.; Hileman, F. D.; Hilker, D.; Jason, C.; Johnson, J. L.; Lamparski, L. L.; Lau, B. P-Y.; McDaniel, D. D.; Mechan, J. L.; Nestrick, T. J.; Nygren, M.; O'Keefe, P.; Peters, T. L.; Rappe, C.; Ryan, J. J.; Smith, L. M.; Stalling, D. L.; Weerasinghe, N. C. A.; Wendling, J. M. Anal. Chem. 1985, 57, 2717-2725.
5. Currie, L. A. In "Trace Residue Analysis, Chemometric Estimations of Sampling, Amount, and Error"; D. A. Kurtz, Ed.; ACS Symposium Series No. 284, American Chemical Society: Washington, D. C., 1985; pp. 50-51.

6. Gray, K. "EPA Says Those Who Don't Trust Well Water Should Buy It in Bottles," Midland Daily News, December 20, 1985: Midland, MI.
7. Long, G. L.; Winefordner, J. D. Anal. Chem. 1983, 55, 712A-724A.
8. Thompson, M.; Ramsey, M. H. Analyst 1985, 110, 1413-1422.
9. Urh, J. J. Amer. Lab. 1986, 18, 105-113.
10. Lovett, R. J.; Welch, D. L.; Parsons, M. L. Appl. Spectros. 1975, 29, 470-477.
11. Koizumi, H. Anal. Chem. 1978, 50, 1101-1105.
12. Herber, R. F. M.; DeBoer, J. L. M. Anal. Chim. Acta 1979, 109, 177-179.
13. Siemer, D. D. Anal. Chim. Acta 1980, 119, 379-382.
14. Reeve, V.; Jeffery, J.; Weihs, D.; Jennings, W. J., Forensic Sci. 1986, 31, 479-488.
15. Currie, L. A., Anal. Chem. 1968, 40, 586-593.
16. Murphy, T. J., In "Accuracy in Trace Analysis: Sampling, Sample Handling, Analysis, NBS Special" Publication 422, Vol. I, pp 509-539.
17. Snyder, S. C., ASTM Standardization News, April 1985, 35.
18. Larson, G. F.; Fassel, V. A., Appl. Spectros. 1979, 33, 592-599.
19. Federal Register 1979 (Dec. 3), 44, 69514-7.
20. Federal Register 1984 (Oct. 26), 49, 43348.
21. Federal Register 1984 (Oct. 26), 49, 43349.
22. "EPA Method Study 20, Method 610-PNA's, Polynuclear Aromatic Hydrocarbons", Environmental Protection Agency, 1984.
23. Federal Register 1985 (Nov. 13), 50, 46880-900.
24. Federal Register 1985 (Nov. 13), 50, 46906-7.
25. Horwitz, W. Anal. Chem. 1982, 54, 67A-76A.
26. Federal Register 1986 (April 2), 51, 11396-11412.
27. Zervos, C. In "Trace Residue Analysis"; D. A. Kurtz, ED.; ACS Symposium Series No. 284, American Chemical Society: Washington, D.C., 1985; pp. 235-252.

RECEIVED September 28, 1987

Chapter 17

Reporting Low-Level Data for Computerized Data Bases

Martin W. Brossman[1], Gerard McKenna[2], Henry Kahn[1], Donald King[3], Robert Kleopfer[4], and John K. Taylor[5]

[1]U.S. Environmental Protection Agency, Washington, DC 20460
[2]U.S. Environmental Protection Agency, Edison, NJ 08837
[3]Ministry of the Environment, P.O. Box 213, Rexsdale, Ontario M9W 5L1, Canada
[4]U.S. Environmental Protection Agency, Kansas City, KS 66115
[5]National Bureau of Standards, Gaithersburg, MD 20899

The problems of effectively describing and qualifying low-level data are compounded by the need for utilizing brief codes in combination with numerical measures of the analyte in large computerized data bases. These codes, typically supported by complex definitions, attempt to caution a wide range of data users with diverse applications. In attempting to address the data description problem issues of limit of detection, limit of quantification (alternately called limit of quantitation), criteria of detection, applicable confidence limits, etc. - must all be addressed. Frequently, statistical rigor is in conflict with the pragmatic use of the data. Furthermore, the difference between the quantity of analyte and results of the measuring system is often confused. The panel is composed of members of a task force formed to resolve data coding policies for one of the largest environmental data systems, STORET. They discuss statistical and analytical approaches and issues involved in coding data with the accompanying advantages and hazards.

Overview and Introduction (Gerard McKenna)

As more and more environmental measurements are generated, there becomes an obvious need to input environmental measurements into computerized databases. This replaces the antiquated and di-appearing practice of stuffing the data into desk drawers and file

0097-6156/88/0361-0317$06.00/0

cabinets. Within the EPA, there have arisen a number of such data systems such as, for example, STORET, SAROAD, and ODES.

One trade-off of these systems, which certainly contain many positive benefits, is that they decrease greatly the ability of a chemist to communicate to a data user all the reservations he has about his measurements. Most of this area of concern surrounds measurements that lie very close to zero. Computer formats try to reduce the chemist's words of warning to alphabetical symbols, used as codes for different classes of limitations.

A number of efforts have taken place to establish conventions of reporting codes. These include, for example, work done by the American Chemical Society Committee on Environmental Improvement, work done by a sub-committee of the American Society for Testing and Materials, the International Joint Commission between the United States and Canada, conventions established by the EPA contract laboratory program and conventions being established by the STORET low level work group. In addition, there have been numerous conventions set at grass-roots levels by individual EPA, state and private water testing laboratories.

Out of these efforts have come a multitude of terms such as Limit of Detection, Method Detection Limit, Instrument Detection Limit, Limit of Quantitation, Criterion of Detection and a multitude of symbols such as the less than sign, ND, TR (for trace), U, M, J, T and W, and K. EPA is now proposing LTL (less than lower limit of detection) and LTC (less than criteria of detection) in computer standards. Some of the conventions come with rigid definitions for prescribed use and others come with vague definitions and allow for "analytical judgment" and flexibility.

Lack of convention and standardization of codes and symbols has caused a great deal of confusion and impatience with analytical chemists from the outside world. A quick, simplistic view of the situation can erroneously lead one to think an agreed upon, standardized convention and one that is universally accepted is not that difficult a task.

To the computer person there probably appears the proliferation of too many codes. To the attorney, it appears that analytical chemists are trying to "waffle" too much. To the statistician, it probably appears that too much tampering of data is going on and to the public it appears that there may be an attempt to mask or conceal information with the use of these codes.

Most analytical chemists will quickly agree on the real solution to the problem. That is to choose up front, or if need be, develop a more sensitive and appropriate methodology for the intended use. Such a method should get you well up on the measurement and quantitation range and leave all the low level coding problems behind. However, this usually is not the solution. First of all, many of our state-of-the art methods which we have now are not more sensitive than are areas of environmental concern and it takes considerable time and effort to drive down the levels of detection. Also, for some parameters, environmental toxicity exposure levels are being determined concurrently with survey measurements being made and, therefore, all positive finding is treated with great interest, even if around zero. Also, the public sometimes perceives any finding, even if qualified, as significant even as Avogadro's number is approached.

Consequently, efforts must continue to establish a universal convention to treat low-level data. Such convention must:

1. be satisfactory to the analytical chemist and not compromise or over-generalize the limitation he is trying to communicate;
2. allow for varying levels of risk for Type I and II errors to accomodate different intended uses of the data and different management philosophies with regard to its use.
3. be consistently understood, even by chemists not well versed in statistics so consistent use is fostered; and,
4. be free from modifying or adulterating the data to make it unuseable to a statistician.

In any event, coming up with such a convention will never preclude the need for a serious investigator or data user from going behind the data in the data base. This includes further communicating with the originating analytical chemist either directly or indirectly through the examination of all supporting data and quality control documentation. Also, it includes understanding the data in the context of the experimental design for which it was generated.

We hope that the following presentations and your input will provide us with some help in resolution of these problems.

A Rethink of the Factors Involved in Reporting Results Below the Method Detection Limit. (Donald E. King)

The application of the concept of "detection limit" as a criterion by the analytical scientist for withholding the results of low-level measurement is not supported. This usage may arise from the mistaken belief that the one-tailed statistical t-test is a test of the quality of the result, rather than the extent to which the result indicates that analyte is present in the sample. It is also the result of confusing the limitations of detection and measurement with the limitations of the analytical process and the impact of sample matrix effects. A further argument against the use of this concept as the basis for not reporting results is the improper hypothesis that a low result is necessarily derived from a population of results with mean zero.

Instrumental detection limit (IDL) protocols are used to determine when an analog signal is sufficiently different from the background noise to conclude that a measureable "real" signal has been observed. But IDL terminology and estimation protocols are not standard.

More recently data reportability has been linked to the repeatability of measurement of samples after a more or less extensive sample preparation and analysis process. Many analytical chemists currently consider it improper to report measurements observed to be below their Method Detection Limit (MDL). They fail to recognize that the measurement was in fact above the IDL and is therefore a real result.

The analyst often tends to over-estimate the "detection limit" because of concern about the accuracy of setting zero, the potential for bias, misidentification, etc., and fear that the data user will draw unsupportable conclusions.

The symbol "<" (less than) has been variously used to indicate a zero result, or to indicate a measurement below the "detection limit" but not necessarily "zero". More recently, STORET has provided the codes W, U, T, and M to indicate when a measurable response was observed and whether or not a result has been reported. "W" signifies a non-response. "T" signifies a measured response below 1.64 SD which has been reported. "U" signifies a result below MDL which has not been reported. "M" signifies a result between MDL and LOQ.

It is presumed that MDL is set at 3SD and LOQ at 10 SD, but the type of data and protocol used to estimate SD is not verifiable. Analysts have a tendency to inflate their estimates of MDL. The use of W and T for coding low-level results was included in standard D4210-83 prepared by the A.S.T.M. D19 Committee (1).

In the Laboratory Services Branch of the Ontario Ministry of the Environment, the codes <W and <T were introduced several years ago. Data coded <W is to be interpreted as a zero measurement. Data coded <T is to be interpreted as "trace" or "tentative" non-zero measurements. Recently the value of W is set equal to the "Logicial Reading Increment" (LRI). The value of T is usually 5 times W. Larger factors of 10 or 20 times W are used for methods subject to greater uncertainty in target analyte identification or quantitation . Because W is usually approximately 2/3 of the standard deviation, T (=5W) will exceed the 95% confidence level for detection of analyte in the sample, and will often be essentially equivalent to the 3s detection limit.

Standard deviation (SD) is the only logical basis for selecting a reading increment (RI). In order to obtain a reliable estimate of SD, the data used for its calculation should have been reported in increments at least smaller than 2/3 the SD. While there is no particular benefit in reading in increments smaller than one-half of SD, the very high risk of error associated with excessive roundoff is readily demonstrated (2).

Thus, the basis for selecting reading increments should be to pick a digit 1, 2, or 5 (with appropriate decimal location), which is just less than the estimated SD. If SD is in the range 0.5 to 1.0, the Logical Reading Increment (LRI) should be no larger than 0.5. But smaller reading and reporting increments are permitted.

The process of statistical inference requires us to select an hypothesis (fancy word for assumption) about the result, and then prove that this hypothesis was incorrect. In the case of Method Detection Limit, we assume that the result belongs to a distribution whose mean is centered on "zero". Based on a one-tailed test, if the analytical result is far enough away from zero we then conclude that such a result must come from some other distribution, in which case there is some likelihood that the sample contains the target analyte. MDL is conventionally set at 3 SD.

If a sample actually contained an amount equal to the MDL, then 50% of the time a result would be obtained which was below MDL. Failure to report that result can induce a "false negative" decision.

There is no logical basis for the initial hypothesis that a low result comes from a population with mean zero. A low result can be obtained from any of the billions of the result populations on the analog number line between zero and twice the detection limit. Before the statistical test is even applied, there is already an immeasurable risk of a "false negative" conclusion, aggravated by the fact that conventional wisdom rejects the reporting of results below MDL.

A more appropriate use of the one-tailed test follows. If a health guideline were set at 10 units, and the SD were about 3.3 (i.e., MDL was 10), then probability tables predict that a result greater than 14 (i.e., 10 + 1.64 x 3.3) would suggest the sample contains more than the guideline (risk of error <5%), while a result of less than 6 would suggest the sample contains less than the guideline. Notice that failure to report these low values prevents decision making.

The analyst does have the right to conclude analyte absence, given a very low result, when previous experience indicates that very few, if any, similar samples from a similar source have yielded any sort of a response, let alone a trace. But if many similar samples yield low but non-zero results, he must admit at some point that a low result may mean "presence". We have found for some parameters that a new more sensitive method still fails to detect analyte, while for others the previous indications of presence are confirmed. This is just the sort of evidence that statisticians look for, and which is denied to the data user when we fail to report low results.

The analyst is obliged to report what was observed as well as his conclusion. If the signal was caused by the targeted analyte, it must be quantitated. An important factor in reviewing our position on the reporting of low data is to recognise the effect of the computerized data bases in separating the data source from the data user. If we do not report the result, we are then reporting an inference based on a single value.

It is certainly important to warn data user's that some of their data was obtained from the bottom of the analytical range. Regardless of the precision of analysis and measurement, the ability of the analyst to define analyte zero and to determine and correctly apply all the corrections for background, reagent, and other lab blanks, is still limited by the analyte, the sample matrix, the analytical method, and his or her skill and experience.

This potential for error does not mean that low results are necessarily in error. The mechanics of estimating and setting zero and digitizing readings have improved dramatically in the last few years. Potentially, in absolute terms, this data is much more accurate than previously considered as long as zero calibration is controlled. Qualifying low results allows the analyst to cross his/her fingers, while allowing the data user full access to all available information.

Reference:

1. ASTM - D19 D4210-83, Intralaboratory Quality Control and a Discussion on Reporting Low-Level Data, A Standard Practice (1983).
2. In prep. D.E. King, A Rethink of the Factors Involved in Measuring, Reporting, and Interpreting Data.

Reporting of Low-Level Data for a Specific Computerized Database
(Robert Kleopfer)

Since 1982, the U.S. Environmental Protection Agency Laboratory in Kansas City has been actively involved in investigating sites contaminated with "dioxin". Over 15,000 samples from some 250 sites have been analyzed for 2,3,7,8-tetrachlorodibenzo-p-dioxin (TCDD). A computer database was developed to track a number of different types of information pertaining to each sample, including sample identification number, analytical results and qualification code.

A "U" code was used to signify that TCDD was not detected in a given sample. The detection limits were calculated for each individual sample based on a 2-1/2 signal to noise definition for the primary ions recorded by the Gas Chromatograph/Mass Spectrometer/Data System (GC/MS/DS). An isotope dilution method based on isotopically labeled dioxin was used to convert signal level to concentration units. Thus the actual reported detection limit varied from sample to sample due to differences in internal standard recovery, chemical noise, and instrumental sensitivity. The "U" code was used when the signal to noise was less than 2-1/2 or the qualitative identification criteria for dioxin were not achieved. For example, the "U" code was used in instances where signal was present in excess of 2-1/2 times the noise, but the isotope ratios (due to naturally occurring chlorine 35 and chlorine 37 in TCDD) were outside of acceptable ranges.

Data quality was indicated with three different codes. A "V" code was used to indicate completely valid data. Those data met stringent EPA criteria including the correct measurement of performance samples. An "I" code was used to indicate invalid data. Those data were considered completely unusable and indicated that a "critical" control feature was not achieved. For example, the reporting of a "not detected" for a positive performance sample would invalidate all associated samples. Finally, a "J" code was used to signify data in which "non-critical" controls were not met. For example, if a performance sample was reported to contain TCDD at levels just outside of the acceptance window, then associated samples were coded "J". The data were considered useable from a qualitative aspect (dioxin was present) but the quantitative amount was in question.

By using these simple definitions we were able to maintain a large database which contained the necessary information without confusing the data user.

ACS Principles for Environmental Analysis Recommendations for Reporting Low-Level Data. (John Taylor)

Measurements are made for many purposes but ordinarily with some specific use in mind. Once the use is defined, data quality objectives can be established and the most cost-effective way to achieve them can be sought. Sometimes the data quality objectives are not reached and this can cause utilization problems. When data are used for other than their intended purpose, its quality also can make such usage inadvisable.

Much of measurement data can fall into two distinct classes. In the first case, they are made to describe the characteristics of a population, for example the heavy-metal content of shell-fish. Quantitative values are necessary for computing means and standard deviations and only true outliers are excluded from the data set, often as a final step in the data analysis. In the second case, the question may be whether a specified value has been exceeded or even if a detectable amount of analyte is present. Only the answer yes or no is considered to be important and numerical values may or may not be reported, except for critical samples.

In recognition of the various uses and chances for misinterpretation of data, the ACS Committee on Environmental Improvement (1) made several recommendations for reporting data that may even seem to be contradictory when taken out of context.

The report issued by the committee recommends that all data reported should have clearly defined limits of uncertainty. When this is done, any user may evaluate its usefulness for any purpose. No discouragement to report numerical values for any data was made nor intended as long as its uncertainty is specified.

The question of detection, that is the decision based on data, was addressed as follows. If a measured value is smaller than its uncertainty (3 standard deviations) the measured value is not considered to be significant at that level of confidence, hence the analyte sought is reported as "not detected" (ND) with the limit of detection (3 sigma) given in parentheses. This means that by chance alone, the value obtained could have been found for a sample not containing any measurable amount of the analyte of concern.

The word measurable is a keyword. Ordinarily some value is obtained when any sample is analyzed. But if the result is not statistically significant, one can have little confidence in it. The "3 sigma" limit is arbitrarily recommended as the decision level.

Because of the large uncertainty of data exceeding but close to the above limit, the report recommends that such individual values be reported as detected (D) with limits of detection indicated rather than reported as numbers. The reason for this is to emphasize the semi-quantitative nature of such data and to discourage over-interpretation by unsophisticated users. For example, the result 5 micrograms ± 3 micrograms hardly can be considered as reliable to one significant figure in that the probable range of 2 micrograms to 8 micrograms limits its quantitative usefulness severely.

As a measured value increases, its relative uncertainty decreases so that at some point it becomes quantitatively useful. The lower limit of quantitation (LOQ) was arbitrarily defined as "10 sigma". Accordingly, the report recommends reporting numerical values greater than the LOQ as numbers together with their associated uncertainties.

The above considerations are based on interpreting a single measurement. If the mean of a set of n measurements is reported or used for decision, then the limits are multiplied by the factor $1/\sqrt{n}$.

The LOQ is useful for describing the capability of a methodology since it represents a practical lower level for obtaining quantitative single measurements. In recognition of the uncertainty of measurement (often unappreciated by laymen and even regulators) the report stresses that regulations should not be set below a measurable level, namely the LOQ of the methodology recommended for monitoring compliance. Below such a level, regulations hardly can be enforceable due to measurement uncertainty of random error.

In thinking about the above, one must understand the important differences between the two kinds of data described. Unfortunately time and cost considerations influence both the ways data are measured and their quality so that data made for one purpose may not have peer status with similar data for another purpose. This relates to historical data, as well. As methodology improves and as attention to quality assurance increases, data once acceptable may become obsolete, and may need to be discarded, regardless of how painful this may be. This is especially true of low-level data. Data reported as "less-than some value" may have little quantiative value at some later date even if a number had been reported because of measurement uncertainty. When data are properly weighted by the factor $1/\sigma^2$, as is necessary when combining it with other data, the rationale for discarding is clarified.

One can always make quantitative use of any reliable data, even if below the detection level, but only when its uncertainty is known and a sufficient number is available. As detection levels are lowered, levels of concern are lowered as well, so that analysts are continually asked to "measure the unmeasurable". What can be done in research situations often is infeasible for routine decisions. Accordingly, there will always be a gulf between "research" and "practical" data. The ACS recommendations were directed largely at the latter type. While the recommendations may appear to encourage loss of quantitative information, the actual loss may have little significance in most cases due to quality considerations. However, no data should be edited in this or any other manner if so-doing would preclude its future use for an other-wise useful purpose.

When evaluating data sets, objectively, one is forced to conclude that much of it is being produced by laboratories with limited competence to do so. In fact, it would be better not to report some data. Contracts to provide data often are let on the basis of price which has no relation to quality. Too many data are accepted on the basis of face value. There is a need to filter data for its quality before use and to reject it if need be. A means to qualify laboratories and to monitor continuing performance is necessary if this problem is to be solved.

Data, just as any commodity, must be bought on the basis of quality. If it does not meet pre-described quality standards, it is not paid for and rejected, just as in the case of goods and products (2).

References

1. Principles of Environmental Analysis, Anal. Chem. Vol. 55, 2210-18 (1983).
2. J. K. Taylor, Quality Assurance for a Measurement Program, in "Environmental Sampling for Hazardous Wastes", G. E. Schweitzer and J. A. Santolucito, Eds., ACS Symposium Series 267, A.C.S. Washington, D.C. 20036 (1985).

Summary and Closure (Martin W. Brossman)

Our panelists have discussed a range of approaches and issues related to the coding and reporting of low-level data for computerized data bases. The very diversity of the discussion illustrates the difficulty we and others have had in arriving at a consensus on a mutually agreed upon approach. We are faced with a very serious, rather intractable, situation. Many pollutants are hazardous down to levels which tax our ability to measure their presence and quantity with any reliability. However, there are no commonly agreed upon methods to describe the results of these low-level measurements and even less consensus on ways to reduce the descriptions to codes necessary to use with our large computerized data bases. (We even find strong differences in preference or acceptance of the term quantitation for quantification.) A brief summary of our panel discussions will illustrate some of the problems and possible solutions. Gerald McKenna has discussed the basic issues involved with low-level measurement and coding. He has listed four criteria which a universal convention for treating low-level data should meet. While not all inclusive, these very criteria illustrate the immensity of the task. The criteria require satisfying both the analytical chemist and a range of intended data users - while simultaneously meeting the rigor of the statistician and be understandable and useable for the non-statistician. Such requirements suggest that at best we may have to be content with a series of conventions of restricted application for some time.

The King, Kleopfer and Taylor papers illustrate a variety of conventions and issues related to describing low-level data. (Henry Kahn's discussion of statistical issues was unavailable for inclusion here.) Don King discusses problems related to terms and coding conventions together with some basic statistical issues. Don King's preference is for a variation of the "W" and "T" codes proposed by ASTM. He argues for the importance of careful establishment of such limits as method detection and instrument detection limits while at the same time accurately reporting findings beyond the limits, with qualification. "Qualifying low results allows the analyst to cross his/her fingers, while allowing the data user full access to all available information."

Robert Kleopfer discusses a different aspect of reporting of low-level data for computerized data bases than that discussed by Don King or John Taylor. Kleopfer discusses a computerized data system and coding approach designed for a single analyte-Dioxin. In the case of Dioxin, a comprehensive set of controls and performance requirements have been established. These controls and performance requirements are included in a data review protocol. Data may be rejected or accepted based on conditions related to a single measurement or a data set. The data reviewer, using the data review protocol, will assign a "V" code to indicate the data meets control and performance requirements as specified in the data review protocol. An "I" code indicates invalid data and a "J" code indicates some noncritical controls were not met. Unlike the codes and conventions discussed in other papers, the Dioxin data approach gets into the issues of data validity both at the "limits of detection" and those related to criteria applicable at ordinary levels of analysis.

John Taylor focuses primarily on issues related to the recommendations of the ACS Committee on Environmental Improvement, which he co-authored. The conventions recommended here are designed to ensure data certainty for use in routine decisions as opposed to research use. Limits of detection are set at 3 sigma limits and limits of quantitation at 10 sigma limits. Taylor notes that a gulf will always exist between research and "practical" data - and we may also add the implied need will therefore exist for different conventions and resulting non-standardization.

The ACS approach discussed by John Taylor and extensions of the ASTM approaches discussed by Don King illustrate some of the practical difficulties faced in attempting to compare and evaluate approaches. The ASTM approach utilizes terms, criteria of detection and limit of detection. The ASC recommendations include terms, limit of detection and limit of quantitation. Both approaches provide a statistical basis for the terms. However, attempts by our task force on low-level data to make a rigorous conceptual and statistical comparision of the approaches have been unsuccessful. Even similar terms are defined in different, non-comparable ways, and additional terms and concepts are used which are unique to each approach.

We have attempted to describe issues related to and approaches for comprehensive and reliable reporting of low-level data for computerized data bases. Hopefully, we have also conveyed some concept of the importance of the problem and the diversity of approaches aimed at problem resolution. However, we find no agreed upon methods to describe the results of low-level analyte measurements or consensus on reducing descriptions to codes for use in large computerized data bases. Part of the difficulty may be ascribed to conflicting demands including meeting the needs of diverse users. However, there also appears to be a lack of rigor in addressing the problem even where a low-level data convention and its codes are designed for a highly specific data base. A systematic examination and specification of the data need and use followed by a careful development of the appropriate "convention" and codes could be helpful.

Certainly, a carefully developed and documented protocol would permit more effective evaluation of the validity and appropriateness of conventions for describing low-level data. A joint effort of ACS and ASTM could, at a minimum, recommend the issues to be addressed and a documentation structure for low-level data conventions. Our task force for one would be pleased to provide inputs ranging from general frustrations to specific recommendations.

RECEIVED March 2, 1987

Author Index

Affiliation Index

Subject Index

Production by Cara Aldridge Young
Indexing by Linda R. Ross
Jacket design by Carla L. Clemens

Elements typeset by Hot Type Ltd., Washington, DC
Printed and bound by Maple Press, York, PA